CW01371007

MALTING AND BREWING SCIENCE

Volume I   Malt and Sweet Wort

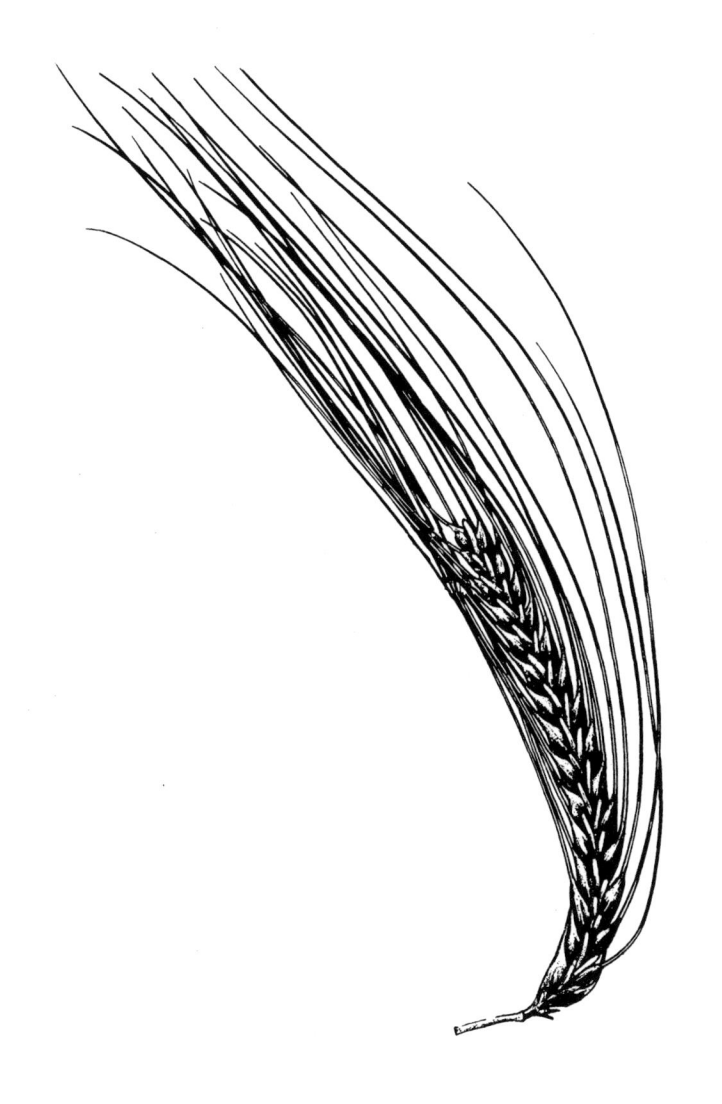

MALTING AND BREWING SCIENCE

Volume I Malt and Sweet Wort

D. E. BRIGGS, J. S. HOUGH

Department of Biochemistry
The University of Birmingham

R. STEVENS

School of Pharmacy
Sunderland Polytechnic

and

T. W. YOUNG

Department of Biochemistry
The University of Birmingham

SECOND EDITION

KLUWER ACADEMIC/PLENUM PUBLISHERS
NEW YORK, BOSTON, DORDRECHT, LONDON, MOSCOW

First edition 1971
Second edition 1981
Reprinted 1986, 1990, 1991, 1994, 1995, 1996

© 1981 D.E. Briggs, J.S. Hough, R. Stevens and T.W. Young

Printed in Great Britain at St Edmundsbury Press, Bury St Edmunds, Suffolk

ISBN 0 412 16580 5

Apart from any fair dealing for the purposes of research or private study, or criticism or review, as permitted under the UK Copyright Designs and Patents Act, 1988, this publication may not be reproduced, stored, or transmitted, in any form or by any means, without the prior permission in writing of the publishers, or in the case of reprographic reproduction only in accordance with the terms of the licences issued by the Copyright Licensing Agency in the UK, or in accordance with the terms of licences issued by the appropriate Reproduction Rights Organization outside the UK. Enquiries concerning reproduction outside the terms stated here should be sent to the publishers at the London address printed on this page.

The publisher makes no representation, express or implied, with regard to the accuracy of the information contained in this book and cannot accept any legal responsibility or liability for any errors or omissions that may be made.

A Catalogue record for this book is available from the British Library

Library of Congress Cataloging-in-Publication Data available

∞Printed on permanent acid-free text paper, manufactured in accordance with ANSI/NISO Z39.48-1992 and ANSI/NISO Z39.48-1984 (Permanence of Paper).

CONTENTS

Preface	page ix
1 Outline of malting and brewing	1
1.1 Malting	1
1.2 Milling and mashing	3
1.3 Boiling with hops	6
1.4 Wort cooling and fermentation	6
1.5 Beer processing	8
1.6 Energy, water, effluent and manpower	9
1.7 Types of beer	10
1.8 Excise	11
1.9 Other beverages and allied products	11
1.10 The brewing industry	12
1.11 List of technical journals	13
References	14
2 Barley	15
2.1 Barley, cereals and brewing	15
2.2 Barley cultivation and harvesting	21
2.3 Drying and storage	29
References	37
3 Some aspects of barley grain physiology	39
3.1 Introduction: an outline of malting	39
3.2 Laboratory tests on grain	40
3.3 Water uptake by barley grain	41
3.4 The penetration of grain by solutes	43
3.5 Dormancy and viability	44
3.6 Respiration and fermentation during steeping	50
3.7 Steep liquor	52
3.8 Choosing commercial steeping schedules	53
References	55

4	The biochemistry of malting grain	57
	4.1 Introduction	57
	4.2 Physical changes in malting barley	57
	4.3 Chemical changes during malting	64
	4.4 Grain organization, growth conditions and the biochemistry of malting	96
	4.5 The chemistry of kilning	104
	References	107
5	Malting conditions and malt types	110
	5.1 Floor malting	110
	5.2 Extraction prediction	113
	5.3 Small-scale malting	115
	5.4 Malting losses	116
	5.5 Variations in steeping conditions	118
	5.6 Pneumatic malting	120
	5.7 Germination temperatures and malting	121
	5.8 Additives in malting and techniques for reducing malting losses	122
	5.9 Kilning	133
	5.10 Different malt types	134
	5.11 Rootlets	141
	References	142
6	The technology of malting and kilning (including the physical principles of barley drying)	145
	6.1 Floor malting	145
	6.2 Trends in malting equipment	146
	6.3 Kilning malt and the physics of barley drying	171
	6.4 Types of malt kiln	183
	6.5 Kiln furnaces	188
	6.6 Separation of culms	191
	6.7 Dispatch and transport of malt	191
	References	192
7	Brewing water	194
	7.1 Geology of water supplies	194
	7.2 Water quality	195
	7.3 The supply of water	195
	7.4 Desalination	197
	7.5 Deionization	198
	7.6 Softening procedures	199
	7.7 Other water treatments	201
	7.8 Water purity	201

7.9 Brewing water	206
7.10 Water economy in maltings and breweries	209
7.11 Effluent treatment	211
7.12 Water in beer	219
References	220
8 Adjuncts, sugars, wort syrups and industrial enzymes	222
8.1 Introduction	222
8.2 The common cereal grains	224
8.3 Analyses of adjuncts	226
8.4 Raw cereal grains	230
8.5 Cooked intact cereal grains	232
8.6 Grits	232
8.7 Flaked cereals	234
8.8 Flours	235
8.9 Refined starches	237
8.10 Sucrose and invert sugars	238
8.11 Sugars and syrups from purified starch	242
8.12 Malt extracts and wort replacement syrups	244
8.13 Industrial enzymes	248
8.14 Future possibilities	251
References	251
9 The chemistry and biochemistry of mashing	254
9.1 Introduction	254
9.2 Hot-water extract	255
9.3 Other analyses	259
9.4 Mashing–some general considerations	262
9.5 Carbohydrates in the wort	268
9.6 Nitrogenous materials in sweet wort	269
9.7 Vitamins	271
9.8 Polyphenols	272
9.9 Organic acids and lipids	273
9.10 Mineral salts in mashing	274
9.11 Water hardness and mash pH	275
9.12 Other substances added to mashes	277
9.13 Some properties of enzymes relevant to mashing	278
9.14 Mashing and pH	279
9.15 Temperature, time and mashing	280
9.16 Mash thickness	288
9.17 Mashing and the dissolution of starch	290
9.18 Sparging	293

9.19 Spent grains	295
9.20 Other grist materials	296
References	300
10 Preparation of the grist	**304**
10.1 Storage, cleaning and weighing	304
10.2 Dry milling of malt	306
10.3 Wet milling	314
10.4 The grist case	316
10.5 Grist composition	317
Examples	318
References	319
11 Mashing	**320**
11.1 Traditional brewing	320
11.2 Traditional infusion mashing	322
11.3 Clarity of worts	330
11.4 Decoction and double-mash systems	330
11.5 Temperature-programmed systems	334
11.6 Modern mashing equipment	336
11.7 Continuous mashing	342
11.8 The theory of mash filtration	344
11.9 Choice of mashing systems	347
11.10 High-gravity brewing	352
11.11 Spent grains	352
References	354
Appendix	355
Index	363

PREFACE

Some ten years have passed since the publication of the first edition of *Malting and Brewing Science*, a period of many changes. As before, this edition is an aid to teaching, particularly the MSc course in Brewing Science at Birmingham University, but it is also aimed at the requirements of other students of the science of malting and brewing throughout the world. In general, technological aspects are covered more fully in this new edition, although not malting and brewing practices that are exclusive to Britain. Nevertheless, the amount of technological information available is too great to be comprehensively covered in one book. Scientific principles and information receive more attention, but for details of analytical procedures reference should be made to the most recently published material of the American Society of Brewing Chemists, the European Brewery Convention and the Institute of Brewing.

The new edition appears as two volumes because a single one would be inconveniently bulky. The first volume outlines the entire process and leads from barley, malting and water to the production of sweet wort. In the second volume there are chapters on hops and hop products, production of hopped wort, fermentation, yeast biology and all aspects of beer quality and treatment.

Decisions about the units of measurement proved difficult; metric units commonly used in the Industry are given and in brackets are equivalents in degrees Fahrenheit, Imperial measures and UK barrels. Considerable information on equivalents is given in a special section in each volume.

References given with each chapter are aimed at offering the reader an entry into the appropriate literature. This restriction offers an economy of space but often fails to mark the special contribution of particular authors. Only for historical reasons or where a method bears the name of its developer or where a diagram or table is acknowledged have authors' names been included in the text.

The authors of the first volume wish to thank all those who have helped in its production, particularly the constructive critics. Within the University, we wish to thank Miss J. Gainham for typing and secretarial work, Mrs P.

Hill for art work, Mrs D. Clarke and Mr J. Redferm for photography. Specialized advice has been received from the following: J. Baxter, P. A. Brookes, O. T. Griffin, D. Howling, P. D. Hull, M. J. Lewis, M. J. W. Lloyd, Miss D. A. Lovett, D. J. Lubert, P. A. Martin, F. Moseley, T. J. Palmer, S. Poole, W. W. Sisler, J. B. Smith, R. Vogel, D. A. Watts and R. Zytniak. Our thanks also go to those who have permitted us to reproduce, in the original or in modified form, various diagrams and tables, acknowledgements for which are given in the legends.

Chapter 1

OUTLINE OF MALTING AND BREWING

For the purposes of this book, beer refers to a beverage which involves in its production (*i*) extracting malted barley (perhaps mixed with other materials) with water, (*ii*) boiling this extract, usually with hops, and (*iii*) cooling the extract and fermenting it using yeast. The fermented beverage is then normally clarified and dispensed in an effervescent condition. In this outline of malting and brewing, each of these stages of production and the choice of raw materials will be briefly considered, prefaced by a short description of the manufacture of malt.

1.1 *Malting*

Malting involves germination of the grain until the food-store (endosperm), which is available to support the development of the germ of the grain, has suffered some degradation from enzymes. The maltster is concerned with both the degradation of the endosperm and the mobilization of the enzymes of the grains. But growth of the germ or embryo is incidental to the making of malt and leads to depletion of the endosperm material through respiration of the embryo and its growth. Thus the maltster's requirements deviate from those of the farmer whose primary interest is in the growth of the embryo to form a mature fruiting plant. When the degradation of the endosperm has progressed to only a limited extent, the maltster terminates both its degradation and the growth of the embryo by drying the grain. The maltster cures the malt by kilning, thus enabling him to store the dry malt for long periods in a stable state. Malt is therefore cereal grain, usually barley, which has germinated for a limited period of time, and then dried. It represents a rich source of carbohydrate, degraded protein, various B vitamins and inorganic material. In addition there is an abundance of enzymes, particularly ones degrading starch, unless the kilning process has been severe for the purpose of making highly coloured malts.

The malting process involves the collection of stocks of suitable barley, the storage of the cereal until it is required, steeping the grain in water, germinating the grain and finally drying and curing it on the kiln. Fig. 1.1 shows a typical maltings in diagrammatic form. Not all barleys yield suitable malts for the brewer, and selection is based on many criteria including

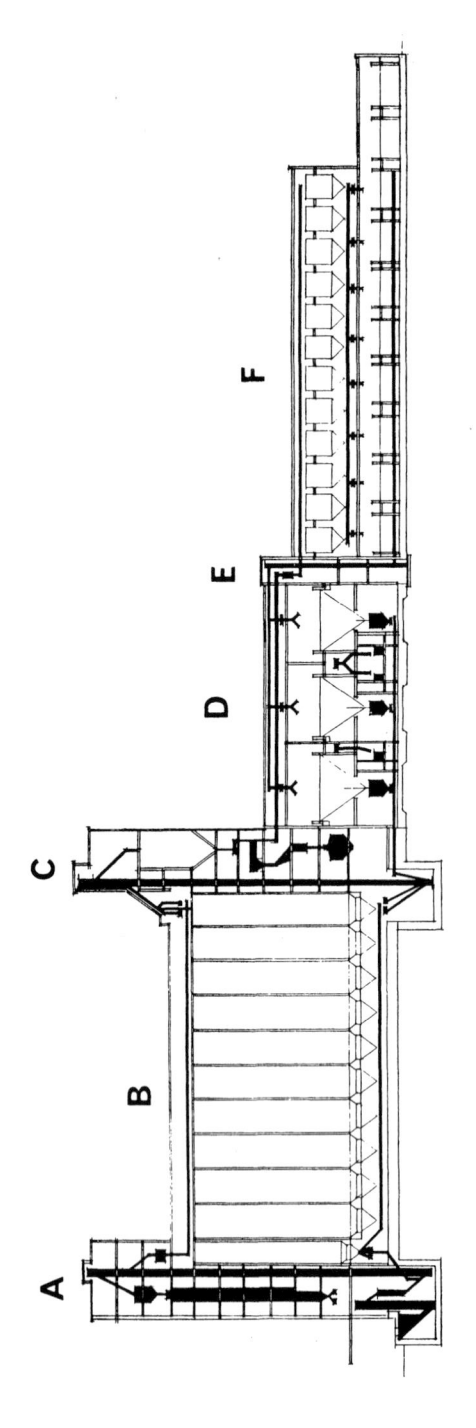

Fig. 1.1 A section of a maltings equipped with pneumatic germination boxes. The sequence of operations is intake (Section A), to barley storage (Section B), followed by steeping and germinating (Section F), kilning (Section D), cleaning (Section C) and malt storage (Section B). Diagram courtesy of ABM Ltd, Newark on Trent.

(*i*) rapid and synchronous germination of the barley grains, (*ii*) even enzymic degradation of the endosperm, (*iii*) an adequate complement of enzymes even after kilning, and (*iv*) low levels of fibrous materials and total nitrogen. The farmer will only grow barleys suitable for malting if such varieties have desirable agricultural properties which include (*i*) good yield, (*ii*) stiff straw, (*iii*) early ripening, and (*iv*) disease-resistance. Some varieties of barley produce grains which will readily germinate in damp weather before the crop is harvested; such barleys are avoided except in arid regions. At the other extreme, some varieties produce grains which will germinate only with difficulty; these grains are dormant. Usually dormancy can be broken by a few months' storage or mild heat treatment. Varieties which do not respond to such mild treatment are avoided by the maltster.

1.2 *Milling and mashing*

The brewer uses as his primary raw materials, malted barley, water and hops. The laws of some European countries, for instance West Germany, preclude the use of any other raw material for beers but elsewhere unmalted cereals and sugars are commonly used in addition. Malted barley is carefully ground in a mill in such a way that the husk of the grain is left substantially intact while the rest becomes a very coarse powder. This flour is particularly rich in starch and also in the enzymes capable of degrading it rapidly when water is added to it.

After milling, there are several methods of mashing to obtain a satisfactory wort (an aqueous extract of the malt). They include:

(1) The infusion mashing system, common in Britain, which is the simplest.
(2) The decoction mashing system, common in Germany and Central Europe.
(3) The double mash system, used in almost all other countries.
(4) A combination of (2) and (3).
(5) Temperature-programmed mashing.

A generalized flow diagram of the brewing process is shown in Table 1.1, and a brewery for carrying out double mashing is shown in Fig. 1.2.

Infusion mashing involves the intimate mixing of the ground material, or grist as it is called, with hot water. Control is exercised on the amount of water used and its temperature so that the mash produced has a final temperature of about 65°C (149°F) and has a thick porridge-like consistency. The temperature is held constant for a period which may be as short as 30 min or may extend for several hours. Enzymes of the malt attack principally starch and its degradation products (amylolysis), but some breakdown of nitrogenous materials (proteolysis) also occurs. The production of sweet wort takes place in a mash tun, a vessel which has a slotted base which acts like a sieve. In order to wash virtually all the sweet wort from the undegraded

TABLE 1.1
Sequence of operations encountered in the brewing process

Operation	Input	Product
1. Milling	Malted barley (plus unmalted cereal in some instances)	*Grist*
2. Mashing-in	Hot water	*Mash*
3. Mash digestion	Enzymes and substrates from grist	*Digested mash*
4. Wort separation	Hot water	*Sweet wort* +spent grains*
5. Wort boiling	Hops, steam (plus wort syrups in some instances)	*Hopped wort* +spent hops* +hot trub*
6. Wort clarification	Sieve or centrifuge or settling tank	*Hopped wort* +hot trub*
7. Wort cooling and aeration	Refrigeration, air or oxygen	*Cold aerated wort* + cold trub
8. Fermentation	Yeast	*'Green' beer* +yeast* +cold trub
9. Conditioning (lagering or aging)	Refrigeration	*Carbonated beer* ready for filtration
10. Filtration	Filter sheets or kieselguhr	*Bright beer*
11. Packaging	Packages such as bottles, cans or kegs	*Packaged beer*

* Spent hops and hot trub are often mixed with spent grains for cattle food. Surplus yeast is used for whisky fermentations and for yeast extract manufacture. Note that a small proportion of the yeast crop passes with the green beer into the conditioning operation.

material (the spent grains), further supplies of hot water at say 79°C (175°F) are sprayed over the surface of the mash, usually by means of revolving sparge-arms. The hot water is commonly referred to as 'hot liquor' by brewers.

Decoction mashing differs from infusion mashing in several respects. The endosperm of the malt used is in some cases less enzymically degraded and requires more extensive enzymic action in mashing. To aid this, the malt is both ground more finely than in the infusion process and is mixed with water of lower temperature, say 35°C (96°F). Extensive proteolysis and solubilization occurs. During the conversion a portion of the mash (often one-third) is boiled in a mash copper or kettle and returned to the rest of the mash in the mash vessel for mixing. The overall temperature rises to about 50°C (122°F). Later, a further third of the mash is withdrawn and boiled. On return to the main mash, the temperature increases to about 65°C (149°F), and amylolysis now occurs rapidly. After a last decoction with one-third of the mash, all enzyme activity ceases in the consolidated mash at say 75°C (167°F). The wort is separated from the grains, usually with a vessel rather like a mash tun in an infusion brewhouse and called a lauter tun. (Lauter is the German for 'clarifying'.) Alternatively a mash filter is used.

The double mash system is a method having features in common with

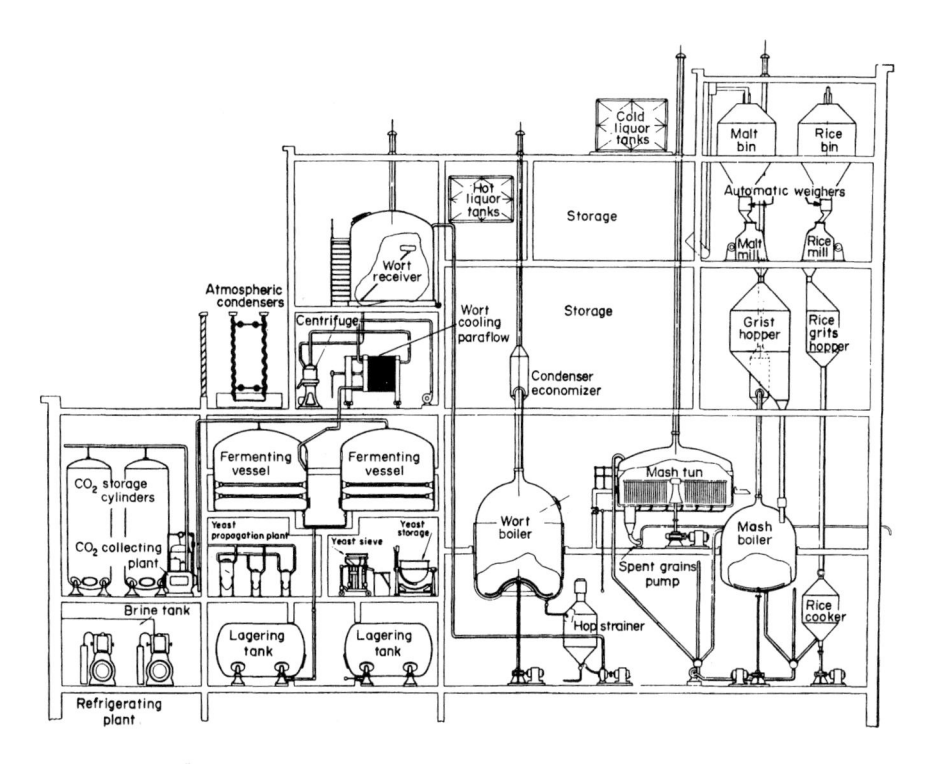

Fig. 1.2 A brewery with equipment for decoction and double mashing. The mash tun is normally called a 'lauter tun'. Diagram by courtesy of APV Ltd., Crawley.

infusion and decoction mashing. Malt with a high content of enzymes is normally used. It is ground and mixed with water to yield a mash at 38°C (100°F). Unmalted cereal in the form of maize (corn) or rice grits is mixed with a little malt, milled and carefully heated in a cereal cooker to 100°C (212°F). The two mashes are then mixed together to achieve about 70°C (158°F), and later the temperature is raised and held at 73°C (163°F) for about 1 hr. The wort is then separated from the grains in a lauter tun. It is not surprising that some breweries adopt mashing methods which are intermediate between decoction mashing and the double mash systems.

Another variant which has achieved popularity in recent years is temperature-programmed mashing in which steam heating (or high-pressure hot-water heating) raises the mash temperature at precise times to predetermined temperatures. In all cases, the choice of method depends on the type of malt available, the proportion of unmalted cereal used, and the type of beer required.

1.3 Boiling with hops

The wort produced by mashing is boiled in a copper or kettle in order to (*i*) arrest further enzyme action, (*ii*) precipitate some of the proteinaceous material of the wort as a coarse coagulum (hot trub), (*iii*) sterilize the wort, and (*iv*) hasten certain desirable chemical changes. Very often excess water is freely evaporated, but in some cases boiling under pressure is carried out. The boiling process is often associated with the addition of (*i*) sugars or syrup (copper or kettle sugars), (*ii*) materials which encourage coagulation of protein and tannin material, and especially (*iii*) hops or bitter products derived from hops. The essential oils of the hops are largely lost by steam-volatilization if steam is permitted to escape from the copper. Indeed, unpleasant flavours are normally imparted to beers if the more volatile essential oils are retained. The bitter resins of the hops are extracted during the boil, and particular attention is paid to levels of humulones (α-acids) and lupulones (β-acids). During the boiling the humulones are substantially isomerized to isohumulones which are more bitter and water-soluble. Some of the lupulone material is oxidized during the boil to yield a range of bitter and non-bitter products. The optimum duration of boiling from the point of view of bitterness is about 90 min.

Hops are selected primarily on the basis of their bitter resin content, and secondarily on other flavour criteria. In recent years, there has been a great increase in the use of hop powders, pellets and extracts because (*i*) these are readily stored, and so full advantage can be taken of the years when good hops are plentiful, (*ii*) they are far less bulky to store than whole hops, (*iii*) they deteriorate less quickly and may require no cold-storage, and (*iv*) boiling time can be reduced.

When whole hops are used rather than extracts, the contents of the copper are run into a hop-back where the spent hops are encouraged to form a filterbed which retains most of the hot trub. Alternatively the hops are separated by a sieving mechanism from which they are removed by a screw-conveyor. The hot trub (and any insoluble material from hop powder or pellets) is taken out of the wort by centrifugation or by a whirlpool tank.

1.4 Wort cooling and fermentation

In order to prepare for fermentation, the clear hopped wort is cooled, usually in a plate-heat exchanger. Further protein and tannin material tends to come out of solution as a fine break or coagulum termed 'cold trub'. This precipitation of hot and cold break is encouraged because it reduces the chance of similar material emerging from packaged beer as a haze. During the cooling, it is usual to aerate or even oxygenate the wort because the next processing step involves the growth of yeast, which is encouraged by the presence of dissolved oxygen even though the concentration is normally only about

7-14 p.p.m. (µg/l) of dissolved oxygen in the wort.

It is traditional practice to ferment infusion-mashed worts with top yeasts, that is strains of yeast which at the end of fermentation tend to rise to the top of the vessel and form a yeast-head. The beers so produced are termed ales. It is normal to ferment other types of wort by bottom yeasts, strains which tend to sediment to the bottom of the vessel towards the end of fermentation. The beers so produced are sometimes called lagers (the meaning in German is 'stored'). This term therefore refers to the secondary fermentation, maturation and storage which these beers traditionally receive after primary fermentation.

The yeast is intimately mixed with the wort, at temperatures around 16°C (61°F) for top yeasts and 10°C (50°F) for bottom yeasts. About 1-2g of pressed yeast is used for every litre of wort. The yeast grows using amino acids of the wort as the main source of nitrogen, and the glucose, fructose, sucrose, maltose and maltotriose of the wort as the main sources of carbon and energy. The yeast multiplies to give some 5 to 10 times the original weight, and its metabolism is overwhelmingly fermentative so that much of the carbohydrate assimilated is degraded to ethanol and carbon dioxide. Fermentation is complete in 2-10 days depending on conditions, particularly temperature.

The flavour of the beer produced depends in part on the quality of the wort used and in part on the metabolism of the yeast. Of particular importance are metabolic reactions which produce substances of strong flavour, such as alcohols, aldehydes, diketones, esters, ketones, mercaptans and other sulphur-containing materials.

It has been traditional to separate the yeast from the beer at the end of the fermentation by skimming or suction in the case of top yeasts and by decantation of beer from bottom yeasts. In more recent times, centrifuges have been used by some breweries. A portion of the recovered yeast is used for pitching (inoculating) further worts. Clearly, the traditional separation of top yeasts is assisted if the cells float very readily to the surface to form a stable foam, assisted by the rising bubbles of carbon dioxide. If, however, this separation occurs too early in the fermentation process, the rate of conversion of carbohydrate to ethanol (termed attenuation) is impaired. Similarly, attenuation is hindered by bottom yeasts settling too early in the fermentation. Yeasts have therefore been selected on the following criteria.

(1) The rate and extent of growth. (Too little growth and slow growth of the yeast hampers the fermentation but too much growth reduces the yield of ethanol for a given amount of assimilable carbohydrate.)

(2) The rate and extent of fermentation. (On the whole, it is desirable that, consistent with reasonable yeast growth, the maximum amount of carbohydrate is converted quickly to ethanol.)

(3) The flavour and aroma of the beer produced. (This depends largely on the complex of metabolic pathways and production of aromatic and flavoured metabolic end products.)

(4) The ability to form a yeast head or to sediment.

The temperature at which fermentation is carried out is important from these points of view. Thus many top yeasts will give satisfactory performance at temperatures in the range 15–20°C (59–68°F) and bottom yeasts in the range 10–15°C (50–59°F), but not necessarily the converse. In recent years, new processing techniques have tended to make the distinction between top and bottom yeasts less obvious. A further development is to dilute strong beers to normal strengths after fermentation. Such high-gravity brewing gives rise to substantial savings in the energy needed for wort boiling and cooling and in the size of vessels required to hold worts and beers.

1.5 Beer processing

Ale is traditionally run from fermenters into casks, and a small amount of sugar and finings is added. The sugar permits a secondary fermentation in the cask and the carbon dioxide produced dissolves in the ale. The finings (usually fish collagen) form a coagulum with suspended yeast and, on settling, leave a clear ale.

There is a tendency in modern breweries for yeasts to be separated at least in part by centrifuges. From this stage onwards there are many possible ways in which the beer may be treated. The simplest in concept is to chill and filter the beer and then, after partly flushing out the volatiles with carbon dioxide, the beer is carbonated with the same gas under pressure. For many beers, the processing is more complex. Secondary fermentation (conditioning) is encouraged by addition of fermentable sugar, and the temperature is gradually reduced to about 2°C (35°F). Various complex biochemical reactions take place which modify and improve the flavour of the beer during storage (maturation) at this low temperature. Eventually the beer is filtered and dispensed into bottles, cans, kegs or bulk tanks. It is usual to filter sufficiently finely to eliminate any yeasts and bacteria present in the beer, but in order to ensure that the beer is sterile, it is often subjected to a pasteurization. This treatment may take place before the beer is dispensed into sterile containers or the filled packages may be pasteurized.

For most beers, it is considered desirable that (*i*) they remain brilliantly clear for several months stored at any temperature in the range 0–25°C (32–77°F), (*ii*) they develop a stable white foam on pouring into a glass, and (*iii*) the flavour and gas content remain stable. Brewing experience indicates that intelligent selection of raw materials and process conditions enable the brewer to achieve these goals substantially. Complementary post-fermentation treatments are nevertheless often adopted, such as addition of proteolytic enzymes, in order to degrade those proteinaceous materials which would otherwise separate as haze.

Alternatively, adsorbants such as certain silica gels or insoluble polyvinyl pyrrolidone may be added so that the content of protein or polyphenol precursors of haze may be reduced. Again, materials may be added which enhance the beer foam, or reduce the dissolved air content of the beer so that the flavour and colour is more likely to be stable.

1.6 *Energy, water, effluent and manpower*

In recent years the cost of energy has increased substantially (Table 1.2), an indication of the short space of time before the world's oil supplies are exhausted. The malting and brewing industries have responded by making their present plants more efficient in the use of power and designing new equipment with this requirement particularly in mind. Savings arise from (*i*) good 'housekeeping', especially by individual process workers, (*ii*) short to medium developments such as increased insulation, energy load shedding and recycling of hot water, and finally (*iii*) long-term capital investment involving for instance steam and electricity co-generation. There is clearly room for these savings because most breweries use twice as much energy as is theoretically required.

TABLE 1.2
Energy costs from 1972 to 1978 in the U.K.
Pence per 10^5 B.Th.U. (25.2 Mcal or 105.5 MJ)

	1972	1973	1974	1975	1976	1977	1978
Electricity	19·7	22·0	24·5	42·5	50·2	56·3	65·0
Coal	2·8	3·2	3·9	7·0	8·5	9·8	10·8
Gas	4·2	6·5	8·9	10·8	12·8	18·2	20·5
Oil	3·0	5·8	8·8	11·3	13·2	17·5	18·5

In parallel with fuel costs, the costs of water and effluent disposal have also increased so that there is a continuing need for good housekeeping and for new equipment to utilize less water and produce less effluent or alternatively reutilize water. Representative figures are provided in Chapter 7.

The position with regard to manpower is more complex. There has been a clear tendency for maltings and breweries to automate their processes, especially using microprocessors, and to reduce the labour force needed. However, in some countries a substantial proportion of the population is unemployed, causing political pressure to be applied to industry to maintain or increase its workforce. As a consequence, the speed of introducing automation has in some instances slackened, working hours have been reduced and more shifts per day introduced.

Other political pressures have arisen in recent years connected with environmental pollution and the use of additives in foodstuffs. Measures to combat environmental pollution have restricted or prevented the use of certain fertilizers, fungicides and pesticides in the production of cereals, including barley. Such measures have also had bearing on the discharge of fumes from maltings and breweries and on the release of effluents and solid wastes. In the case of additives, in many countries, there have been attempts by central government and consumer groups, to restrict the use of chemical additives in foodstuffs and to insist on such additives being listed on package labels.

1.7 *Types of beer*

Pale-coloured top-fermented beers are often referred to as 'pale ales' or, in the case of bottom-fermented beers, as 'Pilsner lagers'. 'Mild ales', which have a lower content of bitter substances than pale ales, are also usually sweeter and darker in colour but not invariably so. 'Brown ales' may be identical with dark mild ales. Dark bottom-fermented beers are also produced, and are associated particularly with Munich. The darkest ales are 'stouts' and 'porters', but the last term is now rarely used. There has been very great interest in recent years in beers with a lower calorific content. Dilution of beers achieves this objective but, if the alcohol concentration is to be maintained, it is necessary instead to reduce the level of unfermented carbohydrates. Either the mashing process is modified to produce a higher proportion of fermentable carbohydrates or enzymes such as amyloglucosidase (glucoamylase) are added to the fermenter to have the same effect. Such beers may be referred to as 'light', 'lite' or dietetic. There are also demands in certain countries for non-alcoholic beers. They are either carbonated worts or beers from which the alcohol has been stripped.

The strength of a beer is not easily defined because 'strength' may refer to alcoholic content, to the specific gravity of the wort, or even to the content of bitter substances. In most cases, however, the reference point is the specific gravity of the wort because Excise Duty is paid in many countries on this value. In Britain, the specific gravity is measured in the usual way and multiplied by a thousand; thus a wort of specific gravity of 1.040 at 20°C (68°F) is referred to as SG 1040. Until recently, brewers pounds were used. Thus a barrel of water (36 gallon) weighs 360 lbs, but a barrel of wort of SG 1040 would weigh 374·4 lbs. This wort is therefore said to be 14·4 brewers' pounds gravity. Outside Britain, the specific gravity has been expressed in terms of sucrose solutions of the same specific gravity. In von Balling's Tables of 1843, wort of SG 1040 is equivalent to 9·95% wt./wt. sucrose at 17·5°C. Improvements and corrections were made to von Balling's Tables by Dr Plato in 1918 and in many European Countries specific gravity is expressed in degrees Plato.

A difficulty arises in describing a beer by the specific gravity of the wort from which it is produced. Thus a beer may have a specific gravity of 1.008 (SG 1008) and be derived from a wort of SG 1040. This is resolved by referring to the wort gravity as 'original gravity' so that the beer in question is said to have an OG 1040 and SG (or present gravity) of 1008. With specific gravity expressed in terms of Balling or Plato, the beer may be said to have an 'original extract' of say 10°P.

1.8 Excise

Excise considerations have had important consequences in the choice of processing methods. Once a process has been modified for these considerations, it often so remains because further change in technique might alter the character of the beer produced. In Britain, imposition of duty on malt led to the development of certain methods of malting. The impost was lifted long ago but still is reflected in the type of malt used, the details of milling and mashing. Duty is now paid on the wort collected after boiling and cooling. This has had some influence on the design of fermentation vessels and yeast propagators, on the recovery of beer from harvested yeast, from filters and so on. In other countries excise is levied on the alcohol content of the beer and, in some, beers are classified in alcohol bands and are charged excise accordingly. Finally, there are countries where excise is paid solely on the volume of beer leaving the brewery.

1.9 Other beverages and allied products

There are many other products, beverages and others, which have a relationship to beer. Thus malted barley is the basis of malt vinegar. In the manufacture of malt vinegar, the milling and mashing processes are similar to beer production. The sweet wort, is however, not boiled; it is fermented with yeast and the unhopped beer produced is acetified by acetic acid bacteria. In the UK whisky is produced from malted barley alone (malt whisky) or a mixture of malted barley and maize (grain whisky). The derived sweet wort is pitched with a mixture of distillers' yeast and brewers' yeast; at the completion of fermentation, it is distilled. Certain still fractions are collected and matured for several years in special casks and blended before bottling. No yeast is reused.

In cider manufacture, the fermentable material is derived from apples, and in wine production, grape juice is fermented. Traditionally, the yeast present on the fruit during processing, or on the equipment, was responsible for fermentation. As with brewing, more sophisticated methods of production entail the selection and propagation of suitable yeast strains so that the quality of the product is more uniform. Because the juice is often treated with sulphur dioxide to preserve it, yeast strains which are resistant to sulphur dioxide are selected; most wild yeasts are sensitive to the preservative and therefore die. Cider may be distilled to give calvados or apple-jack, and wines to give brandies or cognacs.

Some African beer is based on malted sorghum and millet corns which are milled and mashed and yeast added. The beverage is sometimes drunk while it is still fermenting.

Bakers used brewers' yeast up till about 80–100 years ago, but its performance in the bakery was low and variable. Subsequently, bakers' yeast was produced on unhopped malt wort but yields of yeast were low, even when the wort was aerated. Modern methods involve substantial aeration of the yeast as it grows in a molasses and salts medium; the molasses are added at such a rate that the yeast is never starved of sugar but the sugar level is always low. This method of production greatly improves yields of yeast and prevents the formation of alcohol.

1.10 *The Brewing Industry*

The Brewing Industry is not large in terms of its contribution to industrial production or total employment. Probably in no country does it employ more than 0·5% of the working population, nor does the value of its output rise to more than 1·5% of the Gross National Product. World production of beer is estimated as 870 million hl/annum. In traditional beer-drinking countries the increase in beer consumption rises roughly in step with the increase in population, with per capita consumption ranging roughly from 30 to 148 litres/annum. In many traditional wine-making countries, there has been a much greater increase in beer-consumption, for instance in Italy and Spain. Except for Denmark, Eire and the Netherlands, only a small proportion of the beer produced in any country is exported. Britain is unique in having wholesale and retail outlets substantially owned and controlled by the brewers, certainly over 48%.

The Brewing Industry is well organized into national and international associations. Of particular note are: (*i*) the European Brewery Convention which organizes regular congresses, conducts collaborative research and develops analytical methods (published in Analytica EBC); (*ii*) the American Society of Brewing Chemists which has similar functions; (*iii*) the Master Brewers' Association of America which is particularly concerned with brewing technology; (*iv*) the Institute of Brewing which exists in Britain and Australasia and is concerned with collaborative research, development of analytical methods, technical instruction and publication.

There are many technical journals concerned primarily with malting and brewing and a list is supplied in the next section. A list of textbooks on brewing and allied subjects, written in English, is also given.

1.11 List of technical journals (including those that have terminated in recent years)

Alcohol Beverage Abstracts
American Brewer
Association of Growers of the new Varieties of Hops, Annual Booklets
Australian Wine, Brewing and Spirit Review
Beverages
Bios
Birra e Malto
Bottlers' Year Book
Bottling
Brasserie
Brasserie et Malterie de Belgique
Brasserie Malterie Europe
Brauer und Mälzer
Brauereibesitzer und Braumeister
Brauereitechniker
Brauwelt
Brauwissenschaft
Brewers' Almanack
Brewers' Digest
Brewers' Guardian
Brewers' Guild Journal (now, The Brewer)
Brewers' Journal
Brewing Trade Review
Brygmesteren
Bulletin des Anciens Élèves de l'École de Brasserie de Nancy
Bulletin de l'Association des Anciens Élèves de l'Institut des industries de Fermentation de Bruxelles
Bulletin de l'Association des Anciens Étudiants en Brasserie de l'Université de Louvain
Bulletin of Brewing Science (Tokyo)
Cerveja
Cerveza y Malta
Communications of the Master Brewers' Association of America
Deutsche Brauwirtschaft
Echo de la Brasserie
Fermentatio
Hopfen Rundschau
International Bottler & Packer
International Brewer and Distiller
International Brewers Journal
Internationale Fachzeitschrift für Brau-Garungs-und Kaltetechnik
Internationaal Tijdschrift voor Brouwerij on Mouterij
Journal of the Institute of Brewing
Malting Barley Improvement Association. Annual Report
Mälzer
Mitteilungen der Versuchsstation für das Gärungsgewerbe in Wien
Modern Brewery Age
Monatsschrift für Brauerei
Petit Journal de Brasseur
Proceedings. American Society of Brewing Chemists (now, Journal of the American Society of Brewing Chemists)
Proceedings of the Conventions of the Institute of Brewing
Proceedings. European Brewery Convention
Report of the Fermentation Research Institute (Japan)
Report of the Research Laboratories of Kirin Brewery Co. Ltd.
Schweizer Brauerei Rundschau
Société d'Encouragement de la Culture des Orges de Brasserie et des Houblons en France. Rapport sur la campagne
Svensk Bryggeritidskrift
Tageszeitung für Brauerei
Technical Proceedings Annual Con-

vention of the Master Brewers' Association of America
Technical Quarterly of the Master Brewers' Association of America
Versuchsstation Schweizerischen Brauereien. Jahresbericht
Versuchs—und Lehranstalt für Brauerei in Berlin. Jahresbericht
Wallerstein Laboratories Communications
Wye College Department of Hop Research. Annual Report

REFERENCES

[1] BRIGGS, D. E. (1978). *Barley*, Chapman & Hall, London.
[2] BRITTON, D. K. (1969). *Cereals in the United Kingdom*, Pergamon Press, Oxford.
[3] BURGESS, A. H. (1964). *Hops: Botany, Cultivation and Utilization*, Leonard Hill, London.
[4] CHARM, S. E. (1963). *Fundamentals of Food Engineering*, Avi, Westport, CT.
[5] DE CLERK, J. (1958). *A Textbook of Brewing* (2 vols.), translated by K. BARTON-WRIGHT, Chapman & Hall, London
[6] COOK, A. H. (ed.) (1958). *Chemistry and Biology of Yeasts*, Academic Press, New York.
[7] COOK, A. H. (ed.) (1962). *Barley and Malt*, Academic Press, New York.
[8] EARLE, R. L. (1966). *Unit Operations in Food Processing*, Pergamon Press, Oxford.
[9] FINDLAY, W. P. K. (ed.) (1971). *Modern Brewing Technology*, Macmillan, London.
[10] HIND, H. LLOYD (1950). *Brewing Science and Practice* (2 vols.), Chapman & Hall, London.
[11] HOPKINS, R. H. and KRAUSE, B. (1947), *Biochemistry Applied to Malting and Brewing*, Allen & Unwin, London.
[12] JEFFREY, E. J. (1956). *Brewing Theory and Practice*, Kaye, London.
[13] POLLOCK, J. R. A. (ed.) (1979). *Brewing Science*, Three Volumes, Academic Press, London.
[14] PREECE, I. A. (1954). *The Biochemistry of Brewing*, Oliver & Boyd, Edinburgh.
[15] ROSE, A. H. (1961). *Industrial Microbiology*, Butterworths, London.
[16] ROSE, A. H. and HARRISON, J. S. (eds.) (1969). *The Yeasts* (3 vols.), Academic Press, London.
[17] ROSE, A. H. (ed.) (1977). *Economic Microbiology*, Vol. 1, Academic Press, London.
[18] WHITE, J. (1954). *Yeast Technology*, Chapman & Hall, London.

Chapter 2

BARLEY

2.1 *Barley, cereals and brewing*

The barley plant belongs to the Gramineae, the family of grasses. This group is of outstanding importance in man's economy, since it contains the species which yield food grains such as wheat, barley, rye, oats, rice, maize, millets and sorghum, as well as other important plants, such as bamboos, sugar cane and the pasture grasses which support grazing animals.

At present, malted or raw cereals in milled, refined, flaked or cooked states provide virtually all the grist used in the brewery mash tun. In the past 'malted' peas and beans were used, as were potato flakes, potato starch and couch grass rhizomes. Small quantities of wheat and rye are malted, and are used in confectionery or in making special beers such as the German Weissbier (*syn.* Weizenbier). Most brewing malt is prepared from barley. However, malted sorghum is used to make African Kaffir beer, malted rice has been used, and the early American settlers used malted maize. Many brewing sugars and syrups are prepared from maize (Chapter 8).

2.1.1 THE BARLEY PLANT [1–3]

Of the cereals, only the botany of barley is considered here. The first sign of germination of seed grain in the field is the appearance of the small white chit (coleorhiza; root sheath) at the basal end of the grain. This splits and several (often five) seminal roots grow into the soil. Meanwhile the coleoptile, or leaf sheath, grows and emerges from beneath the husk near the apex of the grain. When the coleoptile has cleared the soil surface the first leaf grows out through a pore at the tip. Eventually several leaves appear. They are numbered in order of appearance to define the crop's developmental stage, which is used as a guide in timing applications of fertilizers and sprays of pesticides or selective herbicides. In older plants it can be seen that each leaf arises at a swollen node or 'joint' and consists of a sheath, rolled round the stem, and a blade which extends away from it. Gradually the seminal roots extend into the soil and branch, and secondary 'coronal' or 'adventitious' roots form, and grow from the base of the plants. Roots penetrate the soil to a very varied extent, depending on soil conditions, sometimes reaching a depth of about 1·8 m (6 ft). Several stems, or 'tillers', may form at the base of the

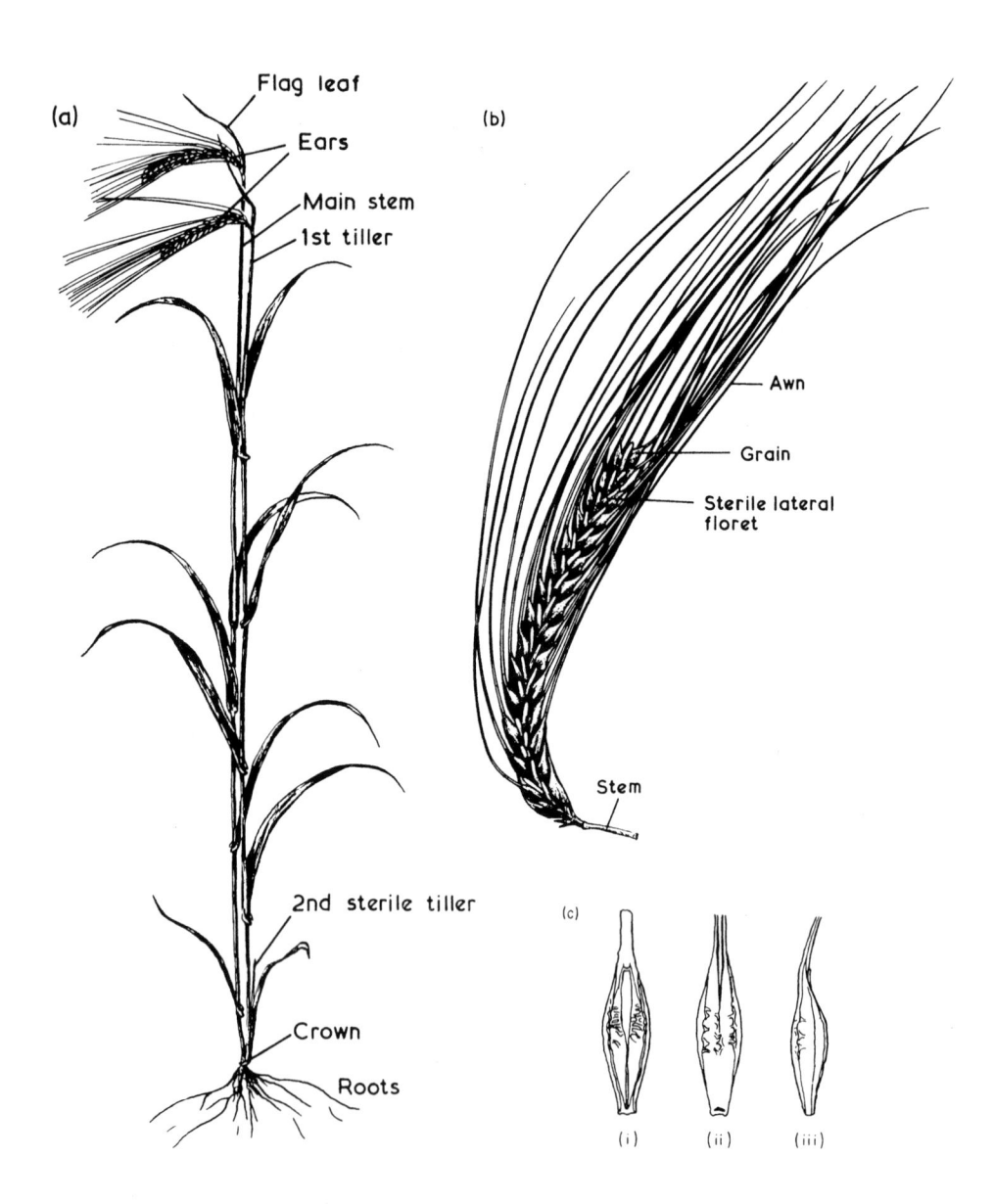

Fig. 2.1 (*a*) A fully grown plant of a two-rowed barley, Irish Archer I, with two fertile tillers. After Hunter (1952) [17]. (*b*) A more detailed view of a two-rowed ear of barley. (*c*) Three views of a barley grain, with the base of the awn in place; (*i*) the ventral furrowed side; (*ii*) the rounded dorsal side, and (*iii*) a lateral view.

main stem. Initially the plant forms a rosette close to the ground, but just before the onset of flowering the stems elongate and the heads or ears (inflorescences) appear at the apices from within the upmost leaf sheaths (Fig. 2.1). Older barleys often reached 150 cm (approx. 59 in) in height, but now shorter varieties of 60–90 cm (approx. 24–36 in) are more common. In autumn-sown or 'winter' barleys, growth is checked at the rosette stage by the winter conditions. In spring-sown or 'spring' barleys, growth is continuous to maturity. The inflorescence or ear has an axis, the rachis, which is a modified extension of the stem having short internodes. A group of three simple flowers, a triad of spikelets, occurs at each node, and these alternate along the rachis. The rachis must be tough to prevent ear shattering and grain loss during harvesting. Where all the spikelets are fertile, six rows of grain are formed on each ear. The central spikelets produce symmetrical grains, and the lateral spikelets produce grains skewed in two senses. When the lateral spikelets are sterile only the central florets of the triads form grains, and the ears are two-rowed (Figs. 2.1 and 2.2). When the rachis internodes are short the grains are pushed together and the ear is 'dense'. When the internodes are longer the grains are less tightly packed along the ear, which is lax. When six-rowed barleys have lax ears the rows of lateral florets may overlap, giving rise to so-called 'four-rowed' barleys (Fig. 2.2). The traditional malting barleys of Europe are two-rowed and have plump symmetrical grains. Six-rowed barleys are more widely grown, worldwide, and some varieties of them are malted.

At flowering each spikelet consists of two small sterile glumes and two conspicuous green bracts (flowering glumes) coming from an inconspicuous pedicel on the rachis. The pedicel is extended as the rachilla or 'basal bristle' which occurs in the furrow of the mature grain. The conspicuous bracts are (*i*) the *palea inferior*, or lemma, which is furthest from the rachis and is normally extended into a long thin awn (or rarely an appendage called a 'hood') and (*ii*) the *palea superior* or palea which is nearest to the rachis. The awns, when they are present, give barley ears their 'beard' or brush-like appearance (Fig. 2.1b). The palea and lemma come to form the husk of the grain on the ventral (furrowed) and dorsal sides respectively. The flowering glumes enclose a simple unilocular ovary equipped with two feathery stigmas, three anthers and two small hairy scales – the lodicules.

The barley inflorescence is adapted for wind pollination but usually flowers are more than 99% self-pollinated. Two fertilizations occur. The egg, fertilized by one pollen nucleus, grows to form the embryo plant. A triploid nucleus, resulting from a fusion between a second male pollen nucleus and two maternal nuclei, multiplies and finally gives rise to the starchy endosperm and the aleurone layer. These tissues grow and crush the nucellus almost beyond recognition, and they become invested in the testa (seed coat;

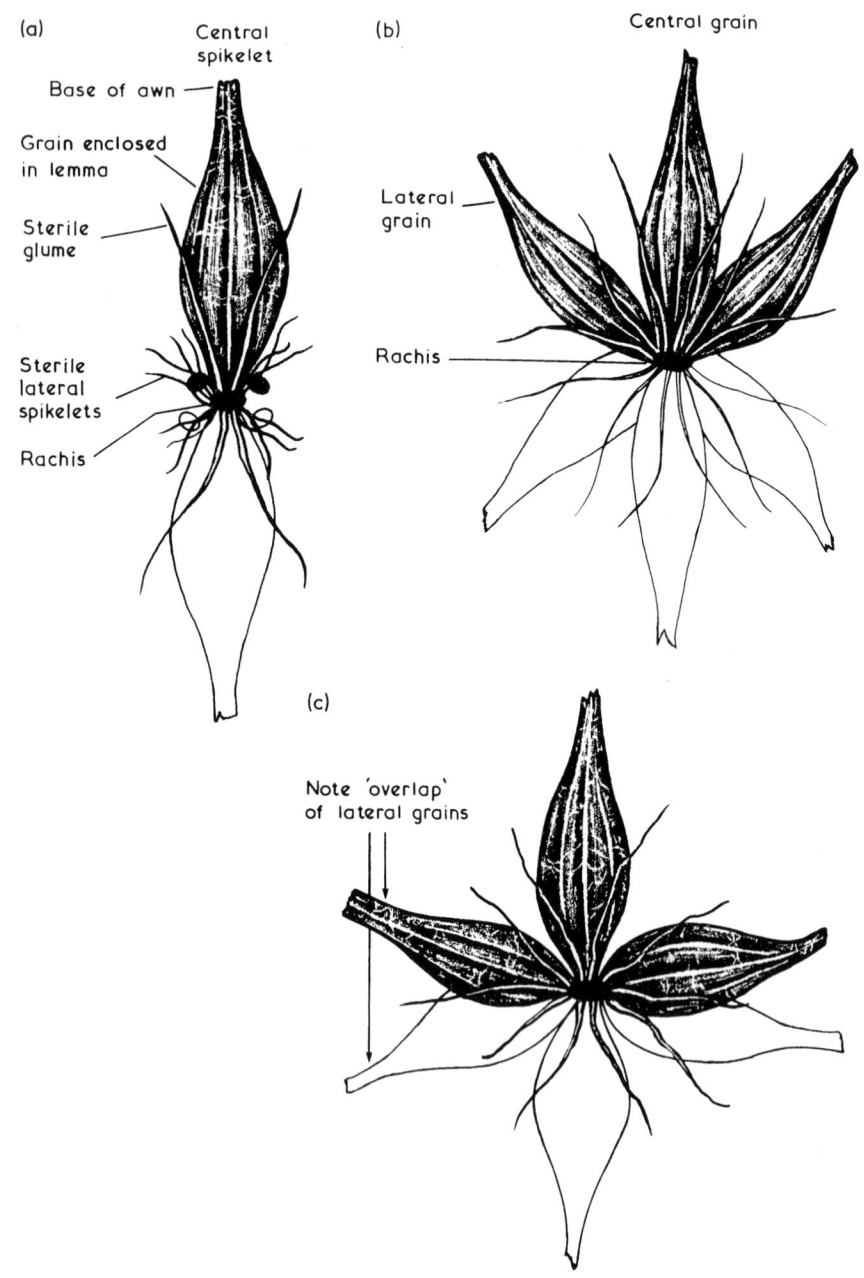

Fig. 2.2 The general arrangement of grains in barley ears, viewed from below. The grains shown in most detail arise at one node; the grains shown in outline arise at the node above. (*a*) Two-rowed barley. (*b*) Six-rowed barley. (*c*) Lax-eared six-rowed barley, a so-called 'four-rowed' barley.

tegmen). The testa, which is derived from the inner integument of the ovary and the integument of the nucellus, becomes fused to the ovary wall, the crushed remains of which form the pericarp (the fruit coat), which in turn usually fuses to the palea and lemma, which together form the husk. In some varieties the husk does not adhere to the grain, which is said to be naked (like wheat). Naked barleys appear unsuitable for malting, but in some areas, e.g. Japan, they are used as human food.

2.1.2 THE BARLEY CORN [1, 3, 4]

Because the seed coat is fused to the pericarp, the mature barley corn is an indehiscent fruit, a caryopsis. The grain is elongated, slightly compressed, but approximately cigar-shaped (Figs. 2.3 and 2.4). The proximal, basal or

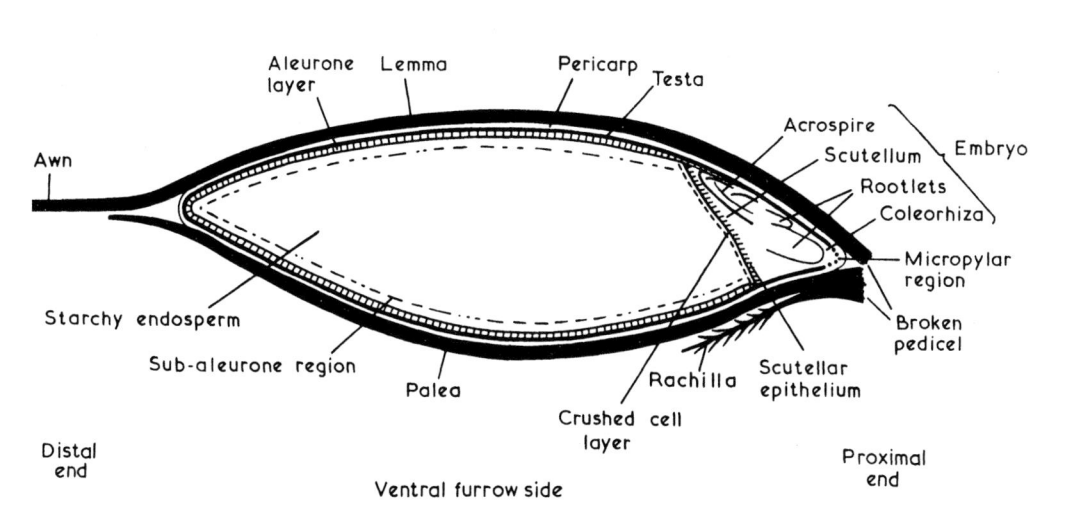

Fig. 2.3 Schematic longitudinal section of a barley grain, taken to one side of the furrow, showing the disposition of the parts.

embryo-containing end of the grain is attached to the rachis. In threshed grain this attachment is usually broken, as is the base of the awn at the apical or distal end of each corn. The small embryo is disposed towards the dorsal or rounded side of the grain, the two dried lodicules fitting between it and the husk. The ventral side of the grain has a furrow, or crease, running along its length. The testa and pigment strand, which bridges the gap between the edges of the testa, are largely or wholly impermeable to water. Water gains entry at the base of the grain, near the tip of the root sheath, probably via the micropyle, which traverses the testa.

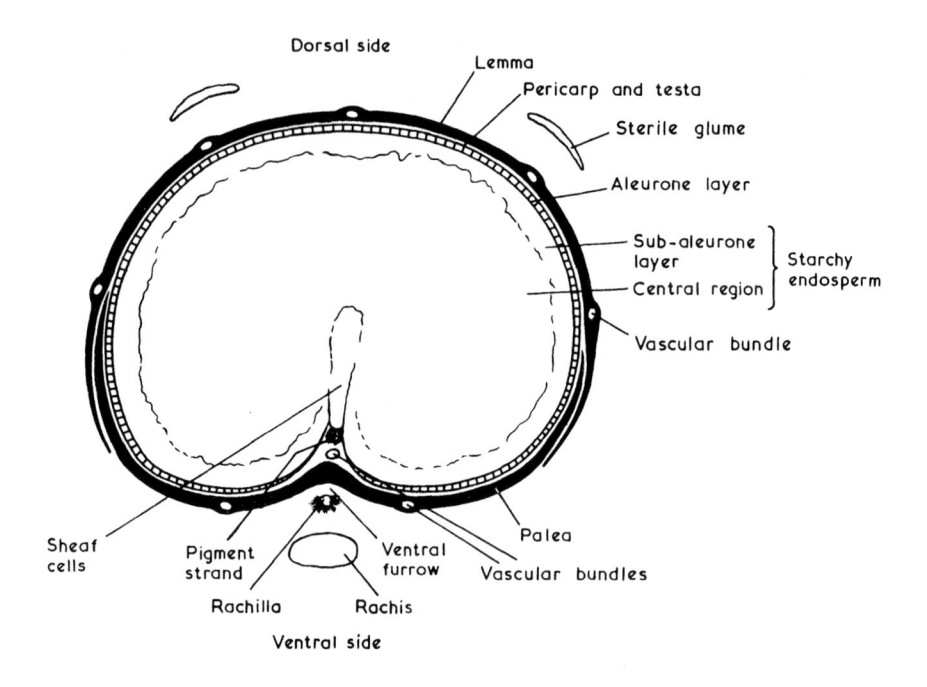

Fig. 2.4 Diagram of a transverse section through the broadest part of a plump barley grain.

The main tissue of the grain, and its main foodstore, is the dead and non-respiring starchy endosperm. This tissue is made of thin-walled cells, packed with starch grains embedded in a protein matrix. In mature grains all traces of the cell nuclei, and many other organelles, have disappeared. This tissue is not homogeneous, for example the protein and β-amylase contents are highest in the sub-aleurone layer, where the cells are comparatively small, and average size of starch grains is smallest. In transverse sections of the endosperm the cell walls radiate from a vertical crest of starch-free 'sheaf cells' which run the length of the endosperm above the furrow and are probably residual nucellar tissue. In section the starchy endosperm of good malting barley appears white and opaque, 'floury' or 'mealy' owing to numerous small air-filled cracks in the materials around the starch grains. In contrast, poor quality grain, often relatively rich in protein, has a greyish and translucent, 'vitreous' 'glassy' or 'steely' appearance. Neither is the starchy endosperm homogeneous away from the furrow. Adjacent to the embryo, abutting the scutellum, the starchy endosperm cells of the crushed cell layer are devoid of starch, or indeed any obvious cellular contents, and the layer seems to consist of cell walls closely pressed together. The aleurone layer, usually about three cells thick, invests the starchy endosperm, except at the furrow and at the interface with the scutellum, and as a single layer of reduced cells overlaps the edges of the embryo. The roughly cubical cells have thick cell walls, they respire, and have prominent nuclei and cytoplasm densely packed with organelles, oil drops and complex aleurone grains containing reserve proteins, minerals and possibly carbohydrate. Starch is absent. The aleurone layer plays a major role in producing hydrolytic enzymes, which are essential during malting and subsequently during mashing, as well as being a specialized reserve tissue.

The quiescent embryo contains an axis made up of the initials of the coleoptile, the first leaves and rootlet initials which are surrounded by a coleorhiza or root sheath. The axis is attached to and partly recessed into a flattened expanded embryonic tissue, the scutellum, which itself abuts the starchy endosperm. The scutellum adjacent to the crushed-cell layer of the endosperm is faced with a specialized columnar epithelium, one cell thick. The scutellum has absorbtive and secretory functions, and contains some reserve substances.

The husk, lodicules and pericarp ensure that water is rapidly distributed over the surface of the grain by capillarity, so that the embryo easily becomes hydrated and may germinate. The husk and pericarp provide mechanical protection for the grain. They almost invariably have a large population of micro-organisms which particularly occur with dust on the grain's surface, between the husk and pericarp, and within the pericarp itself. Micro-organisms apparently do not occur within the testae of healthy undamaged grains. The testa prevents the outward diffusion of sugars, amino acids and other low-molecular-weight nutrients from the interior of the grain. Damage of any sort, breakage, chipping or bruising which readily occurs, during harvesting or handling, may reduce the viability of the grain, or otherwise reduce its suitability for malting.

2.2 *Barley cultivation and harvesting* [3, 5, 6]

About 20% of the British barley crop is used in malting. The remainder is mainly utilized in animal feedstuffs, in which a high 'crude protein' content is desirable. This requirement, combined with the need to use generous quantities of nitrogenous fertilizers to obtain high yields, has resulted in a rise in average nitrogen contents of grain used in malting. Maltsters prefer 'low-nitrogen' grain and pay more for it, but the farmers are unwilling to accept the loss in yields associated with deliberately producing 'low-nitrogen' material, since a crop intended for malting may be found not to have reached malting quality when harvested. Thus maltsters must often buy the best samples from a 'general purpose' crop.

Barley for malting should be grown from pure seed of a variety that maltsters have found to be suitable, and which meets the farmer's agronomic requirements, such as being high yielding and disease-resistant. Modern varieties often have short and stiff straw, are resistant to lodging and ear shedding or shattering, and are suitable for mechanical harvesting. Seed grain may be dressed, e.g. with organic mercurial or other compounds, to make it resistant to soil-born pests and diseases, or with systemic fungicides to make the seedlings resistant to mildew, for example. In most countries lists of recommended varieties are available [e.g. 7]. In Britain the best current varieties have much greater yields per unit field area and produce grain which malts easily to give malt with 3–5% more extract (i.e. the material that brewers need – Chapter 9) than varieties grown before 1939, but having similar nitrogen contents. This may be partly due to grain of newer varieties having lower percentages of husk. Grain for malting should be harvested 'dead-ripe', in dry sunny weather. By sowing varieties that ripen at different times, or by planting fields of both winter barleys and spring barleys (which usually ripen later) it is possible to stagger the harvest. Traditionally British malting barleys are spring-sown, but malting-quality winter barleys are becoming more common (e.g. Maris Otter). Also, within limits and if excessive winter-killing of the seedlings does not occur, the earlier the seed is sown the lower the grain nitrogen, and the higher the grain yield, the thousand-corn weight, and the potential for making high-extract malt (Table 2.1). The longer growing period allows a maximal accumulation of starch in the grains, which in turn gives rise to most of the brewer's extract. Thus seed is sown as early as weather permits, and indeed 'alternative varieties', which are not fully winter-hardy and which do not need an extended cold period for vernalization, may be spring- or autumn-sown. However, in severe winters the proportion of plants killed may be very large or, rarely in the U.K., complete.

Seed rate (grain number/unit area) is critical. If seed is sown too thinly, ground is wasted, the barley plants tiller profusely and may produce some ears which are unripe at harvest. Furthermore weeds develop with little initial competition. If sown too thickly, expensive seed is wasted, the barley stands are too dense, the grain yield is not enhanced, and the plants are more prone to fungal diseases. Usually seed is planted at a depth of about 4 cm (1½ in). The soil of a traditional seed bed should have a good tilth and be damp, but never waterlogged, to ensure even germination and steady growth. Recently 'direct drilling' into the unploughed stubble of previous crops, in which chemical sprays have been used to control weeds, has become an established practice in some areas. For malting barley, the soil should be light and have a neutral, or slightly alkaline reaction. For example, shallow soil with a chalky subsoil is suitable. For many years crops have been rotated to 'rest' the soil and help prevent rises in populations of soil-born pests and

TABLE 2.1
The effect of variety and time of sowing on the yield, thousand corn weight and nitrogen content of the harvested barley
From BELL and KIRBY (1966) [19]

Varieties Date of sowing	Carlsberg	Proctor	Provost	Spratt-Archer
		Yield of grain (kg/ha)*		
16.3.55	4374	4161	3746	3658
5.4.55	3620	3482	3256	3143
27.4.55	1911	2124	1571	1094
		Nitrogen content of grain (%)		
16.3.55	1·46	1·64	1·83	1·63
5.4.55	1·61	1·68	1·78	1·69
27.4.55	1·98	2·26	2·08	2·25
		Thousand corn weight (g)		
16.3.55	42·5	35·7	31·9	38·2
5.4.55	37·7	32·9	31·4	34·6
27.4.55	35·9	32·5	34·6	33·3

* To convert kg/ha yield to cwt/acre, divide by 125·7.

cereal weeds. In continuous cereal cropping grain quality and yields tend to be reduced, but the practice can be profitable and is now widespread following the development of effective methods for controlling most weeds, pests and diseases. Nitrogenous fertilizer, applied at too high a dose rate or late in the growing season, produces high nitrogen grain. By drilling some fertilizer with the seed, and applying a carefully chosen top dressing at about the three-leaf stage, near-maximal yields of malting-quality grain can be obtained. The fertilizer used usually contains nitrogen, phosphate and potassium and supplementary materials, e.g. trace elements, to complement the supplies available from the soil. Nitrogenous fertilizers usually have the most dramatic effects on yield.

Barley yields are usually best when a cool damp growing season, which maintains steady growth, is followed by fine dry weather at harvest, to ripen and dry the grain, facilitate harvesting and reduce subsequent problems caused by excessive dormancy, or the occurrence of split or pregerminated grains. Very hot dry summers with associated moisture lack and induced premature grain ripening give rise to thin, incompletely filled grain, with a high nitrogen content. Weeds compete with barley, reducing the efficiency of harvesters and grain yield. In addition, weed seeds in the grain reduce its value. Weeds are controlled by crop rotations, good cultivation techniques and applications of selective herbicides. Barley is attacked by many pests (including nematodes and insects) and diseases, particularly those caused by fungi, which can drastically reduce yields. For choice weed-seed-free and disease-free seed grain of disease-resistant varieties is dressed and sown on cleaned land. The growing crop is regularly inspected and may be sprayed to combat incipient outbreaks of pests or diseases. Sometimes systemic fungicides are applied almost as a routine. Some diseases, e.g. mildew, caused by *Erysiphe graminis* or eyespot (*Cercosporella herpotrichoides*), may attack the ear itself, or the whole plant, so reducing the flow of nutrients to the grain and reducing the yield. Others, e.g. take-all (*Ophiobolus graminis*), may kill the whole plant.

2.2.1 HARVESTING [3]
In the past, near-ripe barley plants were cut and bound into sheaves. The sheaves were arranged as stooks in the fields and were allowed to dry before being built into stacks. Threshing occurred later, at the farmer's convenience. This system sometimes gave rise to microbial attacks, giving 'mow-burnt' or slimy grains – conditions that are now almost unknown. Now almost universally harvesting and threshing are carried out using combine harvesters. The dead-ripe plants are cut, the grain is threshed, separated and retained while the straw is discarded, to be burnt or baled and collected later. The slightest fault in adjusting the threshing mechanism may give rise to incomplete threshing and grain losses or to too close threshing and bruised and broken grains. Thus, at harvest time, large quantities of fresh grain (15–25% moisture, fresh wt. basis) come from the fields in a short period, and must be processed and stored in such a way that viability and quality are maintained. Usually in Britain it must be dried and stored cool, usually with ventilation and possibly with treatments to combat insect infestation.

Drying grain with hot air is usual, but care is needed to ensure that viability is not reduced. Farmers store feed grain in other ways, e.g. in hermetically sealed silos, in chilled stores, or after treatment with chemical preservatives, but grain treated in these ways is unsuitable for malting [3].

2.2.2 BARLEY SELECTION AND BREEDING [3, 4, 8, 9]
Programmes of selection and breeding have produced new barley varieties with: (*i*) greater grain yields/hectare; (*ii*) shorter and stiffer straw that is resistant to lodging; (*iii*) ears that do not shatter, so they are more suitable for mechanical harvesting; (*iv*) earlier ripening; (*v*) greater disease-resistance; (*vi*) greater uniformity; (*vii*) grains able to yield more brewers extract after malting. The modern breeder has available many thousands of strains of cultivated and wild barleys from which he can try to derive new desirable varieties. The cultivated barleys and a group of wild barleys (*Hordeum spontaneum;* possibly *Hordeum agriocrithon*) are in fact one species, although misleadingly, strains or cultivars are often given Latin binomial 'specific' names in the 'keys' used in reference classification systems. The cultivated barleys are often grouped together as the Cerealia. Of these four common types called by their common, but incorrect, binomial names are:

(*i*) *Hordeum vulgare* L. Six-rowed barleys with all the grains of approximately equal size, with all the lemmae awned or hooded. Many malting barleys occur in this group outside the UK.

(*ii*) *Hordeum intermedium* Kcke. Six-rowed barleys with small lateral grains, the lemmae of the lateral florets having neither awns nor hoods.

(*iii*) *Hordeum distichon* L. Two-rowed barleys, with the reduced lateral florets having rudimentary non-functional sexual organs. Traditional European malting barleys fall into this group.

(*iv*) *Hordeum deficiens* Steud. Two-rowed barleys with the lateral florets so reduced that they totally lack sexual organs.

Each of these groups is composed of numerous varieties (cultivars) that have names such as Golden Promise, Maris Otter and Ark Royal (UK) or Olli, Klages and Piroline (N. America). Grain of different cultivars may be distinguished by minor morphological characteristics, such as the dimensions and hairiness of the lodicules or the rachilla. However, identification by simple inspection is not easy, and increasingly other techniques such as staining or the patterns given when particular groups of enzymes or other proteins are separated by electrophoresis are being introduced to aid in the identification of threshed grain. Maltsters have strong preferences for particular varieties found by trial to give grain samples of good malting quality. Until about 1800 land races comprised the barley crop. These were mixtures of types grown from 'saved' seed, and known by the name of their districts of origin. Increasingly seed was saved from robust attractive high-yielding plants, the seed was multiplied in isolation to avoid contamination and the performance of the selected strain, grown in bulk, was evaluated. Superior strains were retained. This process was helped by the nearly complete self-pollination of barley. A case in which the performance of one selected Archer type is compared with another Danish selection is shown in Table 2.2.

TABLE 2.2
A comparison of two selections of 'Archer' barley
Based on HUNTER (1952) [17]

Barley variety	Yield (kg/ha)	Yield (cwt/acre)	Total nitrogen content (% dry wt.)	Thousand corn weight (g)
Irish Archer	2950	23·5	1·51	37·9
Tystofte Prentice (selection)	3155	25·1	1·47	39·5

Selections from Spratt-Archer hybrids gave some undesirable late-ripening and coarse-grained types, as well as superior lines with useful characteristics, such as Earl, which ripened 7–10 days earlier than the parents. Barley varieties are not genetically 'pure', and in the past they have been notably variable. Grain with one name is not identical at all times, and in all places. To combat increasing variability reselection is practised, to recover the original 'type'. The barleys favoured by maltsters alter with time. In Britain Spratt-Archer and Plumage-Archer, typical of pre-1939 varieties, were replaced chiefly by Proctor in the early 1950s. At present, a wide range of malting barleys are grown, for example Imber and the winter variety Maris Otter.

Selection is indispensable for plant improvement, but it does not allow the deliberate recombination of characters from different strains. This is achieved by hybridization, artificial cross-pollination between strains. Desirable types are sought in the descendants derived from the cross, as genetic segregation allows the appearance of recombinants in subsequent generations which are allowed to self-pollinate. New lines are accepted when trials have shown them to be superior to current varieties in terms of yield, disease-resistance, superior agronomic characteristics, or grain quality. The malting quality of a new line is assessed by chemical analyses, micro-malting trials, and finally full-scale malting trials. The barley should germinate and malt ('modify') readily, and give a malt in good yield. The malt must give a brewers extract of good quality and in good yield (Chapters 5 and 9). In Europe national and international barley trials are carried out annually under the auspices of national bodies and the European Brewery Convention (EBC). The publication *Barley Varieties EBC* gives definitive descriptions of European malting varieties [9]. In England the National Institute of Agricultural Botany (the NIAB) publishes a 'List of Recommended Varieties' each year [7].

2.2.3 THE EVALUATION AND PURCHASE OF MALTING BARLEY [3, 10–15]

In Britain barley is grown under widely varying conditions, and the average quality of grain varies widely. Also the grain is offered for sale in small quantities, relative to maltsters total requirements. Small samples taken from a bulk (e.g. 100–500 g) of grain are shown, and the maltsters choose which to buy after inspection. To be truly representative the sample must have been taken from the bulk in a statistically adequate manner. Grain is purchased on the understanding that the bulk delivered to the maltings matches the sample, and is of equal or superior quality to it. If it is not, the grain may be accepted at a lower price, or the delivery may be rejected.

Samples should be withdrawn from many widely spread locations and different depths from a heterogeneous bulk of grain, for example with a sampling spear. These samples must be combined, well-mixed and divided through a mechanical sample-divider. Subsamples should weigh 0·5–1 kg and be stored in well-labelled, water-proof and air-tight containers, which prevent the gain of moisture from or loss to the air. Barley may be sold with 9–25% moisture, which can be determined on a sample. Buying water is expensive and wet grain must be dried, at extra expense, before it can be stored safely. Generally a lower price is given if the grain moisture exceeds 16%. In market conditions the moisture content may be determined approximately and rapidly by one of several methods using, for example, electrical meters, infra-red driers, or rapid-moisture ovens. It is likely that infra-red reflectance meters will come to be used on ground samples of grain. However, the primary method, against which the others are calibrated, is a slow oven-drying technique [12–14].

An experienced buyer, who has become familiar with the season's grain and the varieties for sale, can learn much from 'hand evaluation'. However, as the appearance of samples of one variety differs with the season, and many new varieties are being introduced and because not all forms of grain deterioration can be detected by inspection, hand evaluation alone is insufficient. Thus of the grain selected by hand as of top quality at one British site in 1956–1957 only 33% malted well and no less than 39% was finally rejected [16]. Grain should be of even appearance; the size (width) assortment may be determined using sets of mechanically shaken slotted sieves. A sample should contain few damaged, cracked or half grains, or grains that have split along the ventral furrow. Signs of germination in the ear (pregermination), such as the coleoptile having started to grow up under the husk, should be absent. Grain should have a delicately wrinkled husk of fine appearance, and not be stained or weathered. Discoloration is often the sign of a fungal attack. Grain must smell and taste 'clean', having no trace of bitterness, sweetness or mustiness or other character that indicates heavy microbial contamination. A sample should contain not more than small quantities of contaminants – weed seeds, other cereal grains, stones, nails, straw or other trash.

If the outward appearance of the grain is satisfactory, the maltster may cut a number of grains (often 50) transversely in half, using a corn-cutter or farinator. The proportions of grains having hard, steely endosperms and those having the desirable 'mealy' appearance are noted. Steely grains are often higher in nitrogen, and malt less readily than their mealy counterparts. Sometimes a particular thousand corn dry weight (TCW, g) is preferred. However, even samples of 5000 grains (a number too large to count easily) are statistically likely to give estimates of TCW that differ by up to 4% from the true average of the batch. Also as the exact moisture content of the grain is not known at the time of purchase, the TCW is of limited value. In some instances bulk density of grain (kg/hl; lb/bushel) may be determined. Again this value varies with moisture content.

Grain is purchased subject to (*i*) the bulk matching the sample, (*ii*) the total nitrogen content [or percentage crude protein = N(%) × 6·25], and (*iii*) the viability (%) proving satisfactory. Rapid methods for measuring the nitrogen content of grain, such as those based on the binding of dyes to the basic residues in the proteins present in finely ground grain samples, give adequate results when calibrated against samples of known composition, and of the same grain type. Possibly infra-red reflectance spectrometers will come to be widely used to determine rapidly the moisture, lipid, nitrogen and carbohydrate (even specifically starch and β-glucan) contents of grains ground to flours. However, the standard technique for determining grain nitrogen is the Kjeldahl digestion method, and this is used to check the results of other rapid analytical techniques. The nitrogen ranges of barleys accepted for making malts have risen markedly in the UK in recent years. At present (1979) the generally accepted upper limits are probably: pale ale, 1·65% of N; mild ale, 1·75% of N; lager malt, 1·9% of N; distillery malts, up to 2·5% of N. Fewer grades of malt are made than was formerly the case, and barley for traditional pale ale malt (N 1·3–1·4%) is very rare.

In contrast, in N. America some six-rowed barleys (N 2·0–2·2%) are made into malts with very high enzyme contents. However, malts made from two-rowed barleys seem to be increasingly common. The highly enzymic six-rowed malts are used to convert large proportions of unmalted adjuncts in the mashing process.

At least 95% (preferably 99–100%) of grains in a batch intended for malting should be capable of germinating. Germination tests usually take 3 days, a long period, and do not give a good estimate of viability when grain is dormant. Thus grain is purchased 'subject to adequate germination', and viability is estimated indirectly by scoring the ability of the embryos of longitudinally split grains to reduce solutions of a colourless tetrazolium salt to an insoluble red formazan dye. This red staining, which also occurs in the aleurone layer, depends on the living tissue containing (*i*) functional reducing enzymes, and (*ii*) substrates to act as hydrogen donors. It is considered safe to reject grain when this test gives a poor result. However, it can give misleading results with heat-damaged grain when, for example, 95% of the embryos stain but only 40–50% of the grains germinate.

2.2.4 BARLEY RECEPTION [3, 15]

The deliveries of grain to a reception area are programmed to minimize delays in unloading, to save the time of the delivery vehicles and their drivers and to create a steady workload for the staff. It is best if grain arrives in bulk, e.g. in railway wagons or lorries, that can be emptied by suction tubes, or by tipping or from hopper bottoms into gratings set at ground level. There is no uniformity in the way grain is transported. It may even be moved in sacks. On arrival, and before unloading, samples are taken from the grain at various places and from different depths, the samples are mixed, a subsample from the mixture is compared with the purchase sample, paying particular attention to the appearance, and the moisture and nitrogen contents. If it is substandard, the bulk may be rejected, or may be accepted at a reduced price. The barley is weighed into the 'green grain' store, being bulked with other loads of similar character. Depending on its moisture content, fresh grain may be dried immediately, or may be held in well-ventilated bins until the drier is free to process it. Grain may be stored on farms or at merchant's premises and delivery taken at the maltings at a later time, so spreading the workload and reducing the required grain-storage capacity. However, many maltsters prefer to receive grain as soon as possible after harvest, then to dry, clean and store it themselves to ensure that its quality is maintained. Initially grain may only be roughly 'precleaned' by sieving and aspiration, to remove dust and gross impurities before it is dried and sent to store. More complete cleaning and grading, comparatively slow processes, are deferred to a convenient time (see Section 2.3.3).

2.3 *Drying and storage* [3, 15]

Wet barley respires more rapidly than dry barley, using oxygen and producing heat, water and carbon dioxide. The rate of respiration increases with increasing moisture content, and goes up markedly more rapidly with moisture contents in excess of 13%. Provided that insect pests are controlled, barley may be stored in bulk without loss of viability when the moisture content is around 13–15%, if it is ventilated to dissipate heat and prevent suffocation. However, grain with a moisture level of 10–12%, held at 15°C (59°F) or less, may be stored for about a year without ventilation or cooling, since its respiration is negligible. Thus the cost of storage is reduced. However, it may be desirable to move the grain from silo to silo ('turn it over'), so it may be inspected, to mix it, ventilate it and to dissipate any heat. This creates dust from grain abrasion and is costly. Forced ventilation may be achieved by using fans to drive or draw air through the grain, often by a system of perforated ducts.

Moist grain is susceptible to attack by many insect pests and spoilage fungi, particularly at elevated temperatures (Fig. 2.5). Foci of insect infestations may give rise to 'hot-spots' in which the grain respires faster than the bulk, the temperature rises, the moisture content of the grain is increased by its own metabolism and that of the insects, so enhancing respiration further and accelerating the processes of deterioration. Hot damp air rises from the 'hot spot' through the grain and moisture condenses in, and is taken up by the cooler grain above. These conditions favour the rapid upward spread of

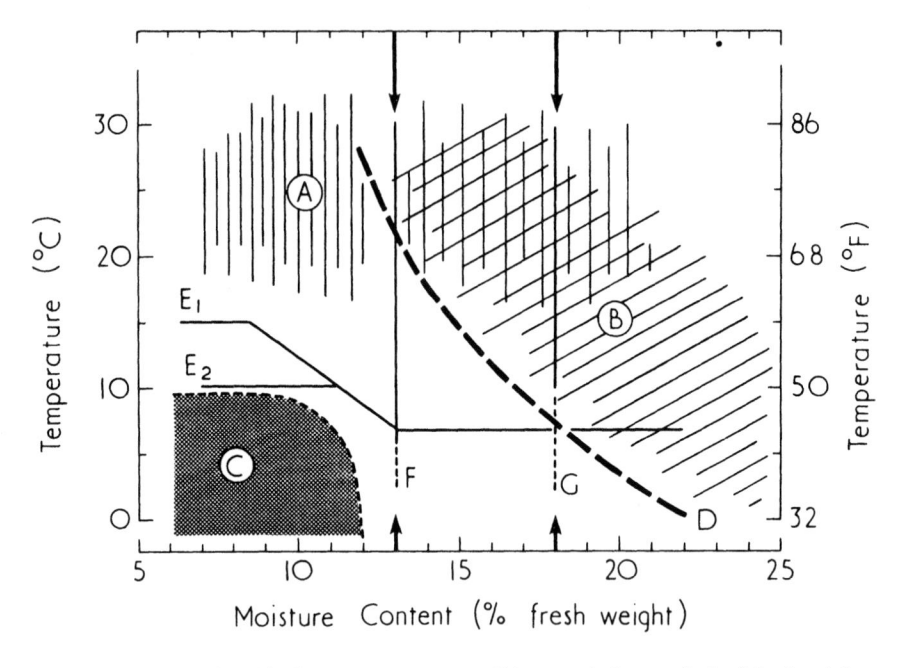

Fig. 2.5 Some relationship between storage conditions and the survival of barley (after Briggs [3], based on various sources). Region A, with vertical shading; conditions under which insect heating is likely to occur. E, line below which most insects (E$_1$) and moths (E$_2$) do not multiply. Line D (- - - -), conditions under which germination is likely to fall by 5% in 35 weeks, a rapid rate of deterioration. Region B, defined by inclined shading and broken line, conditions under which microbial heating of grain is likely to occur. G, line at 18% m.c. (moisture content) below which bacterial multiplication on the grain ceases. F, line at 13% m.c. below which fungi neither grow on, nor damage grains. Region C, generally 'safe' storage conditions. It will be seen that barley stored at 12% m.c. and 15°C (59°F) is not completely 'safe', particularly if insecticides are not used.

heating, and insect and fungal activity, and may lead to the quick spoilage of a large amount of grain. Thus in stores the grain temperature should be ascertained frequently (e.g. weekly) and, since bulk grain is a good insulator, at many locations in a large mass. Any appreciable rise in temperature over the previous values may indicate the onset of deterioration. Suspect grain should be inspected at once and, if necessary, remedial actions should be taken. Moist grain may be chilled in a stream of refrigerated air [e.g. to 4°C (39°F)] and held in store, in a viable state, for extended periods. However, mite infestations may occur, the barley does not 'after-ripen' (lose dormancy), and therefore this method has not found favour with maltsters. The preferred practice with grain intended for malting is to dry it carefully to 11–12% moisture, store it warm at 25–30°C (77–86°F) for 7–14 days to reduce dormancy, then to cool it to less than 17°C (63°F) to minimize chances of deterioration before running it to store. This final cooling can be achieved conveniently when the grain is subjected to a sequence of cleaning and grading operations. Cooling grain to low temperatures immediately after drying can result in it retaining its dormancy, to the maltsters disadvantage. The grain may be treated with an approved insecticide as it is conveyed to store.

The physics involved in barley drying is considered in more detail in connection with malt kilning (Chapter 6). Warm dry air, having a water vapour pressure below the equilibrium water vapour pressure of the grain, is forced through a bed of grain, and removes water from it. The moisture seems to leave the grain via the micropyle, since during drying the germ loses water faster than the rest of the grain, the converse of what happens in steeping. The equilibrium relative humidity of grain increases with grain moisture content (Chapter 6). On British farms many types of driers are used, but in maltings grain is commonly dried on malt kilns, in 'germination-kilning plants', or in continuous-flow tower driers. In these, grain is fed into the top of a tower. As it descends at a controlled rate it is successively warmed (by radiators or hot damp air), then it meets a current of warm dry air which removes water, and finally it is cooled in a current of air at ambient temperature.

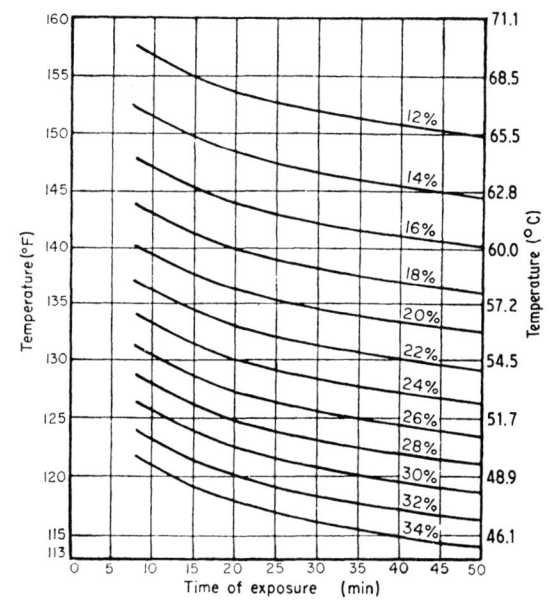

Fig. 2.6 Curves showing combinations of temperatures and times of exposure and grain moisture contents that just cause a detectable delay in germination in barley. After Oxley (1959) [18].

Water is removed faster at higher temperatures, but moist grain is more readily damaged by heat than is dry grain (Fig. 2.6). Adequate safety margins have to be set, since a bulk of grain may be inhomogeneous with regard to moisture content, and machines may not heat all parts of a grain stream equally. Grain may take 2 hr to pass through a tower drier, during which time the 'air-on' temperatures that it encounters may rise from 54°C (about 130°F) to 66°C (about 150°F). The air is cooled by evaporation from the grain, so that the 'air-off' temperatures may be 43°C (110°F) and 52°C (125°F) respectively. In the cooling section the ambient air may reduce the grain temperature to about 30°C (86°F) or less if it is not to be stored warm initially. Grain cooling may be achieved in a special 'flow-through' heat-exchanger. More usually grain is cooled by the aspirating air used to remove dust during the cleaning and grading procedures. In store the grain bulk may be cooled by the fan-driven passage of cold dry ambient air (as occurs on frosty nights). This system is feasible in the colder drier eastern parts of Britain. The principle might be extended to milder more humid areas if air were refrigerated, and possibly predried chemically. Systems for drying grain with cool air streams dried with silica gel, or under partial vacuum, have been tested, but are not in use in the UK.

2.3.1 BARLEY STORES [3, 15]

Malting barley is usually stored in bulks, made up of admixtures of various deliveries. The different lots are usually separated on the basis of age, germination rate, barley variety, nitrogen content, water content and grade. Grain stores range in complexity from heaps of sacks or loose grain held in any odd corner of dry buildings to large purpose-built stores with built-in grain-handling machinery. Perhaps the most convenient stores in which grain movement can be controlled from a central console are groups of concrete bins or silos, with self-emptying hopper bottoms, associated grain-handling and cleaning equipment, and with electrical temperature probes positioned within the storage space, and having facilities for forced aeration. Such stores (*i*) do not readily become infected with insects, rats and mice, (*ii*) can easily be cleaned and fumigated, (*iii*) take up only a limited ground space, and (*iv*) readily allow the storage of a range of grain types. Large 'flat-bed' stores are cheaper to construct, but occupy a large ground area, and are unsuitable for holding a large number of separate grain classes. In these stores the grain is heaped on large flat concrete floors, criss-crossed with ventilation ducts. Inner walls withstand the side thrust of the grain, which is protected by a comparatively inexpensive, but fully weatherproof building. Partitions may be used to keep different grain-lots apart. In some of these stores the bulk of grain (e.g. 30 000 tons) is so large that the surface to volume ratio of the grain heap is small, and the mass of grain 'conditions' the atmosphere above it, so that the grain does not pick up significant quantities of moisture and deteriorate.

2.3.2 PESTS IN STORED BARLEY AND MALT [3]

Old grain and malt stores are usually infested with rats and mice. In properly designed and managed stores rodents never become established. Systematic poisoning with 'warfarin' (a substituted coumarin that causes fatal haemorrhages) or related compounds has been a routine and comparatively safe and effective method of control. However, warfarin-resistant strains now occur, so that it is necessary to use much more dangerous poisons for rodent control. Cats, living half-wild in maltings, are the traditional but ineffective regulators of rodent numbers.

Even healthy grain carries a population of micro-organisms, including fungi, yeasts and bacteria. These occur in the dust, on the grains surface, and between the husk and pericarp and within the pericarp structures. During storage the population changes in composition, 'field fungi' declining and 'storage fungi' increasing in numbers. The size and nature of the population of microbes depends strongly on storage conditions. Harvest- or insect-damaged grains are more readily harmed by microbes. If the grain is stored cool ($\leqslant 15°C$; $59°F$) and dry ($\leqslant 13\%$ moisture), microbes cannot harm grain (Fig. 2.5). However, grain stored under moist and warm conditions may be damaged and killed by microbes, which may also (*i*) produce mycotoxins, (*ii*) interfere with the malting process, (*iii*) give rise to malt with off-flavours, and (*iv*) cause 'gushing' in beer. Damaged barley releases large quantities of soluble nutrients into the steep liquor, allowing micro-organisms to thrive. The steep liquor becomes foul and must be changed more often than is necessary with undamaged grain to keep the grain viable. Thus the volume and 'foulness' (biological oxygen demand; BOD) of the steep liquor is increased, with a consequent increase in the cost of water supply and of effluent treatment.

Insects are the most troublesome pests in maltings. Some species are able to survive and multiply in grain (and malt) of widely differing moisture contents, over a range of temperatures. Of the many insects that occur in maltings in the UK, about 30 species are pests. Common serious pests in maltings are the saw-toothed grain beetle (*Oryzaephilus surinamensis*), the grain weevil (*Sitophilus granarius*) and the rust-red or flat grain beetle (*Cryptolestes ferrugineus*). In addition to the loss in grain weight consumed by mature insects or their larvae, much other grain is damaged, and the percentage germination may be seriously reduced by those insects which selectively eat the embryos. In malt stores, where the relative humidity is very low and insecticides may not be used, the saw-toothed grain beetle, and the Khapra beetle (*Trogoderma granarium*), the rust-red flour beetle (*Tribolium castaneum*) and the confused flour beetle (*Tribolium confusum*) may occur. The Khapra beetle is particularly troublesome, as it can breed in malt containing as little as 2% moisture. Some insects, such as mealworms, may live in damp ('slack') grain dust, while fruit flies may occur in the damp germination compartments. The caterpillars of some moths spin webs which bind grains together, and may prevent a bulk of grain flowing. Customers frequently refuse delivery of insect-infested grain or grain products.

The best safeguards against pest infestations are properly designed stores, holding the grain under optimal storage conditions, exclusion of insect-contaminated grain, sacks, farmfeed or vehicles from the neighbourhood of the maltings, vigorous applications of 'good housekeeping' rules, and the use of approved poisons. Stores and conveyor systems should be fully emptied after use. All grain spillages and dust should be cleared regularly. When emptied, stores should be cleaned and treated with insecticides or fumigants before being refilled with grain. Fresh grain should never be mixed with old stocks. Stocks should be used in rotation, so that storage times are minimized. Storing grain cool and dry with a 'preventive' dose of a nationally approved insecticide (e.g. malathion, pyrethrins or γ-benzene hexachloride) sprayed on to the grain or mixed in as a powder formulation is usually successful. Most insect populations are checked at 15°C (59°F) or less, and at 13% moisture or less. However, mites may breed in damp grain at 7°C (about 45°F). Grain bulks usually contain some insects, and treatments are designed to prevent their multiplication and spread. When an infestation occurs, this may be met with insecticides or by fumigating the grain (e.g. with hydrocyanic acid gas, methyl bromide or chlorinated hydrocarbons such as ethylene dibromide). These dangerous treatments must be carried out by trained personnel, wearing fully protective clothing.

2.3.3 PRECLEANING, CLEANING AND GRADING GRAIN [3, 15]
When barley arrives at a malting, it is rapidly precleaned to remove gross impurities before it is passed to the drier. After these treatments it may be stored safely. Later the grain will be cleaned more thoroughly, by machinery with a lower rate of working, to remove damaged grains, or impurities of about grain size, and will perhaps be graded into different size-classes (Fig. 2.7). Thorough cleaning is slow, so it is postponed to avoid a 'bottleneck' in the barley intake. This treatment may also be used to cool grain after a period of warm storage.

Precleaning consists of a rough screening (sieving) process combined with, or followed by, aspiration of the grain stream with a flow of air. The grain passes through a coarse sieve (screen; rubble-remover) that retains large impurities such as clods of earth, stones, string, snail-shells and straw. However, it is retained by a fine screen which allows small impurities, such as sand and some weed seeds, to fall through. The different fractions are collected separately. The screens may be in the form of concentric rotating cylinders or may be flat and arranged in sets, which oscillate. The ventilating air stream is

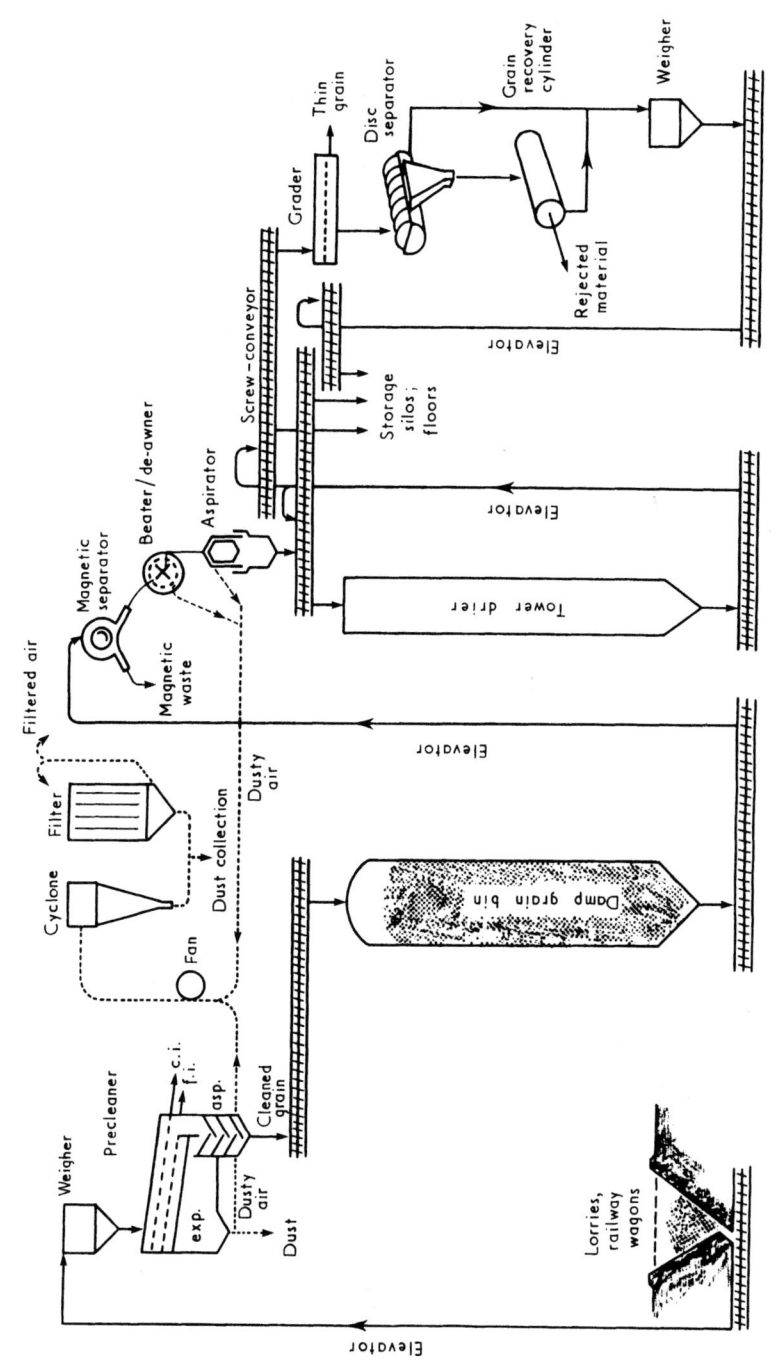

Fig. 2.7 A scheme for grain reception, cleaning, drying and storage. Precleaner: c.i., coarse impurities; f.i., fine impurities; asp., grain aspiration section, with baffles; exp., air-expansion chamber, where dust settling occurs. At present, half-grain separators are little used in the UK.

often drawn across a stream of grain falling between baffles. The air, which is often recirculated, carries away dust and light impurities, which are removed in settling chambers, cyclones or textile-sleeve filters. The dust may be sold for inclusion in cattle-foods. The grain leaving the precleaner is passed through a magnetic separator to remove iron and steel objects and fragments.

Cleaning has to remove impurities of about the same size of barley, such as weed-seeds or grains of other cereals, as well as broken grains or 'half-corns'. This is often combined with treatments to 'grade' the grain, that is, separate it into different classes by width. These machines often have a low rate of working, and their efficiencies fall sharply if they are overloaded. In simple cylindrical separators grain is fed into the upper end of a slightly inclined, slowly rotating metal cylinder, having many sharply cut indentations on the inner face. As the grain stream works its way down the cylinder it is mixed and half-corns and weed-seeds, which are trapped in the indentations, are lifted out of the grain mass and are caught in a catch-trough as they drop from the pockets, as these move into an inverted position. Most of the whole grains, being larger, fall out of the indentations sooner and back into the main grain stream. The 'rejected' material is carried out of the catch-trough by an auger. This stream is often re-treated in a smaller 'recovery' cylinder to reclaim any acceptable grain that has accidentally entered the reject stream. In other more efficient separators the 'lifting' indentations are on both faces of metal discs which rotate in a vertical plane and dip into the grain stream. The liftings are dropped into catch-troughs and are collected separately. Many discs are mounted on one axis, so the indented working surface per unit volume of grain is large, and the rate of grain treatment is high. Again the 'reject' stream may be re-treated to recover useful grain. Reject streams may be graded to recover materials, half grains, wheat grains etc., that are suitable for incorporation into cattle food.

Barley for malting is graded by size (width), because, within one bulk of grain, nitrogen content, rate of water uptake, rate of germination and malt characteristics differ with corn size. Thus optimal malting conditions will differ for each size-class. However, it is expensive to separate barley into many grades and malt each grade separately, according to different schedules, and for moderate-quality malts this is uneconomic. Usually grading now consists of removing 'screenings', 'thins', or 'needles', i.e. grain of less than 2·2 or 2·3 mm width. As much as 20% of the grain may be removed as screenings. The 'bold' or larger grain is malted. The thin grain is used for cattlefeed or, less usually, for making enzyme-rich malt for manufacturing foodstuffs or vinegar.

The sieves (screens) used in grading barley are of two main types. Cylindrical screens rotate and are slightly inclined from the horizontal, so that, as the grain moves through the cylinder, corns narrower than the slots pass through, while wider corns are retained. Often baffles are present to mix the grain and ensure even treatment, while the screen is automatically brushed to dislodge corns wedged in the slots, and dust is removed in an air stream. Such screens are slow working, easily overloaded and are inefficient in that at any time only a fraction of the sieve surface is beneath the grain layer and working. Flat-bed sieves which oscillate horizontally, either to and fro, or round and back around a central axis work more quickly and efficiently. Often their motion has a slight vertical vibration included. Such sieves also incorporate dust aspiration and self-brushing and cleaning and are often arranged in sets so that several grades of grain are separated in one machine.

Grain, and malt, are weighed on receipt, on dispatch and at various handling stages. They may be moved either in small lots contained in sacks, an expensive method incurring high handling costs, or as samples weighing hundreds of tons which are moved by bulk-handling methods. A wide range of conveyors is in use, including helical screws working in cylindrical or half-cylindrical casings, moving belts, chain and flight drag conveyors, and jog-conveyors which, by jerking, convey green malt forward in a series of small hops. Vertical movement is often achieved by bucket elevators. Pneumatic systems, in which the grain is entrained in a rapidly moving air stream, easily move grain in any direction in a 'fluidized' state. The costs of installing and running these systems vary. Pneumatic systems are expensive, but they are flexible, self-cleaning, and pick up all grain and dust and contain them.

REFERENCES

[1] COOK, A. H. (ed.) (1962). *Barley and Malt. Biology, Biochemistry, Technology*, Academic Press, London, 740 pp.
[2] BERGAL, P. and CLEMENCET, M. (1962). p. 1 in [1].
[3] BRIGGS, D. E. (1978). *Barley*, Chapman & Hall, London, 612 pp.
[4] CARSON, G. P. and HORNE, F. R. (1962). p. 101 in [1].
[5] EMPSON, D. W. (1965). *Cereal Pests. Ministry of Agriculture, Fisheries and Food, Bulletin No. 186*, H.M.S.O., London.
[6] MOORE, W. C. and MOORE, F. J. (1961). *Cereal Diseases. Ministry of Agriculture, Fisheries and Food, Bulletin No. 129*, H.M.S.O., London.
[7] National Institute of Agricultural Botany (revised annually), *N.I.A.B. Farmer's leaflet No. 8: Recommended varieties of cereals*.
[8] BELL, G. D. H. and LUPTON, F. G. H. (1962). p. 45 in [1].
[9] AUFHAMMER, G., BERGAL, P., EADE, A. J., HAGBERG, A., REINER, L. and WILTEN, W. (1976). *European Brewery Convention Barley Committee – Barley varieties EBC* (4th edn., loose-leaf, with occasional additions).
[10] HUDSON, J. R. (1960). *Development of Brewing Analysis, A Historical Review*, Institute of Brewing, London, 102 pp.
[11] WAINWRIGHT, T. and BUCKEE, G. K. (1977). *J. Inst. Brewing*, **83**, 325–347.

[12] Analysis Committee of the Institute of Brewing (1977). *Recommended Methods of Analysis of the I.o.B.*, Institute of Brewing, London.
[13] Analysis Committee of the European Brewery Convention (1975). *Analytica EBC* (3rd edn.), Schweizer Brauerei-Rundschau, Zurich.
[14] Technical and Editorial Committee of the American Society of Brewing Chemists (1976). *Methods of Analysis of the ASBC* (7th edn. revised), American Society of Brewing Chemists, St Paul, Minnesota.
[15] NARZISS, L. (1976). *Die Bierbrauerei 1. Die Technologie der Malzbereitung*, Ferdinand Enke, Stuttgart, 322 pp.
[16] MACEY, A. (1957). *Irish Maltsters' Conference, 1952–1961, Selected papers*, Arthur Guinness, Dublin, p. 226.
[17] HUNTER, H. (1952). *The Barley Crop*, Crosby, Lockwood & Son, London, 187 pp.
[18] OXLEY, J. A. (1959). *Irish Maltsters' Conference, 1952–1961, Selected papers*, Arthur Guinness, Dublin, p. 101.
[19] BELL, G. D. H. and KIRBY, E. J. M. (1966), *The Growth of Cereals and Grasses* (eds. MILTHORPE, F. L. and IVINS, J. D.), Butterworths, London, p. 315.

Chapter 3

SOME ASPECTS OF BARLEY GRAIN PHYSIOLOGY

3.1 Introduction: An outline of malting
The simplest traditional malting process is divided into three stages. (*i*) Steeping. Selected barley is immersed in water until it has imbibed a suitable amount, then the excess water is drained away. (*ii*) Germination. The moist barley is allowed to grow under controlled cool conditions in a humid atmosphere, but with no further additions of water. (*iii*) Kilning. When biochemical and physical changes (modification) in the grain have advanced sufficiently the 'green malt' is dried and partly cooked in a flow of hot air in a kiln. Afterwards the brittle dried rootlets, or culms, are separated and collected. The finished malt is sent to store. Rootlets are sold for cattle food.

At the end of steeping the chosen moisture content of the grain (within the range 42–48% fresh weight basis) is sufficient to support modification, but will not allow excessive growth. This is important as growth is associated with a decline in dry matter, lost as carbon dioxide and water (respiration) and rootlets, so there is a 'malting loss', the dry weight loss in converting barley into malt. Thus a balance must be struck – enough growth must be allowed to produce well-modified malt, without allowing malting losses to become excessive. In newer malting processes water may be drained from steeping grain, which is then ventilated, and later re-covered or sprayed with water. This may be repeated. Germination may begin in the steeping vessel, and so the steeping and germination stages tend to merge.

Traditionally germination was carried out in darkness, at comparatively low temperatures (12–15°C; 53–59°F) for choice. Malting was not carried on in the summer, to avoid high temperatures. Modern maltings use refrigeration to cool the air in the germination compartments, and malting proceeds for the whole year. Germination rate and modification intensity are controlled by regulating the moisture content and temperature of the grain. The essential changes of modification are: (*i*) a large increase in the quantities of some of the hydrolytic enzymes present in the grain; (*ii*) there is a partial degradation (hydrolysis, catalysed by enzymes) of reserve substances (cell walls, gums, proteins, starch) in the starchy endosperm. Consequently there is a reduction in the strength of this tissue, and dry malt, in contrast with dry barley, is friable and readily crushed. Gums are degraded so that aqueous extracts of malt are less viscous than those of barley. During germination some soluble hydrolysis products are lost through respiration, while others are used to synthesize other molecules in the embryo. However, the quantities of low-molecular-weight substances (the cold-water extract) increase during malting. Malt from the kiln is comparatively dry (1·5–6·0% fresh weight) and is stable on storage. Kilning not only dries the malt, it also removes unpleasantly flavoured compounds, partially or wholly destroys some hydrolytic and other enzymes, and develops colour and flavour in the final product.

During mashing, crushed malt is mixed with warm water. The preformed soluble substances (cold-water extract) dissolve, and hydrolytic enzymes degrade many of the insoluble components of the malt, including nearly all the residual starch, to soluble products. The solution of all these substances, the sweet wort, is separated from the insoluble residue – the spent grains or draff. The quantity and quality of the mixed dissolved solids, the 'extract' that may be obtained from a malt, are keys to the nature of the beer later made from it, and so are important components of malt quality. The units used to express the amount of extract obtainable from malt vary from place to place (Chapter 9). In the United Kingdom new units are litre-degrees of extract/kilogram of malt (litre°/kg). Previously the units were brewer's pounds of extract per quarter (336 lb) of malt (i.e. lb/Qr). In other areas extract yield is expressed as a percentage of malt weight (i.e. %). Where the extract yield from a malt is determined on a small scale, in the laboratory, using a closely specified analytical technique, the result is called the hot-water extract (HWE). In all cases the extract should specifically be referred to the fresh weight or to the dry weight of the malt from which it was obtained.

3.2 *Laboratory tests on grain* [1–5]

Laboratory tests are made to determine whether a grain sample is of an approved variety, and to ascertain its viability and physiological state, and hence its suitability for malting. Grain will be analysed for its nitrogen and water content, and its ability to germinate. At least 96% of the grains should germinate; those that do not are either dead or dormant, and detract from the quality of the malt. Small-scale malting, micromalting, may be carried out on small samples to find the best conditions for processing the grain on the production scale. If the small samples of grain used in the laboratory are not truly (statistically) representative of the bulk, then tests made on them are useless. Rigid standards of sampling and sample preparation should be used [1, 2, 5]. Samples drawn from all parts of a bulk of grain, using an approved sampling pattern, are thoroughly mixed and split into smaller lots using a sample divider. Samples may be drawn from different depths using grain spears which have pockets that can be opened and closed again when filled with grain. Flowing grain may be sampled with hand-held scoops, or preferably, by automatic devices. Each method must avoid accidental bias, by selecting corns of a particular size, weight or shape.

3.3 Water uptake by barley grain [3, 4]

When grain is immersed in water its moisture content increases rapidly at first but progressively slows down, and, in the absence of germination, effectively ceases (Fig. 3.1). Water uptake rates vary between different varieties

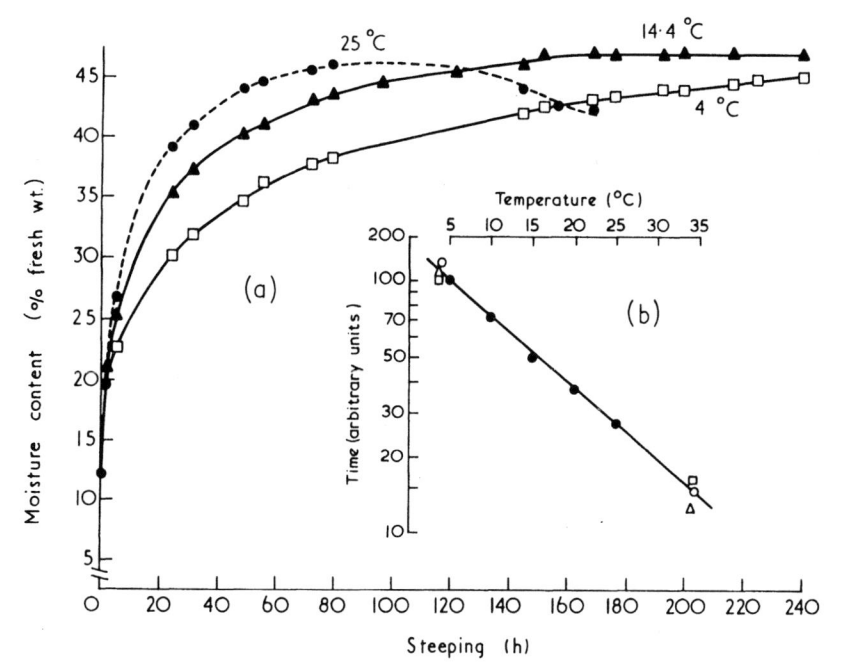

Fig. 3.1 (a) Increase in apparent moisture contents of ungraded samples of Proctor barley steeped at three different temperatures. Samples were rapidly centrifuged, to remove the film of surface water, before the increase in weight was noted. Samples were then returned immediately to the steep. The apparent decline in moisture content in the samples held at 25°C was due to the loss in dry matter caused by the microbial decomposition of the surface layers of the grain [Briggs, unpublished]. (*b; inset*) The relationship between the relative times taken for grain to reach any particular moisture content (after the removal of the surface moisture) at different temperatures. After Briggs (1967) [6].

and different samples of grain, and with the year of harvest and the temperature. Thus for British barleys from a particular region in 1954 it took grain 10–16 hr to reach 30–35% moisture in the steep, but 16–24 hr in 1955, and 6–9 hr in 1956 [7]. Water uptake by ungerminated barley is a physical process, which is independent of the grain's viability, but which is accelerated if the grains are so badly damaged that their surface layers and testae are broken. Experimentally, water uptake is accelerated by incorporating ethyl acetate or chloroform, both lipid solvents, into the steep liquor. Presumably those agents act by increasing the permeability of the surface layers to water [8]. Dissolving appropriate salts or sugars in steep liquor slows water uptake by grain, and reduces the final moisture content achieved. Each grain acts like a small osmometer, and sugars and many salts cannot enter ungerminated and undamaged grain. This effect is not significant with the concentrations of salts found in natural waters. Mealy corns take up water faster than steely corns, and a correlation is said to exist between starch content and the time taken to reach a given moisture level. Water uptake, under standardized conditions, is sometimes used as a criterion of barley quality. Thin grains take up moisture more rapidly, and reach a higher final water content than do wider 'bold' grain (Fig. 3.2). Even moisture uptake is important in malting, and usually corns of 2·3 mm width or less ('thins') are removed. Grains of widths above 2·3 mm vary relatively little in their rates of water uptake (Fig. 3.2).

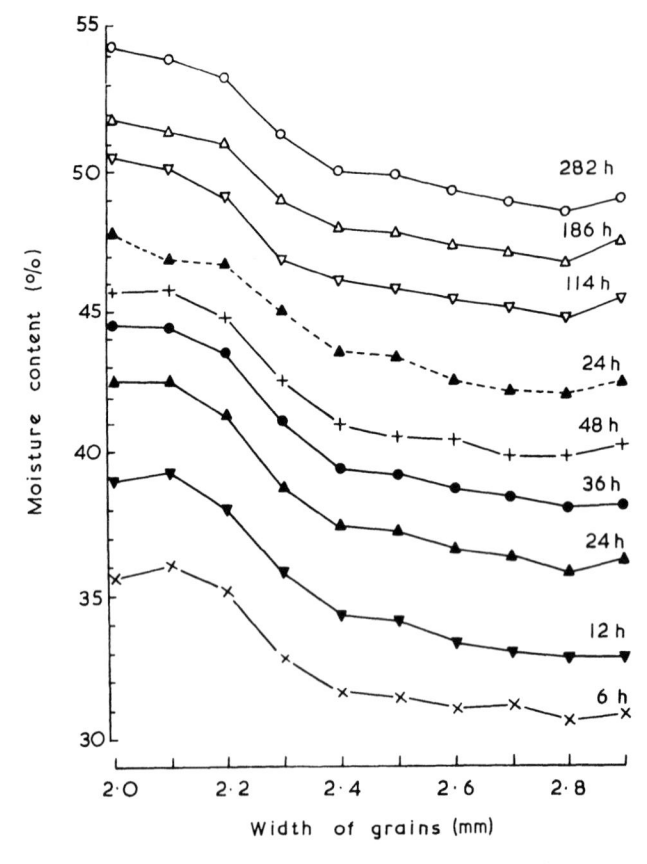

Fig. 3.2 The moisture content of different sizes of grains, separated by sieving one grain lot, after steeping for various periods at one fixed temperature. After Žila *et al.* (1942) [9].

Although the moisture content of steeped grain approaches a limiting value it is doubtful if this is ever reached in malting. If grain has begun to germinate and chit, water penetrates the grain more rapidly (a fact used in some rapid steeping schedules), and furthermore the moisture content readily exceeds the apparent limiting value found for ungerminated grain. Water uptake is more rapid at higher temperatures (Fig. 3.1b). For a particular sample of grain, the logarithm of the time taken to reach any given moisture content is inversely proportional to the steeping temperature (Fig. 3.1b). Thus, at least for a range of two-rowed British barleys, the temperature coefficient for the hydration process is the same. Within limits the duration of the steeping process may be controlled by regulating the temperature. Many empirical 'rules of thumb' that were used to regulate steeping times, when temperatures changed, were unsound. There are upper limits to the temperatures at which grain samples can be steeped without causing damage or death.

When barley is wetted moisture rapidly distributes itself throughout the surface layers. The husk, pericarp, furrow structures, rachilla and lodicules are quickly hydrated, and provide a network of tiny crannies and passages into which water is drawn by capillarity. It gains entry to the internal tissues at, or near, the embryo, probably through the micropyle. In undamaged grain the testa and pigment strand are largely or entirely impermeable. The embryo hydrates rapidly, and swells, followed more slowly by the endosperm. The moisture content of the embryo (%) exceeds that of the husk, or the whole grain [10]. When water is sprayed on to growing grain during malting, it probably remains mainly in the embryo, accelerating growth, respiration and hence the accumulation of malting losses. However, little reaches the endosperm to assist its modification [11, 12]. Total malting loss is the decline in dry matter (%) which occurs during the conversion of barley into malt (Chapter 5). At the end of steeping, grain moisture may be considered as that which is within the grain, together with that which remains as the surface film when the steep liquor is drained away. The surface film (2–5% moisture) may be removed by agitating, blotting or centrifuging the grain. It is absorbed by the grain early in germination. However, while it is present, the surface film impedes respiration, so slowing the rate at which grain chits, or even preventing the germination of water-sensitive grain.

In deep commercial steeps, extremes of hydrostatic pressure can cause grain damage, as can the strongly fluctuating pressures encountered when grain/water mixtures are pumped from one place to another. Grain damaged in these ways grows less well than undamaged control grain, possibly owing to selective damage to grain embryos [13].

3.4 *The penetration of grain by solutes* [3, 4]
The route(s) by which various substances, including the malting additives

gibberellic acid and potassium bromate, gain entry into grain are becoming clearer. In general, strong acids, fully ionized salts and sugars are excluded by ungerminated grains. Nitric acid and nitrates may penetrate slowly, as do weak organic acids and iodine in an excess of potassium iodide, which penetrates most rapidly at the embryo end. Many organic solvents enter readily, perhaps over the whole grain surface. The testa is at most only very slightly permeable to water, and so the semipermeable properties of the grain must reside in the micropyle, or the embryonic structures in contact with it. When the testa has been ruptured by the chit or by mechanical damage, access by water or solutions to the interior of the grain is less limited. Gibberellic acid is excluded by the testa, but can penetrate the micropyle [14]. Sodium bromate probably penetrates the micropyle comparatively rapidly, but also enters over the whole grain surface [15].

3.5 *Dormancy and viability* [1–5, 16]

In any sample a proportion of the grains will not germinate because they are dead. Of those remaining some will not germinate when tested, although they are alive. Such grains are 'immature' and dormant ('sleeping'). Dormancy is common after a cool damp season, but occurs less after a hot dry harvesting period. Varieties differ in the readiness with which they become dormant. Some dormancy is needed to prevent grain 'pregerminating', that is germinating in the ear, during damp weather before harvest. Such pregerminated grain does not keep well in store, and is unacceptable to maltsters. On the other hand, malt cannot be made from dormant barley and so excessively dormant varieties are not used by maltsters. However, moderate levels of dormancy can be overcome. The incidence of dormancy is said to be reduced by supplying growing plants with balanced fertilizers.

Estimates of viability and dormancy vary with the methods of assessment. Tests are carried out on small samples of grain, and results are expressed as percentages. In addition to the difficulties of obtaining representative samples, there are absolute statistical limitations on the accuracy of tests made with small numbers of grain (e.g. 300 or 500).

Thus if five grains fail to germinate in a sample of 100 grains (5%), the true value for the bulk will be 2–11% in 19 out of 20 cases (i.e. these are the 95% confidence limits). If 50 grains fail to grow in a sample of 1000 (still 5%), the range in which the true value will occur is reduced to 4–7% (95% confidence limits), a smaller but still important range of uncertainty [16]. For malting, a grain should have a *viability* or *germinative capacity* of at least 96%. This may be determined by using a tetrazolium staining method (Chapter 2), or by germination tests incorporating treatments which 'break' dormancy [4]. Maltsters often use the Thunaeus test, in which grain is immersed in hydrogen peroxide solution (0·75%) [2]. Most viable grains begin to germinate, and the numbers (*a*) chitted after 3 days are counted. After chitting the grains die. The 'non-starters' are peeled, that is the husk, pericarp and testa are removed from over the embryo by hand, then they are set on wet sand or filter-paper. Some of these also chit (*b*) and the total number of chitted grains (*a* + *b*) is used in calculating the viability as a proportion (%) of all the grains used in the test. Other less-reliable methods are sometimes used. In the Eckhardt test, corns are transected and the parts containing the embryos are set to germinate. The treatment largely overcomes moderate dormancy, but problems arise when micro-organisms grow on the cut surfaces and kill the embryo plants. Decortication with sulphuric acid (50%) nearly always breaks dormancy. However, this technique is not widely used because a proportion of grains, probably those that have been damaged during harvesting or by rough handling or insect attack, are killed by the acid treatment. Most techniques for breaking dormancy are only partly successful, and so tests based on them underestimate viability.

Germinative energy is the percentage of grains that will germinate under the conditions of a specified test. The difference between germinative capacity and germinative energy is a measure of the dormancy (%) of a grain sample. Germinative energy is greatly dependent on the conditions used in the test for germination. For reproducible results the germination conditions should be rigidly standardized. In many early tests this was not done. With dormant grain, maximal germination is often attained at about 13°C (approx. 54°F), rather than at higher temperatures, when the test is sufficiently prolonged. However, maltsters need rapid tests, and germination is usually estimated after 3 days. In the Schönjahn and Coldewe methods grains are placed, embryo ends down, in tapered perforations extending through porcelain plates. These are covered with sand and watered; excess water drains into an enclosing container. Germinated grains are counted from the bunches of rootlets growing from beneath the plates into the moist atmosphere beneath. In the Schönfeld method grains are held in a filter funnel and steeped, drained and allowed to germinate (with occasional mixing) in a humid temperature-controlled chamber. This approach can be used to simulate the effects of different steeping schedules.

In more usual tests grains are set to germinate on paper, or graded sand, held in Petri dishes at a controlled temperature, and wetted with known volumes of water [2, 17]. The number of corns germinated after 1, 2 and 3 days should be noted [2]. At least three replicates, each of 100 grains/dish, should be used. The proportion of the grain that germinates alters with its maturity, and the amount of water supplied to it (Fig. 3.3). Freshly harvested grain usually germinates best with 4 ml of water/Petri dish in a standardized test, and substantially less well with 8 ml of water/dish. By convention the difference between the germinative energy (%, 4 ml) and viability (%) is

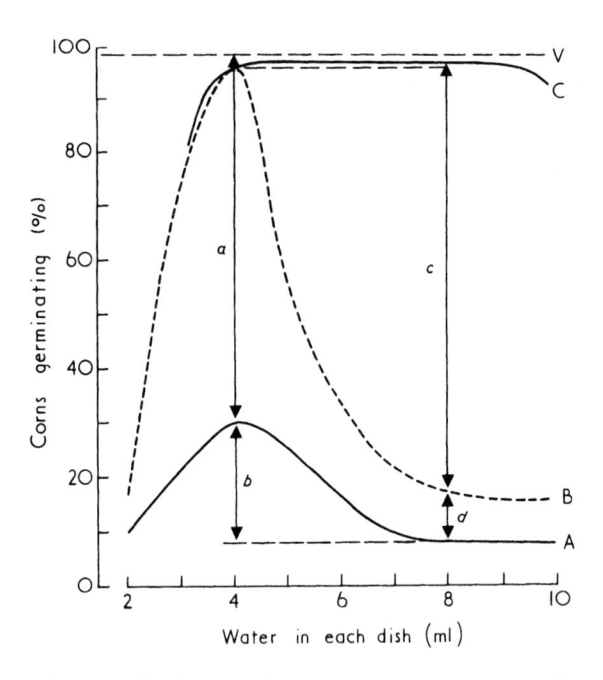

Fig. 3.3 The germination of barley samples at constant temperature on filter paper in Petri dishes, supplied with different quantities of water. The three curves represent grain samples after-ripened to various extents. (A) Freshly harvested barley sample. Profound dormancy, *a*; water-sensitivity, *b*. (B) Partly mature barley, showing little profound dormancy, but considerable water-sensitivity, *c*. Water-sensitivity has declined only by an amount *d* between the samples A and B. (C) Fully mature grain – little dormancy of any kind. The germinative energy (4 ml) is very close to the germinative capacity (viability, V) of the sample.

called the profound dormancy, or just the dormancy, of the grain, while the difference between the germinative energies (%) found with the '4 ml' and '8 ml' tests is called water-sensitivity. The distinction between dormancy and water-sensitivity is useful to maltsters, but really they are two measures of dormancy tested under different levels of stress. In the 8 ml test, grain is covered by a film of moisture which limits respiration. Grains placed dorsal- (embryo-) side downwards germinate less well than those with the furrow-side down, and grains crowded together germinate less well than those spaced out. Presumably micro-organisms in the water compete for oxygen and produce toxic metabolites, and so contribute to these results. When grain is properly stored, dormancy declines. Germinative energy (GE, 4 ml) initially increases more rapidly than germination (GE, 8 ml), so that, although germinability is improving, 'water-sensitivity' (GE, 4 ml minus GE 8 ml) actually increases for a period (Fig. 3.3), so the grain quality appears to deteriorate. In mature grain germinative energy is close to the viability of the sample. On extended storage, 2–3 years under 'normal' conditions, the aging grains begin to weaken and die. Various observations led to ways of malting grain that was partly immature, and showed some water-sensitivity. In well-aerated running water mature grain will germinate while water-sensitive grain will not. A surface film of water prevents the germination of steeped water-sensitive grain, and displacement of the film favours germination. However, if such grain is steeped to about 35% moisture, about the level attained in the '4 ml test', then when the surface film of water has been absorbed by the grain, germination processes will begin, and, after an air rest of 8–24 hr before any visible germination has occurred, re-immersing the grain to hydrate it to a level suitable for malting (e.g. 42–46%) will not prevent normal growth (Fig. 3.4). The concentration of oxygen in water, even when saturated with

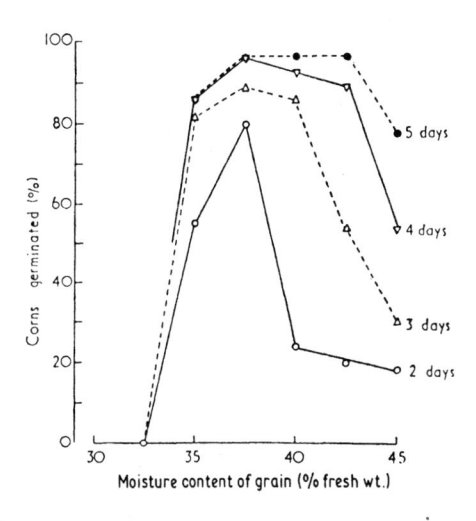

Fig. 3.4 The percentage germination, on successive days, of barley steeped to different moisture contents, then held under malting conditions. After Sims (1959) [18].

air, is comparatively low. Mature and water-sensitive grains respire at about the same rate, which is probably limited by the time taken for oxygen to penetrate the grain tissues. Water-sensitive grain probably requires a higher internal oxygen content than does mature grain before it can germinate. Prolonged steeping of barley at normal, or especially at elevated, temperatures induces a condition resembling water-sensitivity. This is particularly true where a high proportion of damaged grains leads to rapid fouling of the steep water. Germination of such over-steeped grain is favoured by additions of gibberellic acid, and by the removal of the surface film of water (Fig. 3.5).

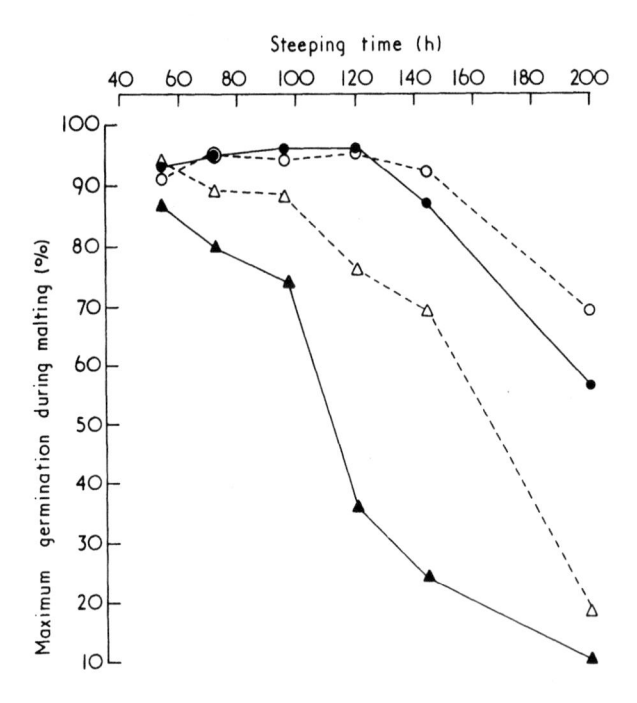

Fig. 3.5 The maximum germination achieved during malting barley samples steeped for various extended periods in water or a solution of gibberellic acid. In some samples the film of surface moisture was removed, by centrifugation, at the end of steeping. After Kirsop (1964) [19]. ▲, steeped in water; ●, steeped in water and centrifuged; △, steeped in a solution of gibberellic acid; ○, steeped in a solution of gibberellic acid and centrifuged.

Many years ago a process was patented for removing the moisture film from the surface of grain using warm air, during its passage from the steep to the germination site. Increasingly maltsters prefer to 'dry-cast' their grain, i.e. convey it dry from the steep, rather than 'wet-cast' it, i.e. move it by pumping it mixed with water. Wet-casting renews the water film on the grain surface, and checks the onset of germination.

Within limits the rate at which grain germinates and grows rises with increasing temperature. However, the maximum germination achieved by partly dormant or immature grains is less at higher temperatures, so that such grain is best malted using long cool (e.g. 12°C; approx. 54°F) germination conditions. Maltsters must, of economic necessity, malt as rapidly as possible and so must use the highest practicable temperature, often 18–20°C (about 64–68°F). However, many attempts have been made to use even higher temperatures.

The causes of dormancy are complex, but seem to involve the metabolic state of the embryo, effects exerted by the grain integuments, possibly growth-inhibiting chemicals, the presence of a water film, the temperature, the availability of oxygen, and the effects of the surface population of microbes. It varies between varieties and is greatly influenced by the weather during the ripening and harvesting period. In the past, when grain had been stored on the ear in stacks, it was sometimes found to be covered with a mucilaginous, microbial slime or to be badly discoloured when threshing began. Such dramatic results of microbial activity are now rare, but the less noticeable populations that invariably occur can still alter the maltability of grain and the quality of the product [4, 20, 21]. Treating grain with steep water made alkaline with lime or sodium hydroxide sometimes enhances subsequent grain germination, presumably by 'cleaning' it and checking microbes. Steep liquor contains growth-inhibitory substances including phenolic acids and ethanol, leached from the grain, and acetic acid which is presumably formed by micro-organisms. Microbes consume much of the oxygen initially dissolved in steep water. These observations at least partly explain the necessity for changing steep liquors, and for separating grains on the wet paper in germination tests. The re-use of steep liquors impairs subsequent grain germination. Barley has been steeped in dilute solutions of formaldehyde, chosen so that germination was only slightly retarded, although microbial growth was prevented. The formaldehyde steep liquor may be re-used, so that microbes may be the major 'inhibitors' in grain leachates, rather than chemicals washed from the grains. Formaldehyde kills *Fusaria* and other groups of fungi on grain that, malted in their presence, would give rise to malt that produces beer with a tendency to gush [22, 23]. At least some agent(s) which cause gushing are mould metabolites. Some moulds, which occur on barley, can produce highly poisonous mycotoxins such as citrinin or zearalinone. Fortunately, these are largely destroyed during malting and brewing, but the risks associated with their possible presence direct attention to the need to avoid grain with heavy mould infestations.

Removing moisture films from grain surfaces, decortication or peeling immature grains favours an increase in germination and indicates that the surface layers strongly influence dormancy. Only rarely are barley grains encountered in which embryos, while viable, do not germinate when separated from the rest of the grain. Increased oxygen pressure, like hydrogen peroxide, tends to overcome dormancy, and the oxygen requirement for germination declines as grain matures. Many chemicals are known to improve the germination of dormant grain, and some are reported to preferentially reverse either profound dormancy or water-sensitivity. For reasons already given this distinction appears misleading. In addition to oxygen and hydrogen peroxide, substances known to improve germination under some conditions include nitrous acid, various nitrates, formaldehyde, thiourea, hydrogen sulphide and other thiols, gibberellic acid, many heavy metals, such as mercuric or ferrous ions, thiol-blocking reagents, such as N-ethylmaleimide or iodoacetic acid, and mixtures of antibiotics. Physical treatments that stimulate germination include decortication (50% H_2SO_4), peeling, cutting the grain in half, puncturing or slitting the integuments near the embryo, warming the grain (40°C; 104°F/3 days), drying the grain (in air is better than in nitrogen) or, better, drying warm (e.g. 35–40°C; 95–104°F), then storing warm (25–30°C; 77–86°F) for several weeks. Taken together these results appear to indicate that dormancy, which declines with time, is due to several factors: (*a*) the physiological state of the embryo; (*b*) the availability of oxygen to the living tissues; (*c*) the density of the microbial population, which will compete strongly for oxygen in the surface layers of the grain, and may produce growth-inhibitory substances; (*d*) the surface layers, and the film of surface moisture, which are barriers to oxygen uptake. Of the dormancy-relieving treatments mentioned, those that do not facilitate oxygen penetration into the grain either stimulate grain metabolism (gibberellic acid) or are highly damaging to micro-organisms (3, 4, 24).

3.6 *Respiration and fermentation during steeping*

By analogy with wheat, up to half of the respiration of dry quiescent grain is probably due to the associated micro-organisms. Of the oxygen uptake of the grain itself about 50–70% is due to the embryo, the rest to the aleurone layer. As the moisture content of the grain increases, so does its metabolic rate, and its heat output. Under anaerobic conditions, as occur for most of the time in traditional steeps, moist grain with its microbial flora rapidly utilizes the available oxygen, but continues to produce carbon dioxide and ethanol from carbohydrates by fermentative processes. Up to half of the ethanol formed may escape into the steep liquor. Prolonged anaerobiosis and the progressive accumulation of ethanol impair germination and may eventually kill grain. High concentrations of carbon dioxide can suppress oxygen uptake and fermentative processes in air, but it is unclear if this occurs during steeping. Experiments in which steeps were sparged with oxygen or carbon dioxide showed that: (*i*) in oxygenated steeps the grains chitted rapidly, but subsequently tended to grow uncontrollably during germination (bolt), and produced malt with very high losses of 20%; (*ii*) grain from steeps saturated with carbon dioxide malted slowly initially, but finished with a normal extract and a malting loss reduced by about 1%. Other experiments indicated that sparging steeps with nitrogen or carbon dioxide improved malt yields by 1% and 2½% respectively [25, 26]. These results appear to cast doubt on the value of the accepted practice of aerating steeps. However, in these trials germination was for fixed periods of time, and not until maximal extracts had been reached – it is likely that peak extracts would be reached sooner by grain aerated in the steeps. Furthermore oxygenation may well have induced chitting in the steep, with a consequent rapid uncontrolled uptake of water which led to bolting and excessive malting losses. In apparent contrast, it was noted that germination was more even if an air rest was given between immersions in water saturated with carbon dioxide. Similarly the traditional English practice of long steeps without aeration tended to stifle the grain, and is said to hold back 'bolters' leading to slower more even germination on the malting floor. However, the practice of near-anaerobic steeping is obsolescent, since in modern malting plants *rapid* even germination is essential. As previously noted prolonged, 'over-steeping' induces a condition resembling 'water-sensitivity' in grain (Fig. 3.5). Excessively long steeping kills grain, which begins to decompose. Steep aeration is used to maintain grain viability and ensure that subsequently rapid even germination occurs. Air rests seem to be more effective than sparging air into the steep liquor while it is covering the grain, but many maltsters use both techniques. If fully mature grain is used, it germinates evenly after aerated steeping. Experimentally, excessively vigorous aeration, e.g. by sparging oxygen-enriched air into a steep, can induce some corns to chit under water. This is undesirable, leading to excessive water uptake in these grains, and uneven germination [27]. Also, such extreme aeration is costly. However, in some newer steeping schedules it is the aim to obtain *all* the grains in a chitted state at the end of an air rest to obtain very rapid and *uniform* water uptake in the final steep (see below). Better results are obtained more cheaply by 'air-resting' grain, that is draining off the water and ventilating the damp grain bed (see below). In floor maltings vigorous growth with the associated high heat output, as induced by steep aeration or applications of gibberellic acid, is likely to get out of control, because the heat cannot be dissipated. Consequently high malting losses and alterations in malt quality ensue. To overcome this problem grain may be cast with a lower moisture content than normal, or alkali-metal bromates may be applied (Chapter 6). In pneumatic maltings, where the refrigerators for cooling the conditioning air have adequate capacity for dissipating the heat, rapid and vigorous germination can be controlled by reducing the 'air-on' temperature, and good malt can be made in a short time. In high-capacity conical steeps it is difficult to mix grain or otherwise ensure that it is equally treated during aeration treatments, consequently steeps of different designs are being used (Chapter 6). Even so, it is doubtful if under-water steep aeration is ever adequate to induce chitting, or even maintain the liquid half-saturated with oxygen from the air, about the minimum requirement for suppressing ethanol formation. It is easier to obtain more uniform aeration with air rests, in properly designed steeps, but even here films of moisture over the grains roughly halve the oxygen uptake, and uneven drying in the grain bed during air rests can lead to uneven germination, and should be avoided. During air rests it is best to suck air downwards through the grain bed to disturb the moisture film, aid water and carbon dioxide drainage (carbon dioxide is a heavy gas), and to replace the interstitial gas between the grains with fresh air.

3.7 *Steep liquor*

The nature of steep liquor affects malt quality. Chalky alkaline waters check microbial growth in steeps, and leach a mixture of acidic, proteinaceous and phenolic materials (so-called 'testinic acid') from grain, leading, it is said, to beers with improved non-biological stability and less astringent flavours. However, lime water (saturated calcium hydroxide solution) is a comparatively inefficient extractor of 'testinic acid'. Much greater quantities are extracted by dilute solutions (0·05–0·1 %) of sodium carbonate or sodium hydroxide [28] used in short immersions at the start of steeping. Sometimes (rarely in England) such preliminary alkaline steeps are used to 'clean' grain and remove 'musty' smells. Caution is needed, since over-treatment with these agents can kill grain. Sodium bicarbonate is less harmful to grain, but is a weaker alkali and might be a less effective cleaning agent. The use of hypochlorite solutions (e.g. calcium hypochlorite or 'bleaching powder') or chlorinated water has been proposed to combat microbes, but these treatments are avoided because of the 'disinfectant' flavour imparted to the malt. Ferrous waters are also avoided, because the iron salts interact with phenolic substances in the husk, imparting a dull greyish unattractive appearance to the malt. The uses of other additives, such as hydrogen peroxide, gibberellic acid and bromates are summarized in Chapter 5.

During steeping and steep aeration, barley is washed, and dust and light corns ('floaters') come to the surface, and 'weir-over' into overflow pipes. They are collected on screens and, after drying, are sold for cattlefood. During steeping many substances, including amino acids, hexose and pentose sugars, phenolic and simple carboxylic acids and phosphates, are dissolved from the surface layers of all grains and from the interiors of damaged or broken grains. Thus, in steeping, grain loses 0·5–1·5 % of its dry weight as dust and dissolved materials. The dissolved materials are utilized by microbes to support growth with the consumption of oxygen and the production of acids; consequently the pH falls, and possibly other growth-inhibitory substances are formed. Consequently it is necessary to change the liquors at least once during steeping, and the large volumes of waste, with their high levels of suspended solids and dissolved materials and consequent high biochemical oxygen demands (BOD) make them costly to purify or discharge to public sewers (Chapter 7). No practical use has been found for this material. It has been sprayed over grassland as an irrigant and a dilute fertilizer, and it has been proposed to use it as a supplementary source of yeast nutrients by adding it to mashing liquor in breweries. However, because of its large microbial population and large volume, and the distance between maltings and breweries, this last proposal is impractical. Both obtaining water and disposing of effluent (residual steep liquors and cleaning water) are becoming increasingly costly, and so maltsters are concerned to reduce the volumes employed. Different conventional plants are said to use 27–90 hl of water/tonne of malt produced (90–300 gallons/quarter; 336 lb).

3.8 Choosing commercial steeping schedules

Petri-dish ('4 ml' and '8 ml') tests are widely used to estimate the maturity of barley, and its readiness for malting. Originally grain was not malted until 'profound dormancy' had nearly gone and 'water-sensitivity' had reached a low figure. It was unexpectedly found that grain giving low values in germination tests could often be malted on a pilot scale. To devise a test closer to commercial practice, small-scale steeps and germinations were carried out in test tubes (micro-malting) and were found to give better estimates of the germination percentage found in commercial maltings. Bag tests in which grain samples (0·5–1·0 kg) were held in nylon net bags and steeped and germinated in contact with large bulks of malting grain gave even better estimates. Yet in every case discrepancies were found between the results of small-scale tests and full-scale malting trials. Although small-scale tests are indispensible for screening barley lots, and they are 'safe' in that they usually give pessimistic estimates of the readiness of a sample for malting, yet the lack of explanations for the observed discrepancies is unsatisfactory [29].

In traditional British steeping, barley was immersed, without aeration, in water at well temperatures, often 10–14°C (50–57°F). The water might be changed 3–4 times, and the grain would be immersed for a total of 50–80 hr until it reached 42–46% moisture. Subsequently many steeps were equipped with aeration devices and now, belatedly, newer installations have facilities for warming steep water to a chosen extent, e.g. to 18°C (64·4°F). However, warm steeping should only be used in conjunction with good aeration. 'Water-sensitive' grain steeped to 35% moisture germinates, but this moisture level will not support adequate modification unless the moisture level is subsequently raised [18, 30] (Fig. 3.4). This discovery led to the introduction of incremental steeping and the revival of air-rest steeping. These techniques are now widely used to obtain rapid and vigorous germination in all grain of widely differing degrees of maturity. In incremental steeping grain is steeped to 35–37% moisture, and, after an air-rest period (8–24 hr) chosen by reference to preliminary trials, just sufficient water is sprayed on to renew the film of surface moisture, a process repeated at intervals as the grain takes up the moisture and the surface dries, until the desired final moisture content of 43–47% is reached. This attractive approach ensures that the grain is well-aerated, and, because there is little effluent from the spray-steeping, little waste water need be produced at this stage with a consequent saving in effluent costs. However, spraying a large bulk of grain evenly requires specialized steeping equipment, and air-rests and immersion steeping is more usual. Historically air rests were usual in some parts of Bohemia, while in Bavaria grain was sometimes steeped in running water [4]. When the process was revived in Britain, a typical schedule would be a steep of 8–16 hr until the grain moisture was 30–35%, followed by an air-rest with downward ventilation with ambient air for 12–24 hr, period indicated by small-scale trials, followed by one or more immersion steeps, for about 24 hr, or until a moisture content of 43–45% was reached. The need for an extra change of steep water was indicated by the liquor becoming unacceptably dirty. At the end of the air rest the moisture content of the blotted grain is about the same as if steeping had been continuous, but the film of surface moisture had been taken up. Thus during the air–rest period the surface film of water (2–5%) must be absorbed by, and meet the needs of, the grains interior. More complex multiple-steeping schedules involving more than one ventilated air rest are now commonplace, with aerations of the steep, and sometimes additions of additives between the steeps. For example, in the first steeping vessel the grain is held wet for 10 hr, is air-rested for 11 hr, is rewet for 1 hr and then transferred wet to the second vessel. In the second vessel it is drained and air-rested for 10 hr, rewet for 8 hr with vigorous aeration, drained and held dry for a further 4 hr, then dosed with gibberellic acid and a soluble bromate during transfer to the third ventilated steep vessel. In this vessel it may be sprayed with water. The grain is held until it is completely chitted, then it is transferred to the germination vessel.

Another set of steeping conditions, designed to deal with the routine production of different malt types from barleys having a range of qualities, show a flexible approach [31]. The first steep lasts 6–10 hr and is carried out at 15–25°C (59–77°F), until the grain moisture is 36–38%. During this period, vigorous aeration is carried out to maintain the vigour of the grain and ensure even mixing of the steep contents. The steep is drained, and during the transfer of the grain to a special ventilation tank gibberellic acid (e.g. 0·25 mg/kg of grain) may be applied, and bromates may be added at the same time. The grain is ventilated for about 24 hr during which time the temperature is allowed to rise to within the range 25–30°C (77–86°F), higher temperatures being used for poorer-quality barleys. Germination begins, but the high temperatures 'stun' the embryos so that rootlet growth is subsequently reduced, and the development of proteolytic activity is suppressed, checking the formation of soluble nitrogen in the malt. At the end of this period the grain should be evenly chitted. It is then steeped for a further period to 46–48% moisture, the higher values being used for grain of lower quality. Higher moisture contents are achieved by either using longer steeping periods or by increasing the steep temperature. Moisture enters the chitted grain extremely rapidly, and a steep of 12 hr at 20°C (68°F) would be typical. The grain is then transferred to the germination compartment. During the transfer further applications of additives can be made.

In concept the practice of 'flush-steeping' falls between incremental spray-steeping and multiple steeping and is suitable for inducing rapid germination without unusually complex steeping, aeration and spraying facilities. During air-rest periods grain respiration rates and temperature rise, and the rise in temperature accelerates the onset of germination. By briefly covering the grain with water (flushing) at intervals, the surface film of water is renewed around each grain and the temperature is controlled. By maximizing the duration of the air rests between flushes excellent aeration is achieved, and rising temperatures ensure that rapid water uptake and even germination is obtained [29]. However, this process carries a penalty: the production of large volumes of effluent [32].

The newer steeping schedules have been developed to enable 'water-sensitive' grain to be malted and to induce the most rapid and even germination possible of any samples of grain. Grains which are slow to germinate produce less extract after malting (Table 3.1). Furthermore, uneven germinated malt 'takes colour' unevenly on the kiln, so giving an irregular appearance to the final product.

TABLE 3.1
Extracts of small samples of malt separated from commercial batches when they chitted, then allowed to complete their germination period
After KIRSOP (1964) [19]

Time at which the grains begin to chit (days from steep)	Sample ...	Hot-water extract (litre°/kg at 20°C)			
		A	B	C	D
0–1		301·2	309·5	310·5	297·2
1–2		298·3	307·2	307·9	293·8
2–7		296·0	—	—	—

REFERENCES

[1] HUDSON, J. R. (1960). *Development of Brewing Analysis, A Historical Review*, The Institute of Brewing, London, p. 12.

[2] Analytical Committee of the Institute of Brewing (1977). *Recommended Methods of Analysis*, The Institute of Brewing, London.

[3] POLLOCK, J. R. A. (1962). *Barley and Malt. Biology, Biochemistry, Technology* (ed. COOK, A. H.) Academic Press, London, p. 303; p. 399.

[4] BRIGGS, D. E. (1978). *Barley*, Chapman & Hall, London, 612 pp.
[5] Analysis Committee of the European Brewery Convention (1975). *Analytica – EBC*, 3rd edn., Schweizer-Brauerei-Rundschau, Zurich.
[6] BRIGGS, D. E. (1967). *J. Inst. Brewing*, **73**, 33.
[7] MACEY, A. (1957). *J. Inst. Brewing*, **63**, 306.
[8] BROWN, A. J. and WORLEY, F. P. (1912). *Proc. R. Soc. London Ser. B*, **85**, 546.
[9] ŽILA, V. V., TRKAN, M. and ŠKVOR, F. (1942). *Woch. Brau.*, **59**, 63.
[10] REYNOLDS, T. and MACWILLIAM, I. C. (1966). *J. Inst. Brewing*, **72**, 166.
[11] KIRSOP, B. H. (1966). *Proc. Irish Maltsters' Conf.*, p. 38.
[12] KIRSOP, B. H., REYNOLDS, T. and GRIFFITHS, C. M. (1967). *J. Inst. Brewing*, **73**, 182.
[13] YOSHIDA, T., YAMADA, K., FUJINO, S. and KOUMEGAWA, J. (1979). *J. Amer. Soc. Brewing Chemists*, **37**, 77.
[14] SMITH, M. T. and BRIGGS, D. E. (1979). *J. Inst. Brewing*, **85**, 160.
[15] BROOKES, P. A. and MARTIN, P. A. (1974). *J. Inst. Brewing*, **80**, 294.
[16] URION, E. and CHAPON, L. (1955). *Proc. Eur. Brewery Conv. Congr.*, Baden-Baden, p. 172.
[17] BISHOP, L. R. (1936). *J. Inst. Brewing*, **42**, 103.
[18] SIMS, R. C. (1959). *J. Inst. Brewing*, **65**, 46.
[19] KIRSOP, B. H. (1964). *Proc. Irish Maltsters' Conf.*, p. 55.
[20] ETCHEVERS, G. C., BANASIK, O. J. and WATSON, C. A. (1977–Jan.). *Brewer's Digest*, p. 46.
[21] HAIKARA, A., MÄKINEN, V. and HAKULINEN, R. (1977). *Proc. Eur. Brewery Conv. Congr. 16th, Amsterdam*, p. 35.
[22] GJERTSEN, P., TROLLE, B. and ANDERSEN, K. (1963). *Proc. Eur. Brewery Conv. Congr. 9th Brussels*, p. 320.
[23] GJERTSEN, P., TROLLE, B. and ANDERSEN, K. (1965). *Proc. Eur. Brewery Conv. Congr. 10th Stockholm*, p. 428.
[24] GABER, S. D. and ROBERTS, E. H. (1969). *J. Inst. Brewing*, **75**, 303.
[25] CHAMBERS, A. R. and LAMBIE, A. D. B. (1960). *J. Inst. Brewing*, **66**, 159.
[26] ENARI, T. M., LINNAHALME, T. and LINKO, M. (1961). *J. Inst. Brewing*, **67**, 358.
[27] EKSTRÖM, D., CEDERQVIST, B. and SANDEGREN, E. (1959). *Proc. Eur. Brewery Conv. Congr., Rome*, p. 11.
[28] STEVENS, R. (1958). *J. Inst. Brewing*, **64**, 470.
[29] MACEY, A. and STOWELL, K. C. (1959). *Proc. Eur. Brewery Conv. Congr., Rome*, p. 105.
[30] REYNOLDS, T., BUTTON, A. H. and MACWILLIAM, I. C. (1966). *J. Inst. Brewing*, **72**, 282.
[31] MACEY, A. (1977–Sept.). *Brewer's Guard.*, **106**(9), 81.
[32] BROOKES, P. A., LOVETT, D. A. and MACWILLIAM, I. C. (1978). *J. Inst. Brewing*, **82**, 14.

Chapter 4

THE BIOCHEMISTRY OF MALTING GRAIN

4.1 Introduction

A knowledge of some aspects of the biochemistry of germinating barley is indispensible for an understanding of the various malting processes, and the alterations which make malt suitable for use in the mash tun. However, many of the biochemical pathways of intermediary metabolism, such as the glycolytic pathway or the tricarboxylic acid cycle which, of course, occur in barley, are not considered in detail here because up to now the knowledge has not been of immediate value in malting. Much of the information is summarized elsewhere [1–3].

4.2 Physical changes in malting barley

During steeping, grain swells and increases in volume by about a quarter; space is allowed for this when steep tanks are filled. The first indication of germination, which usually occurs after casting, is the appearance of the white 'chit', the coleorhiza or root sheath, which breaks through the testa and pericarp and protrudes from the base of the corn, beyond the husk. Later, seminal roots (maltsters' 'rootlets', 'culms', 'coombes', 'cummins', 'sprouts') break through the root sheath and form a tuft at the end of the grain. Meanwhile the coleoptile with its enclosed first leaf (the maltsters' 'acrospire', 'spire' or 'blade') penetrates the testa on the dorsal side of the grain and grows towards the apex between the testa and the pericarp. The growth of the acrospire is commonly used as a rough guide to the progress of 'malting'. Usually the lengths of the acrospires of a number of grains are noted, as fractions of the grain lengths. Thus when the tip of the acrospire is half way up the grain its length is said to be $\frac{1}{2}$; when the tip reaches the grain's apex its length is 1, and so on. In traditional practice green malt is kilned, and growth terminated, when most acrospires are $\frac{3}{4}$–1 the grain's length. In some grain samples the acrospire cannot be distinguished through the husk, so it is necessary to peel the grain or make the husk transparent (for example by boiling grains in copper sulphate solution) before measurements can be made. Traces of starch appear in the embryo after imbibition [4]. It is probably formed at the expense of sugars stored in the embryo, as it also appears in isolated embryos [1].

Shortly afterwards, in the intact grain, the first indications of the breakdown of the starchy endosperm, the 'cytolysis' or degradation of the cell walls of the compressed cell layer are seen. This process begins adjacent to the scutellum and progressively spreads through the starchy endosperm towards the apex of the grain. It may be followed by the loss of the cell wall's ability to take up stains, such as 'Congo Red'. The extent of cell wall degradation increases with time, and is locally complete in grains grown with a high moisture content (Fig. 4.1). Following cell wall dissolution the proteins of the endosperm cells undergo degradation, and the starch granules are partially degraded, as shown by the microscopic etching and pitting they undergo. All these changes, which are termed 'modification', are catalysed by hydrolytic enzymes.

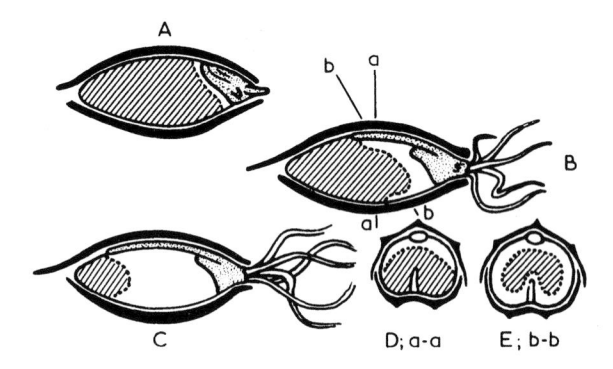

Fig. 4.1 A. B. C, Diagrams of longitudinal sections of barley grains at successive stages in malting. The rounded dorsal side of the grain, along which the acrospire is growing, is uppermost. Unmodified starchy endosperm is shown cross-hatched, while the modified regions are unshaded. D and E, Vertical and inclined transverse sections of grain B, along the axes shown. In vertical section, D (at a–a), modification appears only on the dorsal side. In inclined section, b–b, it occurs completely around the starchy endosperm (cf. [5]).

While endosperm modification is going on, changes are happening in the protoplasm of the cells of the aleurone layer, and the columnar cells of the scutellar epithelium. As discussed later, these changes are associated with the mobilization of the cell's reserves, the manufacture and release of hydrolytic enzymes and (in the case of the scutellar epithelium) the uptake of soluble low-molecular-weight materials from the starchy endosperm and aleurone layer. These soluble substances (sugars, amino acids and minerals) serve as nutrients for the growing embryo. As germination proceeds, the cells of the columnar epithelium separate along their sides and elongate ($\times$ 2–4) forming a fine 'pile' which projects into the dissolved region of the endosperm, greatly increasing the surface area of the cells, and presumably making the epithelium more efficient as an absorbtive or secretory organ.

The modification of the starchy endosperm begins under the entire scutellum, and progresses up the grain parallel to the scutellar face. Gradually the process advances rather faster adjacent to the aleurone layer, but the 'leading edge' of the modified region remains parallel to the scutellar face [5] (Fig. 4.1). Thus the central endosperm is degraded more slowly. Furthermore the 'sheaf cells', situated above the furrow, retain their strength and are little degraded. When wet or dry the endosperm of ungerminated barley is tough and resilient. When germination has occurred, the moist endosperm may be squeezed and rubbed out between the fingers, and when dry it is readily crushed and 'friable'. Thus the processes of modification have reduced the strength of the endosperm and changed its character. Maltsters usually follow the progress of modification by rubbing out samples of grain. In poor-quality 'undermodified' malts, the tips of the grains retain their 'barley character', resulting in hard ends in the kilned product. Undermodified malt is difficult to mill, yields its extract with difficulty, and may give rise to other problems during brewing. Traditionally malt is checked for undermodification by chewing to see if any corns retain hard tips. Hardness is now also tested in less subjective ways. For example, in the Brabender apparatus a coarsely ground sample of malt is reground using a special mill which determines the pressure on the rolls needed to produce a standard fine grind. The result is an 'average value' and does not show the hardness of individual corns, and hence the homogeneity of the sample. Inhomogeneous samples may arise in several ways, for example from uneven germination during malting, or from blending different malts. Other devices measure the hardness of grains individually. Thus the sclerometer measures, and records, the resistance of individual grains to the passage of a cutting knife [6].

In some malting processes, but not all, increasing length of the acrospire roughly indicates the extent of the advance of modification. However, if the husk with the attached pericarp is split so that the acrospire grows away from the grain's surface modification proceeds normally; it is not dependent on the nearness of the acrospire. The changes occurring in modification are mediated by numerous hydrolytic enzymes, many of which increase greatly in amount during malting [1-3]. The soluble products produced by their activity, such as sugars and amino acids, tend to accumulate in the grain, particularly in the endosperm, faster than they are utilized by the embryo. They constitute the cold-water extract (CWE) of the grain. This and other measures, such as the proportion of the nitrogen that is soluble after mashing, are regarded as guides to the extent of modification during malting (Chapter 5).

4.2.1 THE HUSK FRACTION
When malt is crushed and sieved the 'husk fraction' contains the husk proper

(palea and lemma), together with the lodicules, testa, pericarp, some nucellar tissues, aleurone layer, fragments of embryo and acrospire, and some starchy endosperm. The gross composition of such a fraction is shown in Table 4.1.

TABLE 4.1

The approximate composition of a 'husk' fraction of barley malt prepared by sieving. Such fractions contain true husk, together with pericarp, testa, nucellar material, aleurone layers, fragments of embryo and starchy endosperm.

Data of GEYS, cited by HOPKINS and KRAUSE (1947) [7]

Fraction	Proportion (%)
Water	7·4
Protein	7·1
Pentosans	20·0
Fibre	22·6
Fat	2·1
Ash (including silicates)	10·0
Starch	8·2
Other nitrogen-free materials	22·6

Together the true husk and pericarp make up 9–14% of the dry weights of most barleys. Typically the 'husk and pericarp' content of British two-rowed malting barleys is about 10%. Six-rowed barleys usually have a higher husk content, and traditionally 'six-rowed malts' were used to increase the husk content of the brewing grist, to 'open out' the mash, and so obtain more rapid wort separation after mashing. As with all the grain proportions mentioned, the values quoted represent mean values of 'average' samples of grain. There are wide variations in size and composition between grain lots and individual grains within each sample. Estimates of 'husk' content differ depending on the technique of measurement used. Soaking grain in water or a strong solution of ammonia, followed by hand separation and drying and weighing is tedious and imprecise because the samples used are small, and so may be unrepresentative of the bulk, and also materials are lost in solution. 'Pearling' in a special mill, does not remove all the husk, yet removes small amounts of other tissues. Husk, and pericarp, may be rinsed away from grain after soaking in sulphuric acid (50%) or hot sodium hypochlorite solution. The loss in dry weight is used as a measure of husk (with pericarp) content [1, 8], but it includes the weight of materials extracted from damaged corns.

The husk and pericarp may be considered together, since they invest the grain, are usually fused together, appear to serve the same functions, and little is known of their detailed compositions. They do not take an active part in the chemical changes that occur in malting, although in many instances the associated microflora does. The microbes compete for oxygen, and the surface layers restrict its penetration into the grain. Substances leach from the surface layers during steeping, including some phenolic materials [9]. The husk is essential to the formation of filterbeds during wort separation, and it protects the grain against physical damage. Wheat and rye (both naked grains), and naked barleys, are more readily harmed than normal husked barley by mechanical handling. The husk may physically restrict both barley swelling during steeping, and subsequent acrospire growth. The pericarp is of importance in accentuating or causing dormancy (Chapter 3). Grain that has been decorticated with sulphuric acid, and which therefore lacks both husk and pericarp, retains most of its anthocyanogens (proanthocyanins; leucoanthocyanins). However, these rapidly leach from decorticated grains into steep water. The reason for this is unknown. Cellulose comprises about 5% of the whole grain, and most of this is in the husk. Hemicelluloses containing hexose, pentose and uronic acid residues are present, but starch is not. The starch found in 'husk fractions', prepared by sieving milled grain, comes from the starchy endosperm. The polysaccharide cell walls which make up the husk are lignified. This tissue contains traces of proteins, polyphenols, sugars, amino acids and ash. Its surface layers are particularly rich in silica, which gives it its abrasive properties.

It is thought that the 'testinic acid' (mixture of polyphenols and proteins) that is extracted from barley by mildly alkaline steep liquors comes mainly from the husk [9]. The 'husk' fraction of malt has major effects on the flavour and stability of derived beers. Malt grists freed from 'husks' by sieving give bland and insipid beers, while grists enriched with 'husks' produce astringent beers that readily become hazy.

The testa, together with the pigment strand, makes up about 1–3% of the barley corn by weight. This impermeable tissue stains for 'lipidic' materials, and consists of two cuticle-like layers separated by cell walls [1, 10]. It contains a 'cellulosic' polysaccharide component, a chemically resistant 'estolide' of fatty acids and hydroxy fatty acids, and a wax that contains *inter alia* hydrocarbons, and 5-*n*-alkyl resorcinols. It is normally impermeable to water. The effects of organic solvents, such as chloroform or ethyl acetate, in accelerating water uptake may be due to alterations in the testa, or the modified structures which traverse the testa at the micropyle [1].

4.2.2 THE EMBRYO [1–3]
The relatively small embryo (axis *plus* scutellum) is capable of growing into a viable seedling when separated from the grain and cultured on a synthetic nutrient medium. Obviously it has the full potential and metabolic complexity of any small plant. The other parts of the grain either (*i*) protect the inner parts or (*ii*) serve as an elegantly organized food store, and a system for mobilizing these reserves. Neither the embryo, nor any part of the mature grain is green, nor has it any capacity for photosynthesis. Embryos weigh about 1·5 mg,

and represent 2–5% of the grain's dry weight. Ungerminated embryos contain soluble sugars, notably sucrose (14–15% dry weight), raffinose (5–10% dry weight) and higher fructosans. The ash content of the embryo is 5–10% dry weight, about one-fifth of the grain's supply. Cell walls and gums make up a further 16%. The embryo contains a little true cellulose, some pectin (uronic acids) and hemicelluloses. Initially it contains no lignin, but a little forms during germination. The 'crude protein' content is about 34%, but this includes amino acids and other nitrogenous compounds. Most of the embryo lipids (14–17% dry weight) occur in subcellular organelles termed spherosomes. The most abundant lipids are the triacylglycerols (triglycerides) (about one-third of the grain's supply). Thus the embryo itself is well-stocked with easily metabolized reserve substances. In the initial stages of malting the raffinose is fully utilized and it, plus sucrose, fructosan and triacylglycerol provide the initial 'fuel' and 'building blocks' for the growing embryo. Later these are supplemented and replaced by supplies of soluble materials diffusing from the modifying endosperm. During malting the embryo takes up considerable quantities of 'nutrients' and these are respired (so lost as carbon dioxide and water) or converted into the materials of the growing seedling. Of these the rootlets are removed to give finished malt.

4.2.3 THE ALEURONE LAYER AND NUCELLAR TISSUE [1–3, 11]

Residues of the nucellus occur between the aleurone layer and testa, and project into the starchy endosperm as the sheaf cells. It has recently been concluded that the nucellar cuticle becomes fused with the testa as grain matures. Nothing is known specifically of the composition of this tissue, but it is comparatively resistant to enzymic degradation. The aleurone tissue, with the associated nucellar residue and testa, comprises about 8–15% dry weight of the grain. In most malting barleys the aleurone layer is, on average, three cells deep. Each cell has thick cell walls, which are not obviously altered during malting and mashing, although electron microscopy shows that limited wall degradation does occur during malting. Under other conditions of germination wall degradation can be substantial. The cells are filled with cytoplasm, containing a prominent nucleus and densely packed with aleurone grains, which contain proteins, minerals and possibly glycoproteins, and spheroplasts, which contain mainly triacylglycerols. Mitochondria and other organelles are present. During germination the cell contents change drastically in appearance. This tissue is a source of some nutrients for the growing embryo, and it provides many of the hydrolytic enzymes which are important in modifying grain.

In some barley varieties, but not any currently grown in Britain, the cells are coloured blue or red with anthocyanin pigments. Colourless anthocyanogens (proanthocyanins) are present in all commonly grown barleys.

Aleurone cells are wholly devoid of starch, but contain about 20% dry weight triacylglycerol (about 90% of that of the degermed grain), 17–20% 'crude protein' (N × 6·25), and they are rich in minerals (more than 75% of those of the degermed grains), and soluble sugars, particularly sucrose and raffinose, and perhaps fructosans. Some 40–45% of the tissue is cell wall, with a small proportion of associated 'gums', which together contains about 85% arabinoxylan, 8% holocellulose (which contains glucose, mannose and galactose) and 6% protein. Pectins (which contain uronic acids) are virtually absent. Between the interface of the aleurone layer cell walls and the walls of the starchy endosperm deposits of another polysaccharide occur; this material is probably a 'laminarin-like' β-(1→3)-D-glucan.

4.2.4 THE STARCHY ENDOSPERM [1–3, 11]

During malting it is, above all, the accumulation of hydrolytic enzymes in the starchy endosperm and the partial breakdown of its structural components which constitute 'modification'. The material remaining in this tissue at the end of malting provides the bulk of the brewers' extract; consequently malting is carried out in ways designed to minimize excessive losses from this tissue. In contrast, in agriculture, the contents of the endosperm are nearly all used by the seedling in seed grain germinating in the field.

The starchy endosperm is dead, and does not respire. It is not uniform in composition. There is a gradient of increasing starch concentration and decreasing protein and β-amylase concentration moving inwards from beneath the aleurone layer, and there is an invagination of nucellar cell walls as the sheaf cells. Further, the crushed 'cell layer', next to the scutellum, appears to consist of cell walls only, and lacks starch and protein. Thus the composition noted here for this tissue is approximate, and represents an average. The starchy endosperm makes up approximately 75% of the grain's dry weight. The thin cell walls make up about 7% of the tissue and the gums approximately 3% more. The walls contain a holocellulosic 'fibrillar' matrix, about 6% of the whole, containing glucose, mannose and galactose. Of the residual amorphous layer, approximately 23% is arabinoxylan and 71% β-glucan. At most, only small quantities of uronic acids (< 1%) are present. Protein is also present. The insoluble wall components are probably cross-linked by peptides, and/or diferulic acid. Apparently the warm-water-soluble gums are not so linked. 'Protein' (N × 6·25) comprises approximately 9% of the tissue. The protein's strength of adhesion to the cell walls and the starch grains which it invests varies widely between varieties. Where the protein is tough and its adhesion strong it may contribute significantly to the hardness of steely grains. Much of this protein serves as a reserve, and some of it is arranged in small storage granules, of different structure to aleurone grains. All the β-amylase of the grain is located in the starchy endosperm.

In mature grain, starch is practically confined to the starchy endosperm. It makes up about 63–65% of plump two-rowed malting barleys, but less (e.g. 58%) of six-rowed grains. In two-rowed barleys therefore it comprises approximately 85% of the starchy endosperm. During development starch grains are laid down in amyloplasts, and, when mature, retain traces of minerals (including phosphates), lipids, free fatty acids and protein. The polar lipids and free fatty acids may be closely associated with the polysaccharide component [12]. Possibly the residues of the amyloplast membranes still enclose the mature starch granules. The granule's main component is starch. The granules are tightly packed and occur in a wide range of sizes, with diameters of 2 to 40 μm, with two size ranges predominating (1.7–10 μm; 20–48 μm). Large grains are less frequent in the subaleurone layer. The grains have a highly organized structure as shown, for example, by the appearance of a 'dark cross' between crossed Nicol prisms, i.e. in polarized light.

The starchy endosperm contains minerals, some free amino acids and sugars including glucose, fructose, maltose and sucrose. Starch may be dissolved from sections of starchy endosperm, with solutions of salivary amylase, revealing the residual protein and other cell inclusions. Generally cell nuclei cannot be detected in sections of mature grain, probably because they are crushed and destroyed by the starch grains during their development. However, the starchy endosperm probably retains trace amounts of the nucleic acids DNA and RNA.

4.3 *Chemical changes during malting* [1–3]

The gross chemical changes observed during malting are the net result of the degradation of reserve substances, the interconversion of materials in the living tissues, the flux of materials to the embryo, and their incorporation into the growing tissues, the synthesis of new materials and the losses occasioned by (*i*) leaching during steeping, (*ii*) oxidation to carbon dioxide and water, and (*iii*) the separation of the rootlets. Total malting losses usually amount to 6–12% of the original dry weight of the barley. A comparison of the 'average' compositions of two-rowed barleys and malts shows that during malting there is a net degradation of high-molecular-weight polymeric materials, and an accumulation of simpler soluble substances which contribute to the cold-water extract (Table 4.2). Allowing for malting losses it is apparent from Table 4.2 that about 18% of barley starch is degraded during malting. Although this is partly offset by a rise in the levels of simpler soluble sugars, it nevertheless represents a substantial loss of potential brewers' extract. Before going further it must be emphasized that analyses of complex biological materials should be regarded with caution. Some analytical methods are unspecific or imprecise. Often methods capable of collectively estimating all members of a class of compounds, for example polyphenols, are not known. Consequently the proportions shown in Table 4.2 do not add up to 100%.

TABLE 4.2

The chemical composition of European two-rowed barley and malt
Compiled from various sources by HARRIS (1962) [3]

Fraction	Barley	Malt
	\multicolumn{2}{c}{Proportions (% dry weight)}	
Starch	63–65	58–60
Sucrose (16)	1–2	3–5
Reducing sugars	0·1–0·2	3–4
Other sugars	1	2
Soluble gums	1–1·5	2–4
Hemicelluloses	8–10	6–8
Cellulose	4–5	5
Lipids	2–3	2–3
Crude 'protein' (N × 6·25)	8–11	8–11
Albumin ⎫ salt-soluble protein	0·5	2
Globulin ⎭	3	–
Hordein – 'protein'	3–4	2
Glutelin – 'protein'	3–4	3–4
Amino acids and peptides	0·5	1–2
Nucleic acids	0·2–0·3	0·2–0·3
Minerals	2	2·2
Other substances	5–6	6–7

4.3.1 THE CARBOHYDRATES OF BARLEY AND MALT [1–3]

Barley carbohydrates are usually considered in groups defined by the separation techniques used to analyse them. Particular attention will be paid to (*i*) starch, (*ii*) hemicelluloses and gums, and (*iii*) carbohydrates soluble in aqueous alcohol.

Free hexoses, glucose (1, 0·03–0·09%) and fructose (2, 0·03–0·16%) occur in barley and alter in amount during malting (Table 4.3). They, and the pentose sugars arabinose (3) and xylose (4) occur in polysaccharides, as do smaller quantities of galactose (5), mannose (6), glucuronic acid (7) and galacturonic acid (8). The pentose sugars ribose (9) and deoxyribose (10) occur combined in nucleic acids and nucleotides. In addition, fucose occurs in trace amounts, as does glucosamine which is found in glycoproteins.

The soluble diffusible sugars that accumulate during malting originate partly from the degradation of starch, partly from the breakdown of soluble polysaccharides, gums and hemicelluloses, and by synthesis from triacylglycerols. Furthermore sugar interconversions occur in the living tissues. The multiple origins and fates of the soluble sugars, and difficulties encountered in separating them complicate the picture of the alterations that occur during malting.

TABLE 4.3

Changes in the carbohydrate composition of a sample of Carlsberg barley (N, 1·43%) during traditional floor malting. Only traces of free pentoses were ever detected. The disappearance of raffinose was real, but the apparent disappearance of 'glucodifructose' was due to difficulties in separating it from other chromatographic fractions. Carbohydrates are expressed as g/1000 corns. B, Raw barley. M, Finished malt without rootlets. All other samples were analysed with rootlets. Blue value is that of isolated starch; pure amylose, about 1·4; amylopectin, about 0·1. TCW, thousand corn dry weight

HALL et al. (1956) [13]

Carbohydrate	B	Steeping (days) 1	2	3	Germination (days) 1	2	6	9	Kilning (days) 1	2	3	4	M
Soluble carbohydrates													
Fructose	0·027	0·012	0·027	0·042	0·100	0·097	0·132	0·145	0·189	0·267	0·297	0·262	0·194
Glucose	0·016	0·012	0·015	0·031	0·085	0·143	0·173	0·178	0·238	0·312	0·300	0·285	0·462
Sucrose	0·301	0·208	0·298	0·265	0·276	0·370	0·540	0·624	0·519	0·713	0·965	0·755	1·315
Maltose	—	—	0·008	0·012	0·026	0·147	0·179	0·123	0·116	0·109	0·103	0·109	0·257
Maltotriose	—	—	—	—	—	—	0·052	0·063	0·069	0·055	0·055	0·059	0·148
Glucodifructose	0·033	0·040	0·043	0·047	0·029	—	—	—	—	—	—	—	—
Raffinose	0·058	0·054	0·077	0·050	0·014	—	—	—	—	—	—	—	—
Fructosan	0·226	0·584	0·459	0·370	0·401	0·410	0·621	0·842	0·752	0·494	0·477	0·374	0·352
Insoluble carbohydrates													
Glucosan	0·725	0·577	0·711	0·824	0·869	0·876	1·611	2·161	1·396	1·216	1·036	1·129	0·613
Araban	0·047	0·016	0·030	0·039	0·025	0·039	0·120	0·148	0·130	0·141	0·112	0·115	0·110
Xylan	0·050	0·020	0·030	0·039	0·031	0·058	0·139	0·182	0·156	0·175	0·168	0·153	0·142
Starch	25·74	24·72	24·48	25·06	24·63	24·43	22·06	20·70	20·95	20·65	20·90	21·50	20·79
Blue value	0·326	0·331	0·317	0·308	—	0·310	—	0·381	0·371	0·377	0·364	0·374	—
TCW (g)	39·1	39·2	38·8	38·7	38·6	38·9	37·6	37·6	37·2	—	—	36·8	35·3

1 Glucose, β-D-Glucopyranose

2 Fructose, β-D-Fructofuranose

3 Arabinose, α-L-Arabinofuranose

4 Xylose, β-D-Xylopyranose

5 Galactose, α-D-Galactopyranose

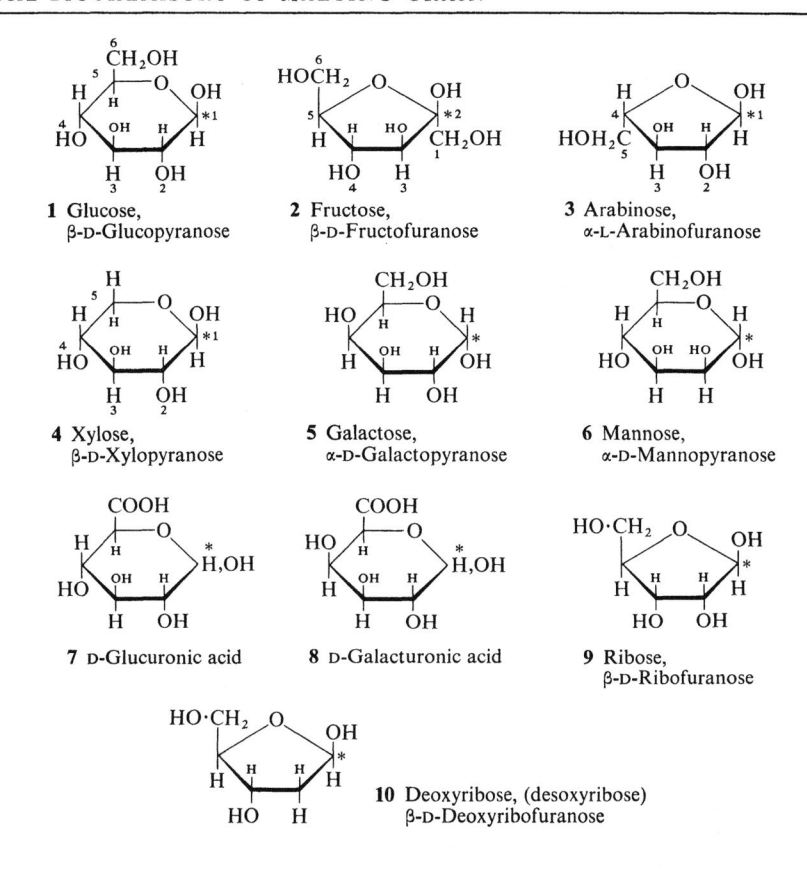

6 Mannose, α-D-Mannopyranose

7 D-Glucuronic acid

8 D-Galacturonic acid

9 Ribose, β-D-Ribofuranose

10 Deoxyribose, (desoxyribose) β-D-Deoxyribofuranose

4.3.2 STARCH AND ITS DEGRADATION [1–3]

Starch, the major component of the barley grain (up to 65%), provides the major part of brewers' extract, either as soluble sugars formed from it during malting or, of more importance, the sugars and dextrins arising from its hydrolysis during mashing. The orderly organization of the starch in the granules is shown by their optical properties, their layered structure detectable under the microscope (particularly after limited amylolysis), and their comparative resistance to enzymic degradation. As noted, the polysaccharide is not the only component of starch grains. The polysaccharide molecules are partly arranged in an ordered 'semi-crystalline' manner, and may be associated with polar lipids. Each grain seems to consist of concentric layers centred on a dark spot, the hilum, and in dried grains or those damaged in milling radial cracks can occur. The orderly structure breaks up when granules are swollen and gelatinized (disrupted and dispersed) in hot water, or when they are battered and 'chipped' in milling operations, or are subjected to other violent physical stresses.

Large and small grains occur, and the latter are thought to be more resistant to amylolysis. However, in normal barleys both large and small granules are usually found to have about the same amylose/amylopectin ratios. The major component of starch grains is a mixture of poly(α-D-glucopyranose) polymers of variable very high molecular weights, called starch. Various techniques separate starch into two fractions, amylose and amylopectin. The starches of most British barleys contain 18–26% amylose (the balance being amylopectin), but varieties are known with as little as 0·3% or as much as 45% amylose. Barleys with 'special' starches have no particular value for the maltster or brewer. Each starch fraction is a mixture of molecules of various sizes differing in detailed structure, and can be fractionated further. 'Pure' amylose consists of unbranched chains of poly [(1→4)-α-D-glucopyranose]. Most preparations are not 'pure' and contain a limited number of α-(1→6) links. Amylose strongly associates with many polar substances, including some lipids, and may be precipitated by them from solution, and form crystalline complexes with them. It gives an intense blue–black colour with iodine. Amylopectin differs from amylose in that its α-(1→4)-linked chains are extensively branched through α-(→6)-linkages, so that these molecules have ramified structures. While amylose molecules have single reducing and non-reducing glucose residues at their chain ends, amylopectin has one reducing chain end, but numerous non-reducing glucose residues which terminate its branches. The ratios of the number of terminal glucose residues to the total number gives an indication of the complexity of the molecules in a fraction. Amylopectin gives a characteristic red tint with iodine solution. The 'blue value' colour given by starch in the presence of iodine compared with the values given by purified amylose and amylopectin provides an estimate of the ratio of these substances to be calculated (Tables 4.3 and 4.4). However, the 'blue value' method gives misleading results when polar lipids are still present. The physical separation of amylose and amylopectin (or fragments of amylopectin obtained by treatment with a debranching enzyme) using exclusion chromatography, seems

TABLE 4.4

Analyses of starches from Carlsberg barley and malt
HARRIS (1962) [3]

	Barley	Malt
Moisture (%)	11·2	14·7
'Protein' content (N × 6·25%)	0·51	2·05
Ratio of glucose residues, non-terminal : terminal	25:1	22:1
'Blue value' with iodine	0·366	0·409
Amylose content (%)	26·2	29·2
Ratio of non-terminal : terminal glucose residues in amylopectin	18–19:1	15–16:1
Amylose (g/1000 corns)	6·72	6·37

a preferable approach. Fractions can be separated which have properties intermediate between those of 'pure' amylose and 'pure' amylopectin so, in fact, starch is a mixture containing a complex range of molecular types.

Barley starch is partly degraded during malting, and the composition of the residual starch progressively alters (Tables 4.3 and 4.4). The changes observed vary a little on different occasions. However, the proportion of amylose in the starch increases, e.g. from 22% (barley) to 26% (malt), although the total amylose content of the grains actually declines, and at the same time the average chain length of the amylopectin is reduced (Table 4.4). It seems that in malting, amylopectin is preferentially degraded. Generally malt starch is more readily degraded than barley starch by standardized mixtures of enzymes.

The mixture of barley enzymes that degrades starch is called diastase. The diastase from ungerminated barley contains a different mixture of enzymes from diastase from malt. The mixture is complex and not all of its component enzymes have been adequately characterized. Although in the mash tun the main characteristics of starch degradation can be explained by the combined activities of α-amylase and β-amylase, at normal temperatures *in vivo* the degradation of starch grains is certainly aided by other enzymes [1-3, 14, 15]. Microscopic evidence shows that the transient starch grains, which appear in the living tissues of the embryo during germination, are degraded in a different way from those of the dead starchy endosperm [1, 4].

Phosphorylase is an enzyme which catalyses the reversible phosphorolysis of the terminal non-reducing residue of starch, releasing glucose 1-phosphate (Fig. 4.2). The location of this enzyme in the barley grain is uncertain, although some occurs in the embryo, where it may usefully catalyse the degradation of transitory starch. The glucose 1-phosphate produced is available for further metabolism. Barley contains many phosphatases which are able to hydrolyse glucose 1-phosphate, phosphorylase and phosphatase, plus inorganic phosphate, are able to act in sequence to degrade starch chains, with the release of glucose. Phosphorylase is not able to hydrolyse α-glucose (1→6) linkages in starch or dextrins, however, or remove substituted glucose residues, so it is only able to degrade the outer chains of amylopectin.

α-Glucosidase, an enzyme concentrated mainly in the embryonic tissues and aleurone layer, increases in amount during germination, and can hydrolyse glucose residues one link at a time from the non-reducing ends of starch chains, dextrins and smaller oligosaccharides, such as maltose (**11**) and isomaltose (**12**). Thus it is able to hydrolyse α-glucosyl-(1→4) *or* (1→6) linkages. Recent evidence suggests that this enzyme is the 'raw starch factor' which accelerates the degradation of starch granules by mixtures of α-amylase and β-amylase [15]. α-Glucosidase has a limited ability to catalyse transglucosidation reactions, and it is likely that other transglucosylases occur in barley. These enzymes are able to transfer sugar moieties or groups of sugar residues from one compound to another with the formation of a similar or a distinct type of linkage. Thus an α-(1→4) link in a chain might be broken and the separated end can be joined to the same or a different chain via either an α-(1→4) link or an α-(1→6) link.

$$\text{G-G-G-(G)}_n\text{-G}\cdots + \text{P}_i \underset{\text{PHOSPHORYLASE}}{\rightleftarrows} \text{G-G-(G)}_n\text{-G}\cdots + \text{G-1-P}$$

Non-reducing chain end — Inorganic phosphate — Shortened chain — Glucose 1-phosphate

$$\text{G-G-G-(G)}_n\text{-G}\cdots + \text{H}_2\text{O} \xrightarrow{\beta\text{-AMYLASE}} \text{G-(G)}_n\text{-G}\cdots + \text{G-G}^*$$

Non-reducing chain end — Shortened chain — Maltose

$$\text{G-G-G}\cdots \text{ or } \text{G}\rightarrow\text{G}^* + \text{H}_2\text{O} \xrightarrow{\alpha\text{-GLUCOSIDASE}} (\text{Substrate-G}) + \text{G}^*$$
or G-G* or G→G-G-···
etc.
Glucose

$$\begin{array}{c}\text{G-G-G-(G)}_n\text{-G-G-(G)}_n\text{-G-G}^* \\ \uparrow \\ \text{G-G-(G)}_p\text{-G}\end{array} + x\text{H}_2\text{O} \xrightarrow{\alpha\text{-AMYLASE}} \begin{array}{c}\text{G}^*, \text{G-G}^*, \text{G-G-G}^*, \\ \text{G-G-G-G-G}^* \\ \uparrow \\ \text{G-G-G}\end{array}$$

Branched and unbranched chains — Mixture of sugars, α-limit dextrins, etc.

$$\begin{array}{c}\text{G-G-G-G-G-G}^* \\ \uparrow \\ \text{G-G-G}\end{array} + \text{H}_2\text{O} \xrightarrow{\text{DEBRANCHING ENZYME}} \begin{array}{c}\text{G-G-G-G-G-G}^* \\ + \\ \text{G-G-G}^*\end{array}$$

Substances with α(1→6) branch points — Products lacking branch points

$$\text{G-G}^* \longrightarrow \text{G}\rightarrow\text{G}^*$$

Maltose — Isomaltose

$$\text{G-G-G-G-(G)}_n\text{-G-G-G}^* \xrightarrow{\text{TRANSGLYCOSYLASE(S)}} \begin{array}{c}\text{G-(G)}_n\text{-G-G-G}^* \\ \uparrow \\ \text{G-G-G}\end{array}$$

Amylose — Amylopectin

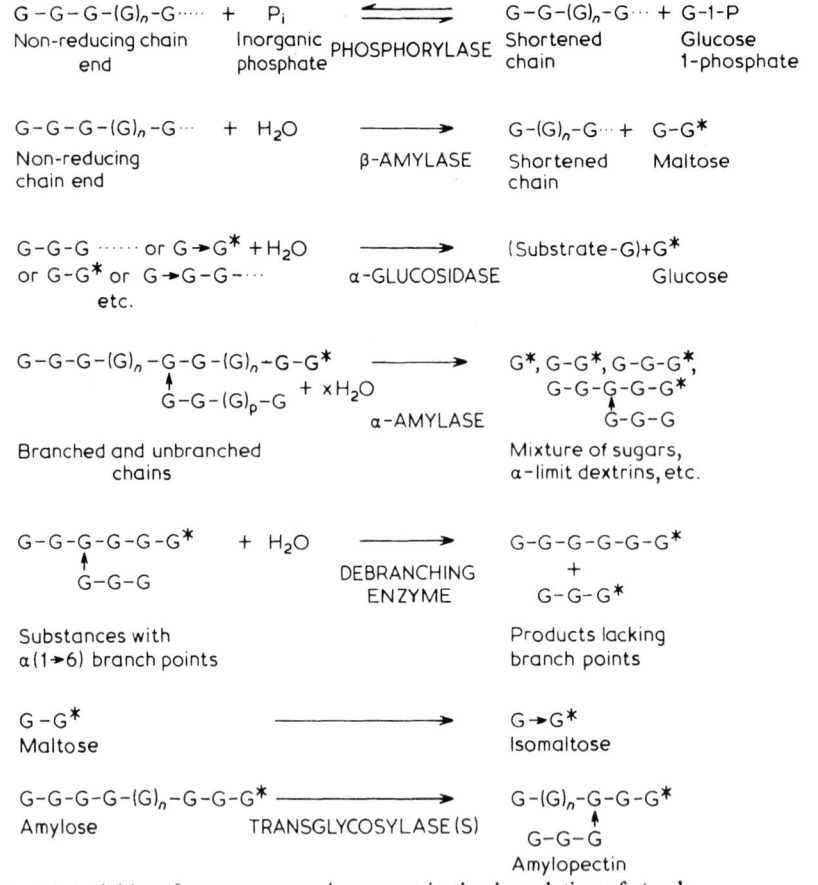

Fig. 4.2 Activities of some enzymes important in the degradation of starch. G = α-D-glucopyranose; - = (1→4) link; → = (1→6) link. * Potential reducing groups.

β-Amylase cannot attack starch granules on its own, but it attacks dextrins and soluble starch at the non-reducing chain ends, hydrolysing the penultimate α-(1→4) links to release the disaccharide maltose (**11**). This enzyme will not attack α-(1→6) linkages, or α-(1→4) links immediately adjacent to them, so while β-amylase acting alone can wholly degrade pure amylose, it degrades amylopectin to maltose and a 'β-limit dextrin' in which all the outer chains have been degraded to stubs, probably two or three glucose residues long, within a short distance of the branch points (Fig. 4.2). β-Amylase occurs in soluble and insoluble forms in barley, but during malting all the enzyme becomes soluble. The bound or latent enzyme may be freed by proteolytic enzymes or agents that can reduce disulphide linkages (such as dithiothreitol). It seems that the bound enzyme is linked to insoluble protein via disulphide bridges, but the salt-soluble enzyme is heterogeneous, being either free or joined by disulphide links to enzymically inactive protein. The 'free enzyme' of malt is known to occur as several isoenzymes, and is slightly smaller than the barley enzyme, so may have undergone limited proteolysis. Apparently no new enzyme protein is synthesized during germination. This enzyme is confined to the starchy endosperm, and is most concentrated in the sub-aleurone layer.

11 Maltose,
α-D-Glucopyranosyl-(1 → 4)-D-glucopyranose

12 Isomaltose,
α-D-Glucopyranosyl-(1→6)-D-glucopyranose

13 Xylobiose,
β-D-Xylopyranosyl-(1→4)-D-xylopyranose

14 Cellobiose,
β-D-Glucopyranosyl-(1→4)-D-glucopyranose

15 Laminaribiose,
β-D-Glucopyranosyl-(1→3)-D-glucopyranose

16 Sucrose,
α-D-Glucopyranosyl-(1→2)-β-D-fructofuranoside

In contrast, α-amylase attacks α-glucosido-(1→4) linkages at random within the starch chains, although attack ceases near to α-(1→6) branches, and is slower near the chain ends. Thus this enzyme acting on its own is able to degrade starch to a complex mixture of sugars (including glucose and maltose) and a wide range of dextrins, some of which contain α-(1→6) linked branches. It is able to slowly initiate an attack on starch granules and then progressively degrade them. This degradation occurs faster if β-amylase is present, and faster still if α-glucosidase is added to the mixture. Wherever a starch chain is split by α-amylase, a new non-reducing chain end is exposed to attack by β-amylase or other enzymes that attack non-reducing chain ends. Thus limited α-amylolysis can enable β-amylase to 'by-pass' α-(1→6) branch points, so that, acting in concert, these enzymes achieve a more rapid and complete degradation of starch (including soluble starch and not only starch granules) than either acting alone. α-Amylase is virtually absent from raw barley, but it is synthesized in considerable amounts in malting, in a manner considered later. This enzyme also occurs as a complex mixture of isoenzymes.

Other enzymes besides α-glucosidase can hydrolyse α-(1→6) linkages. Various enzyme preparations have been called R-enzyme and limit dextrinase, but it now seems that these are really a single entity. 'Debranching' enzyme(s) from germinated barley is able to hydrolyse the α-(1→6) linkages in β-limit dextrin and 'limit dextrins' remaining after the degradation of starch by mixtures of α-amylase and β-amylase.

During kilning diastatic enzymes are degraded to variable extents. It seems that α-glucosidase and 'debranching enzyme(s)' are inactivated more readily than β-amylase, while α-amylase is least readily degraded. Consequently malts intended to have maximal 'diastatic power', such as distillers' malts, are kilned at the lowest possible temperatures.

4.3.3 HEMICELLULOSES AND GUMS [1-3]

True cellulose, pure poly-β-(1→4)-D-glucosan and holocellulose, (a polysaccharide fraction insoluble in sodium hydroxide solution, which contains glucose, mannose and traces of galactose), are confined mainly to the husk, although small quantities do occur in the cell walls of other tissues. Traces of pectin (materials containing uronic acids) also occur in the husk and embryo. Since little is known of them and since they appear not to change during malting, they will not be considered further.

Hemicelluloses and gums are mixtures of polysaccharides having similar chemical compositions. However, gums are extracted by hot water, and raising the temperature (e.g. from 40 to 80°C; 104 to 176°F) extracts more gum and leaves less hemicellulose in the residue. The hemicelluloses require other treatments such as extraction with sodium hydroxide solution to dissolve them. Probably the soluble gums grade imperceptibly into the hemicelluloses, which are made up of 'gum-like' materials, which are less soluble by virtue of having greater molecular weights and/or increasing degrees of cross-linking. The hemicelluloses and gums of most interest to the maltster occur in the starchy endosperm, as components of the cell walls. Undoubtedly they serve as a polysaccharide reserve substance, although they are less important than starch [11]. However, similar classes of material occur in all parts of the grain. As usually prepared, hemicelluloses comprise about 8% and gums 2% of the barley grain. The major components of the gums are (*i*) pentosans containing chiefly arabinose (3) and xylose (4) and (*ii*) β-glucans, made up of glucose (1). However, traces of galactose (5), mannose (6) and uronic acids (7 and 8) have been detected in various gum preparations. Doubtless many minor polysaccharides remain to be characterized.

Pentosans from different parts of the grain vary in their composition. A material from the husk consists of a chain of β-(1→4)-D-xylopyranose residues (4) variously substituted with α-L-arabinofuranose (3) and D-glucuronic acid (7) (Fig. 4.3). Hydrolysates of husk also contain galacturonic acid (8). Pentosan gums from the starchy endosperm are rather simpler in structure, in that the xylan chain is substituted with α-L-arabinofuranose residues only (Fig. 4.4). Similar pentosan gums make up a small part of the polysaccharide of the cell walls of the starchy endosperm, but a major part of the cell walls of the aleurone layer. The ratio of arabinose to xylose residues differs widely between preparation, e.g. from 1 : 1 to very low values, and there are indications that pure arabans and xylans may also occur. Because of their comparatively low viscosity pentosan gums cause few problems in beer manufacture, and so have attracted relatively little interest.

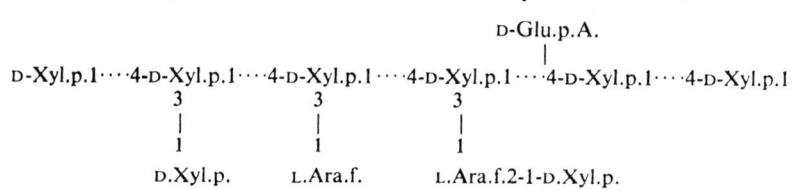

$$\begin{array}{c} \text{D-Glu.p.A.} \\ | \end{array}$$

D-Xyl.p.1···4-D-Xyl.p.1···4-D-Xyl.p.1···4-D-Xyl.p.1···4-D-Xyl.p.1···4-D-Xyl.p.1
       3   3   3
       | | |
       1   1   1
    D.Xyl.p.   L.Ara.f.   L.Ara.f.2-1-D.Xyl.p.

Fig. 4.3 Pentosan hemicellulose of barley husk [3]. D-Xyl.p., β-D-xylopyranose residue. L-Ara.f., α-L-arabinofuranose residue. D-Glu.p.A., D-glucuronic acid (pyranose form) residue. Sugar residues joined by dotted lines may be linked either directly or through a chain of (1→4)-linked D-xylopyranose residues, which probably have the β-configuration.

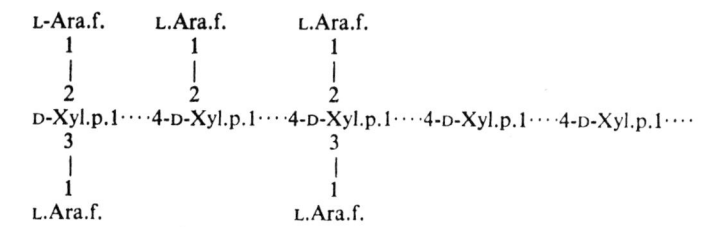

   L-Ara.f.   L.Ara.f.   L.Ara.f.
      1   1   1
      | | |
      2   2   2
D-Xyl.p.1···4-D-Xyl.p.1···4-D-Xyl.p.1···4-D-Xyl.p.1···4-D-Xyl.p.1···
       3   3
       | |
       1   1
    L.Ara.f.   L.Ara.f.

Fig. 4.4 Water-soluble pentosan gum of the barley endosperm [3]. Symbols as for Fig. 4.3.

β-Glucans (*syn.* β-glucosans) are a family of polysaccharides made up of unbranched chains of β-D-glucopyranose residues joined by (1→4) and (1→3) linkages. Again the ratio of the bond types is different in different preparations (e.g. approximately 1 : 1 or 7 : 3), and chain lengths vary. A major proportion of many β-glucan preparations contain repeating units in which single or a small series of (1→4) linkages alternate with (1→3) linkages [16]. However, it seems that in a few areas adjacent (1→3) bonds may occur [17]. Furthermore small quantities of what appears to be a laminarin-like β-glucan, in which most residues are joined by β-(1→3) linkages, has been detected in the cell walls at the interface between the starchy endosperm and the aleurone layer [18]. Hemicellulosic β-glucan may be extracted into dilute sodium hydroxide or into warm water following hydroxylaminolysis or treatment with selected proteolytic enzymes such as a barley acidic carboxypeptidase. Thus it is likely that the less-soluble β-glucans are covalently cross-linked through peptides, and possibly diferulic acid. Cell walls fluoresce blue in ultraviolet light, and other analyses also indicate that they are esterified with ferulic and diferulic acids.

During malting the β-glucans of the cell walls are initially solubilized and the level of β-glucan gums increases, but ultimately the soluble β-glucans are degraded relatively more than the pentosans, so that gum from malt contains 70–90% pentosan, while gum from raw barley contains only 14–20% pentosan. Undegraded β-glucans form highly viscous solutions and can create problems during wort separation, and beer filtration. Furthermore they may form gelatinous precipitates in strong beers. Thus, although β-glucans may conceivably contribute to beer foam and palate fullness they must be adequately degraded during malting or mashing. Although the decreasing viscosity of extracts prepared from barleys at different stages of malting has been a useful guide to following modification in relatively undermodified European lager and North American malts, this index has proved to be an unreliable guide when making well-modified British ale malts. This is because the decline in viscosity stops before modification is complete (see Chapter 5). The alterations in the quantities and qualities of the hemicelluloses and gums which take place during germination are mainly caused by their enzyme-catalysed hydrolytic degradation, although small quantities are synthesized in the growing embryo. In some rapid malting processes extract recovery is maximal before β-glucan degradation is complete, so germination must be prolonged to ensure that worts from the malt are not unacceptably viscous.

Enzyme activity, probably proteolysis, increases the solubility of the hemicelluloses converting them to gums which, in turn, are degraded to oligosaccharides and sugars. Several enzymes are involved in the hydrolysis of the pentosans (Fig. 4.5) [1, 3, 16]. Arabinosidase catalyses the hydrolysis of the α-L-arabinofuranose (3) residues from the xylan chains, or from arabinoxylan oligosaccharides. This xylan chain, in turn, is disrupted, when sufficient arabinose residues have been removed by an endoxylanase, which degrades the molecule to xylan and arabinoxylan oligosaccharides. The existence of an exoxylanase, which is said to attack the non-reducing ends of the xylan chains releasing xylobiose (13), was deduced from inhibitor studies, but this enzyme has not been isolated, and the observed activity may be due to the combined activities of other enzymes. Xylobiose (13) is hydrolysed to xylose (4) by xylobiase, or β-D-xylopyranosidase. It is likely that this enzyme can also hydrolyse single unsubstituted xylose residues from the ends of the xylan chains. As free pentoses do not accumulate to any extent in malting grain, they must be rapidly utilized by the living tissues.

THE BIOCHEMISTRY OF MALTING GRAIN

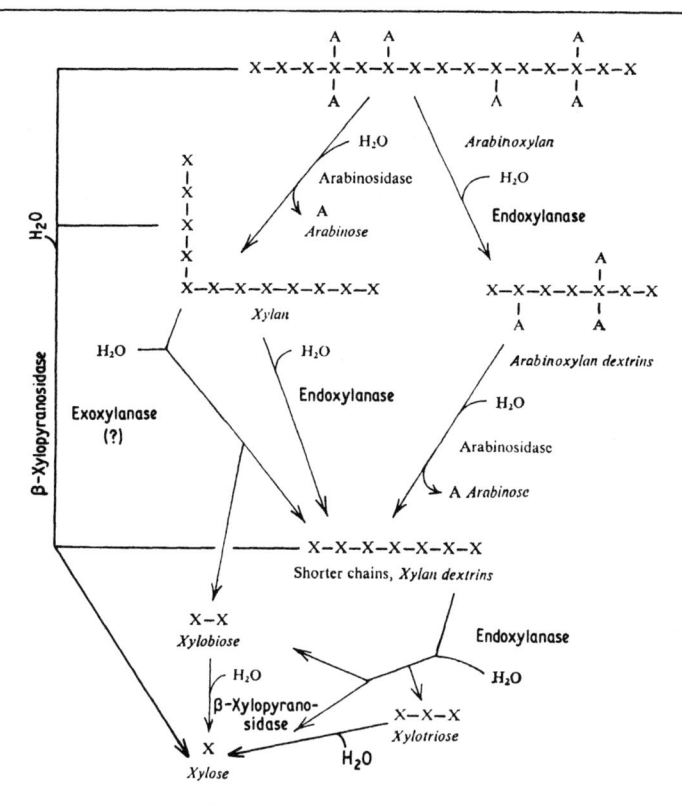

Fig. 4.5 Possible routes by which enzymes degrade arabinoxylan gum. X = β-D-xylopyranose residues. A = α-L-arabinofuranose residues.

The initial increase in β-glucan gum that occurs during malting is at the expense of hemicellulosic β-glucan. It is likely that proteases act to break the peptide cross-links between the hemicellulosic glucan chains [19]. Various β-glucanases and associated enzymes are known in germinating barley [1, 3, 16] (Fig. 4.6). An active 'laminarinase' occurs, which has only limited activity on 'barley β-glucan', in those regions of the molecules where several consecutive β-(1→3) linkages are found. However, the recently discovered 'laminarin-like' β-glucan may be its natural substrate. Two other endo-β-glucanases are present. One, barley endo-β-glucanase, has a high activity on β-glucans having a mixture of β-(1→3) and β-(1→4) linkages, while the other enzyme, a 'cellulase', attacks barley β-glucans and artificial substrates containing only β-(1→4)-linked D-glucopyranase residues. In solution the random hydrolytic breakage of β-glucan chains catalysed by these enzymes causes a rapid fall in viscosity. Some early studies indicated that exoglucanases were present. These were supposed to break the penultimate links at the non-reducing ends of β-glucan chains, releasing disaccharides, either cellobiose (14) or laminaribiose (15). However, these enzymes have not been isolated, and the same disaccharides can be released by the endo-β-glucanases. Two β-glucosidases occur, which are both able to hydrolyse cellobiose and laminaribiose to glucose and hydrolyse the terminal non-reducing β-D-glucopyranose residues from the ends of many β-(1→3)- or β-(1→4)-linked oligosaccharides and polysaccharides. Accumulations of cellobiose or laminaribiose have not been detected in germinating grain, so it is probable that they are rapidly degraded *in vivo*.

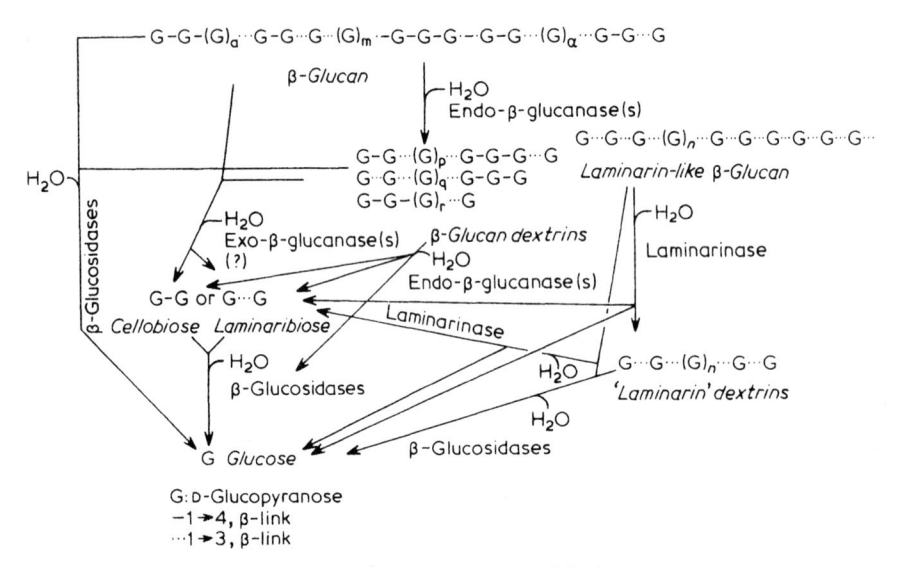

Fig. 4.6 The proposed routes of enzymic degradation of β-glucan gum.
G = β-D-glucopyranose residue. — = (1→4) linkage. ... = (1→3) linkage.

TABLE 4.5

The activities of various enzymes involved in the degradation of gums at different stages in the malting process. Values relative to those of barley, arbitrarily set at 1·0
HARRIS (1962) [3]

	Endo-β-glucanase	Endo-xylanase	Arabino-sidase	Xylo-biase	Exo-xylanase*
Barley	1·0	1·0	1·0	1·0	1·0
Green malt	11	3·1	1·8	2·3	2
'Low-dried' malt	17	2·0	2·1	1·2	0·5
Finished malt	13	1·8	1·1	0·5	0·4

* Approximate values.

The enzymes that catalyse gum degradation change markedly in amount during malting (e.g. Table 4.5). Because of the importance in achieving an adequate breakdown of β-glucan, endo-β-glucanase has been extensively studied. The changes in viscosity that occur when milled barley is mashed have been evaluated as a means for deciding if a barley is of malting quality, as the initial rise in viscosity is due mainly to gums dissolving in solution, and the subsequent decline is due to their hydrolysis. However, the test is without value, since the important factors are the level of endo-β-glucanase and the quantity and quality of the β-glucan remaining in the finished malt.

4.3.4 SOLUBLE CARBOHYDRATES [1-3]

Ungerminated barley contains 1-2% of carbohydrates that are soluble in 70% ethanol (Tables 4.2, 4.3 and 4.6). Sucrose (**16**; 0·34-1·69%) is always the most abundant component, followed by raffinose (**17**; 0·14-0·83%), 'glucodifructose' (**18, 19**; 0·1-0·43%) and higher oligosaccharides and fructosans (**20**; about 0·5-1·0 per cent). Traces of other sugars, including glucose (**1**), fructose (**2**) and maltose (**11**), also occur. These last three occur particularly in the endosperm, the maltose being concentrated in the sub-aleurone layer, where β-amylase is also most abundant.

TABLE 4.6
Sugars, soluble in 90% ethanol, of Ymer barley and malt
MACLEOD, TRAVIS and WREAY (1953) [20]

	Quantities of sugars (mg/1000 corns)	
	Barley	Malt without roots
Fructose, 2	41	221
Glucose, 1	34	629
Sucrose, 16	421	2100
Maltose, 11	22	299
Glucodifructose, 18 and 19	92	180
Raffinose, 17	203	0
Maltotriose, 21	0	74
Other higher oligosaccharides, including fructosans	326	181

Glucodifructose is a mixture of trisaccharides, probably kestose (**18**) and isokestose (**19**). These are the simplest members of two series of fructosans (**20**) in which sucrose is linked to extended chains of D-fructofuranose residues joined mainly by β-(2→1) linkages in the kritesin (inulin) type, and mainly β-(2→6) links in the hordeacin (phlein) type. Sucrose (**16**), raffinose (**17**), 'glucodifructose' (**18 and 19**) and fructosans (**20**) are concentrated in the living grain tissues, the embryo and the aleurone layer.

17 Raffinose,
α-D-Galactopyranosyl-(1→6)-α-D-glucopyranosyl-(1→2)-β-D-fructofuranoside

18 Kestose (6-kestose),
β-D-Fructofuranosyl-(2→6)-β-D-fructofuranosyl-(2→1)-α-D-glucopyranoside

19 Isokestose (1-kestose),
β-D-Fructofuranosyl-(2→1)-β-D-fructofuranosyl-(2→1)-α-D-glucopyranoside

20 Tetrasaccharide 'bifurcose', possibly the basic unit of the fructosans. Extended in direction A with chains of β-fructofuranose units mainly (2→1)-linked, kritesin (inulin) type. Extended in direction B with β-fructofuranose units mainly (2→6)-linked, hordeacin or phlein type

21 Maltotriose, ($n = 1$); Maltotetraose, ($n = 2$); Maltopentaose, ($n = 3$)

During malting the quantities of the simple sugars alter dramatically, the details of the changes depending on the malting conditions (Tables 4.3 and 4.6). Some, like maltose (**11**) and maltotriose (**21**), are formed at the expense of the starch, but sucrose (**11**) is formed in the growing tissues, either from other sugars or from triacylglycerols. Raffinose (**17**), a trisaccharide that contains galactose, seems to disappear completely during malting, and does not reappear. During malting, considerable quantities of sugars are used to support respiration, or they are converted into insoluble structural components of the growing embryo. During kilning many sugars decline in amount, some interacting with nitrogenous materials to form melanoidins, but the quantity of sucrose often increases. The sugars present in malt are the main constituents of the 'preformed' cold-water extract.

4.3.5 REGULARITIES IN THE CARBOHYDRATE COMPOSITION OF BARLEY [1, 21]

The various compositions of different samples of even one variety of barley gave rise to difficulties in assessing the intrinsic malting values of different cultivars. Useful comparisons were more readily made after it had been realized that the composition of grain does not vary in a random manner. Thus within a particular variety there are regular relationships between the total carbohydrate content of a sample and the proportion of various carbohydrate fractions (Fig. 4.7). With increasing total carbohydrate, 'extract' (mainly preformed sugars and starch) increases proportionately more than the 'hemicellulose' or 'insoluble carbohydrate' fractions. Six-rowed barleys generally contain relatively less extract and more of the other fractions than do two-rowed barleys. Barleys having high 'extracts' should give rise to malts with high 'extracts' (Chapter 5). Barley samples that are proportionately rich in carbohydrate (and therefore poor in 'protein', $N \times 6.25$) will make 'extract'-rich malts. If malting is carried out in a standardized way, then there are correlations between barley analyses and malt analyses. Consequently, within limits, barley analyses may be used to predict the gross chemical composition of the malts that are prepared from them [1].

4.3.6 NITROGENOUS MATERIALS IN BARLEY AND MALT [1-3]

The nitrogenous constituents of barley are of great interest to maltsters and brewers. As already noted, a barley sample with a high nitrogen or 'crude protein' content gives a malt with a lower yield of extract than a 'low-nitrogen' sample of the same variety. Barley with a high nitrogen content is

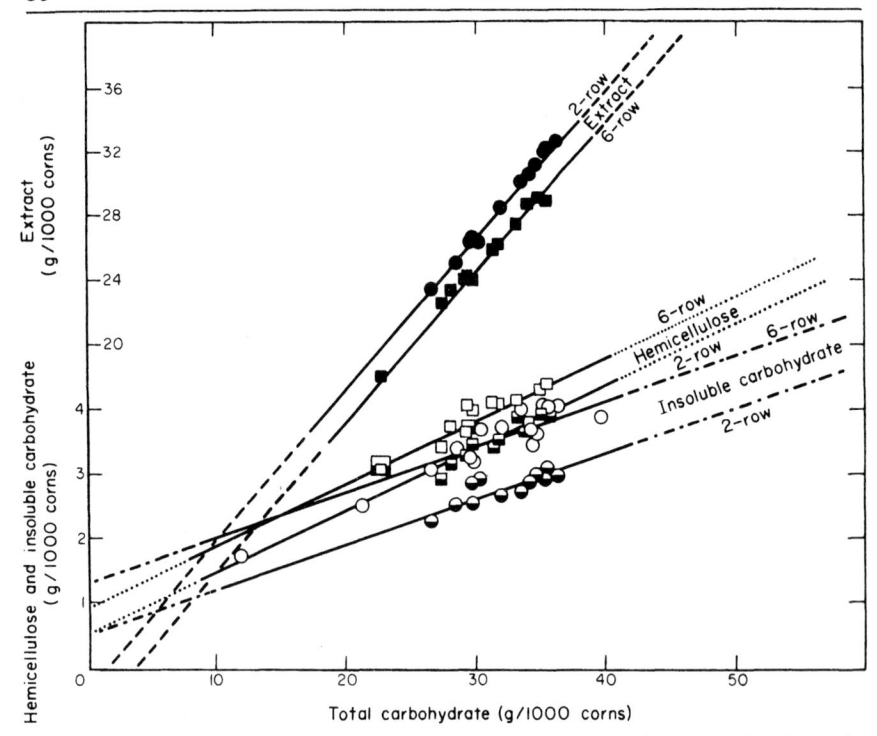

Fig. 4.7 The regular alterations in the proportions of the crude carbohydrate fractions of a two-rowed barley and a six-rowed barley, with alterations in their total carbohydrate contents. After Bishop and Marx (1934) [21].

sometimes slow to 'modify' during malting, has a high respiration rate and its roots grow quickly, so that it makes malt at the expense of a high malting loss. Furthermore the quality of beer (flavour, palate fullness, tendency to form haze) is influenced by the nitrogen content of the malt from which it is made.

By convention the 'crude protein' content of barley is the nitrogen content × 6·25 without regard to the fact that many nitrogenous substances besides true protein are present. In recent years barley samples with nitrogen contents in the range 1·3–3·2% have been on offer in the United Kingdom but in nearly every case samples with nitrogen contents in the range 1·4–1·9% N are used in making malts. Modern methods of analysis (particularly column chromatography, electrophoresis and isoelectric focusing) have shown that, like other living organisms, barley contains numerous true proteins and proteins linked with a wide range of associated substances to yield conjugated proteins such as lipoproteins and glycoproteins. Eventually it should be possible to distinguish several thousand types of protein. Up to now simple fractionations have proved to be most generally useful. Various separation systems have been based on the solubility characteristics and ease of precipitation of various fractions by different reagents [3, 6]. For example Bishop [22–25] extended the classical Osborne method of fractionation, and divided extracts into simple nitrogenous substances, the proteoses and peptones (taken together – non-protein nitrogenous compounds). The proteins soluble in a salt solution are divided into albumins and globulins. Material insoluble in salt solutions, but extractable with warm ethanol is called hordein, while that remaining after the ethanol extraction is called glutelin. Such a separation scheme is empirical. The quantities of the fractions alter during malting (Table 4.7), but it must be noted that the detailed composition of each of these fractions changes with malting time. Still, recent studies have confirmed that these fractions do broadly represent classes of substances with different functions in the grain. Thus the hordein fraction represents mainly storage protein and its subfractions which, when separated by electrophoresis, are characteristic of different varieties. It is clear that each of these fractions is highly heterogeneous. Nevertheless the fractions increase in a regular manner during grain development, and in a series of mature samples of one variety, the amount of nitrogen in each fraction varies in a regular manner with the total nitrogen content (Fig. 4.8). Thus as 'extract', which represents the major storage carbohydrate of the grain, increases most rapidly with increasing grain total carbohydrate, so hordein, the major protein-storage fraction, increases most rapidly with increasing grain total nitrogen (TN) content. The recognition of the 'regularity principle' behind grain composition led to a more rational approach to barley evaluation, and particularly to the development of extract prediction equations (Chapter 5). The details of the relationship between grain TN and the proportions of the different solubility fractions differ in different grain varieties.

TABLE 4.7

Changes in the nitrogen-solubility fractions (N, g/1000 grains) of a Chilian six-rowed barley, steeped at 14·5°C (58°F) during 0–53 hr, germinated between 20°C (68°F) in couch to 120 hr and 14·5°C (58°F) to 290 hr and then to 17·5°C (63°F) to 336 hr. Loaded to kiln at 340 hr. Kilning finished at 460 hr. All samples were analysed with their rootlets. TCW, thousand corn dry weight. NPN, non-protein nitrogen
BISHOP (1929) [24]

Process time (hr)	TCW (g)	Total N (g)	Albumin N (g)	Globulin N (g)	Hordein N (g)	Glutelin N (g)	Proteose N (g)	NPN (g)
0	46.0	0.807	0·099	0·076	0·261	0·267	0·025	0·132
53	45·1	0·795	0·089	0·074	0·262	0·262	0·017	0·091
124	44·5	0·783	0·099	0·071	0·249	0·240	0·008	0·114
170	43·7	0·788	0·120	0·093	0·207	0·207	0·003	0·159
217	44·2	0·804	0·111	0·101	0·199	0·213	0·009	0·168
288	43·4	0·791	0·102	0·111	0·192	0·190	0·023	0·167
337	42·9	0·779	0·085	0·117	0·183	0·195	0·021	0·173
384	42·7	0·782	0·104	0·081	0·191	0·222	0·040	0·150
460	41·4	0·758	0·060	0·114	0·176	0·228	0·033	0·147

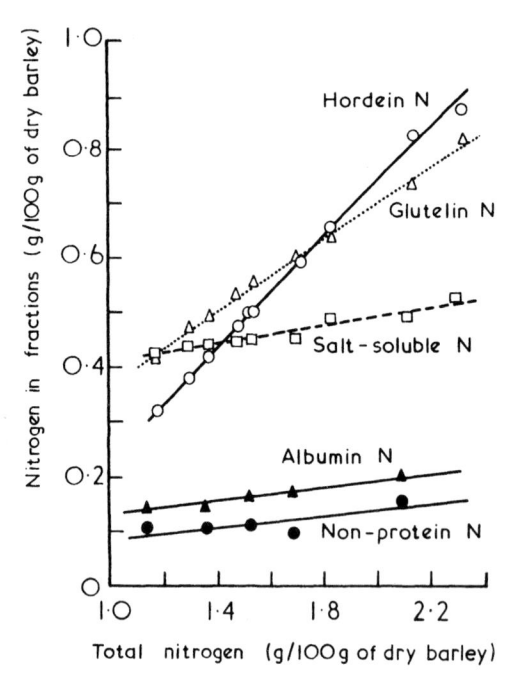

Fig. 4.8 The regular alterations in the proportions of the different crude protein fractions of mature Plumage-Archer barley samples, having different total nitrogen contents. After Bishop (1928) [22].

During malting the proportions of the various nitrogenous fractions alter. There is a net decline in hordein, and a rise in simpler water-soluble substances (Table 4.7). There is a massive breakdown of substances in the endosperm, many hydrolytic enzymes increase in amount, simple nitrogenous

TABLE 4.8

Dry weights and nitrogen contents of parts of barley during malting
After BROWN (1909) [26]

	Dry weight, (g/1000 corns)				Nitrogen, (mg/1000 corns)		
	Total wt grain	Endosperm + integuments	Germs	Roots	Endosperms + integuments	Germs	Roots
Barley, steeped	42·32	40·81	1·51	0	543	83	—
Malt at 5 days	41·70	38·64	1·81	1·26	456	114	57
Malt at 11 days	38·46	31·77	5·00	1·69	323	228	76

substances are taken up in quantity by the embryo, and are used in the synthesis of new tissues. In addition, various changes occur during kilning, for example some nitrogenous substances interact with carbohydrates to form melanoidins, while others, including some enzymes, are denatured. Apart from substances leached during steeping there is no loss of nitrogen from the grain during malting. However, the 'respiratory losses' of dry matter that occur tend to raise the nitrogen content of the grain (Table 4.7), while the substantial quantity of nitrogen that moves into the roots, (Table 4.8) which are eventually removed, tends to reduce the nitrogen percentage in the finished malt. Thus the final nitrogen content of malt depends on these conflicting trends, as well as the nitrogen content of the original barley. When malt is mashed nitrogenous substances come into solution, and are called the total soluble nitrogen (TSN). That part remaining in solution after a standardized period of boiling (or by convention 0·94 × TSN) is called the permanently soluble nitrogen (PSN). These soluble fractions arise partly from preformed soluble substances, and partly from the proteolysis of initially insoluble proteins that occurs during mashing. As malting is extended, so the level of soluble nitrogen rises (both TSN and 'preformed soluble nitrogen'); then eventually it 'peaks' and declines (Figs. 4.9 and 4.10).

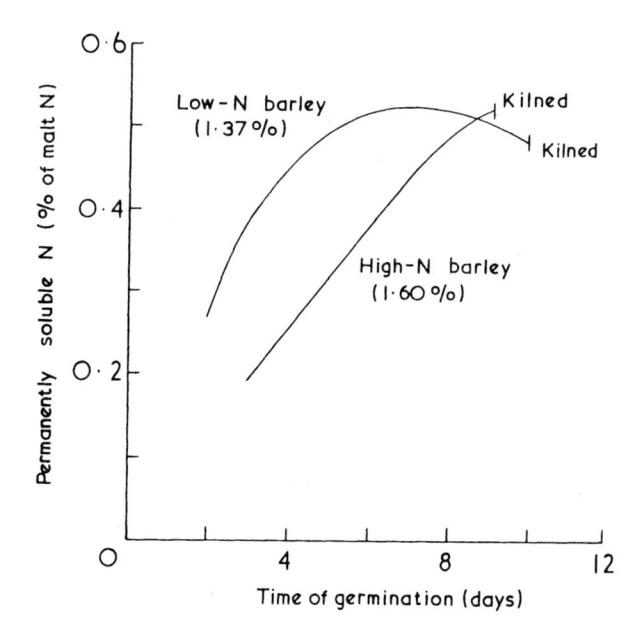

Fig. 4.9 The alterations in the permanently soluble nitrogen extractable from two samples of barley, having different nitrogen contents, during the malting process. After Macey (1957) [27].

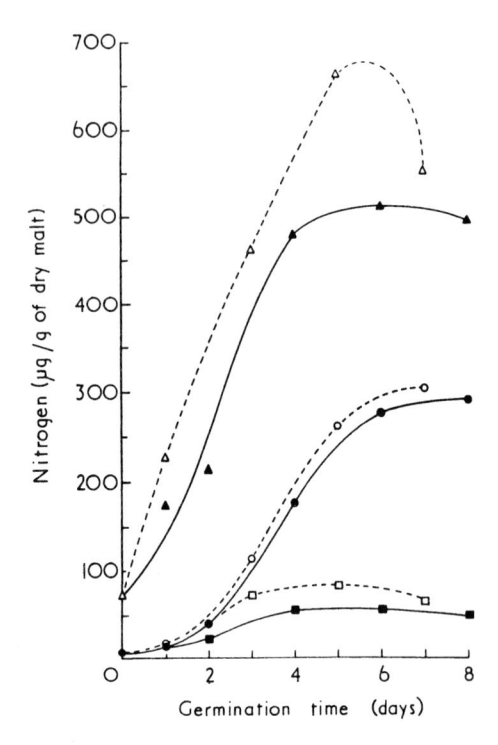

Fig. 4.10 Alterations in the levels of proline nitrogen, total amino acid nitrogen and ammonium nitrogen in Proctor barley, malting with and without an application of gibberellic acid (GA₃). Data of Jones and Pierce (1963) [28]. ▲, Total amino acid N; △, Total amino acid N (GA₃); ●, Proline N; ○, Proline N (GA₃); ■, Ammonium N; □, Ammonium N (GA₃).

The ratio of the soluble nitrogen obtained from malt to the total nitrogen content of the malt is used to follow the progress of malting. The so-called 'nitrogen index of modification', traditionally used in the U.K., is PSN × 100/TN (%) [6], but increasingly the soluble nitrogen ratio (SNR which is TSN × 100/TN (%)) is being used. When a series of malts are made from barley samples (of one variety) of progressively greater nitrogen contents the TSN values of the malts are progressively greater, but they do not increase in direct proportion to the total nitrogen contents of the malts. Consequently the soluble nitrogen ratios decline as the series is ascended.

Using column chromatography as the analytical method, the levels of free amino acids have been followed during malting. Peptides also occur in grain, but analytical methods for these are less satisfactory. In addition to the common amino acids, amides and imino acids that occur in proteins (Table 4.9), barley contains significant quantities of other free amino acids,

TABLE 4.9
Amino acids and related substances of barley and malt

CH$_3$·CH(NH$_2$)·COOH
22* α-Alanine

NH$_2$·CH$_2$·CH$_2$·COOH
23 β-Alanine

HOOC·(CH$_2$)$_3$·CH(NH$_2$)·COOH
24 α-Aminoadipic acid

NH$_2$·(CH$_2$)$_3$·COOH
25 γ-Aminobutyric acid

$\begin{array}{c}NH_2\\HN\end{array}$C·NH·(CH$_2$)$_3$·CH(NH$_2$)·COOH
26* Arginine

HOOC·CH$_2$·CH(NH$_2$)·COOH
27* Aspartic acid

NH$_2$·CO·CH$_2$·CH(NH$_2$)·COOH
28* Asparagine

HS·CH$_2$·CH(NH$_2$)·COOH
29* Cysteine

S·CH$_2$·CH(NH$_2$)·COOH
|
S·CH$_2$·CH(NH$_2$)·COOH
30* Cystine

HOOC·(CH$_2$)$_2$·CH(NH$_2$)·COOH
31* Glutamic acid

NH$_2$·CO(CH$_2$)$_2$·CH(NH$_2$)·COOH
32* Glutamine

NH$_2$·CH$_2$·COOH
33* Glycine

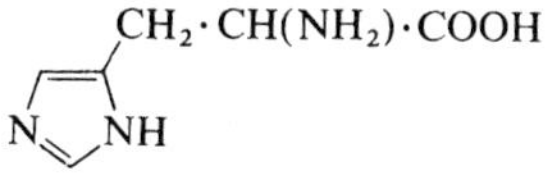

CH$_2$·CH(NH$_2$)·COOH (imidazole ring)
34* Histidine

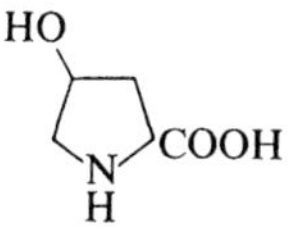

HO- (hydroxypyrrolidine)-COOH
35* Hydroxyproline

$\begin{array}{c}CH_3·CH_2\\CH_3\end{array}$CH·CH(NH$_2$)·COOH
36* Isoleucine

$\begin{array}{c}CH_3\\CH_3\end{array}$CH·CH$_2$·CH(NH$_2$)·COOH
37* Leucine

NH$_2$·(CH$_2$)$_4$·CH(NH$_2$)·COOH
38* Lysine

CH$_3$·S·(CH$_2$)$_2$·CH(NH$_2$)·COOH
39* Methionine

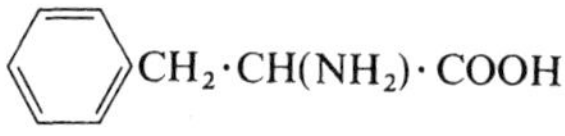

⟨phenyl⟩CH$_2$·CH(NH$_2$)·COOH
40* Phenylalanine

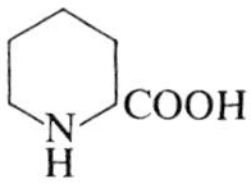

(piperidine)-COOH
41 Pipecolinic acid
(= piperidine-
2-carboxylic acid)

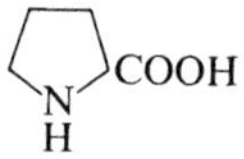

(pyrrolidine)-COOH
42* Proline

HO·CH$_2$·CH(NH$_2$)·COOH
43* Serine

CH$_3$·CH(OH)·CH(NH$_2$)·COOH
44* Threonine

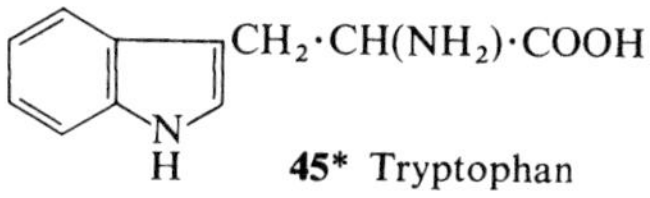

(indole)CH$_2$·CH(NH$_2$)·COOH
45* Tryptophan

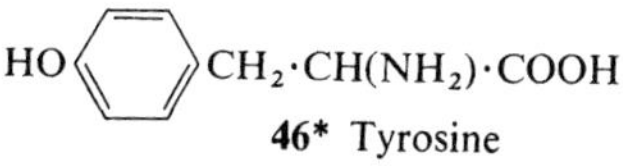

HO-⟨phenyl⟩CH$_2$·CH(NH$_2$)·COOH
46* Tyrosine

$\begin{array}{c}CH_3\\CH_3\end{array}$CH·CH(NH$_2$)·COOH
47* Valine

* Occur in proteins.

such as β-alanine (23), α-aminoadipic acid (24), γ-aminobutyric acid (25), and L(-)-pipecolinic acid (41). The rise in free amino acids that occurs in malting is consequent upon the hydrolysis of reserve proteins in the endosperm (Fig. 4.10). The imino acid proline (42), which forms a high proportion of these substances, is not assimilated by yeast under anaerobic conditions, and so it survives in the finished beer. Proline accumulates in barley plants under conditions of 'water stress' [1], and malting can be regarded as germinating seed under 'water stress'.

$$NH_2 \cdot CH \cdot COOH$$
$$| \quad R$$
48 Generalized formula of an amino acid

$$-CH-CO-NH-CH-$$
49 Peptide link

$$NH_2 \cdot CH \cdot CO \left(.NH.CH \cdot CO \cdot \right)_n NH \cdot CH \cdot CO \cdot NH \cdot CH \cdot COOH$$
$$\quad | \qquad\qquad | \qquad\qquad\qquad | \qquad\qquad | $$
$$\quad R_1 \qquad\quad R \text{ various} \qquad\quad R_2 \qquad\quad R_3$$
50 Generalized peptide chain

A 'generalized' amino acid may be represented by **48**. Peptides and proteins consist of chains of amino acids joined by peptide linkages (**49, 50**). These links are cleaved by a wide range of hydrolytic enzymes. Barley contains several proteases, including at least five endopeptidases, which increase markedly in amount (e.g. × 20) during malting. The thiol-containing proteases have pH optima in the range 3·9–6·0, while the pH optima of the metal-activated enzymes occur in the range 5·5–8·0 [29]. In addition, germinating grain contains three carboxypeptidases (pH optima 4·8–5·6), which are able to attack hordein and increase the solubility of β-glucan, probably by breaking cross-linking peptides [19], and may be responsible for liberating bound β-amylase. Four neutral aminopeptidases and two dipeptidases have also been distinguished. Proteases and carboxypeptidases accumulate in the starchy endosperm during germination, while the other enzymes are concentrated in the scutellum [29]. The marked stimulatory effect of gibberellic acid on the accumulation of free amino acid that occurs in malting (Fig. 4.10), and the fact that levels of endopeptidases and some carboxypeptidases are also greatly enhanced by additions of this hormone, while some of the other enzymes are not, emphasize the importance of these enzymes in degrading the major protein reserves of the starchy endosperm and enhancing the soluble nitrogen fraction [30]. Raw barley also contains proteins that are effective protease inhibitors. These are destroyed during malting [31]. Peptidases are inactivated more readily than proteases during kilning. Some protease activity is insoluble, and this fraction is relatively more stable than the soluble fraction.

4.3.7 NUCLEIC ACIDS AND RELATED SUBSTANCES [1-3]

Nucleic acids represent 0·2–0·3% of the dry weight of the barley grain. They are metabolically active in the living tissues, but it is not clear to what extent they occur in the starchy endosperm and contribute to the nitrogenous reserves of this tissue. However, during malting, hydrolytic enzymes able to degrade deoxyribonucleic acid (DNA; **51**) and ribonucleic acid (RNA; **52**) increase, and nucleosidases and nucleotidases are also active because (*i*) 6% of the soluble nitrogen of decoction wort is contributed by the free bases adenine (**53**) and guanine (**54**), and (*ii*) nucleic acids are totally degraded to soluble materials during mashing.

TABLE 4.10

Constituents of the nucleic acids and their degradation products

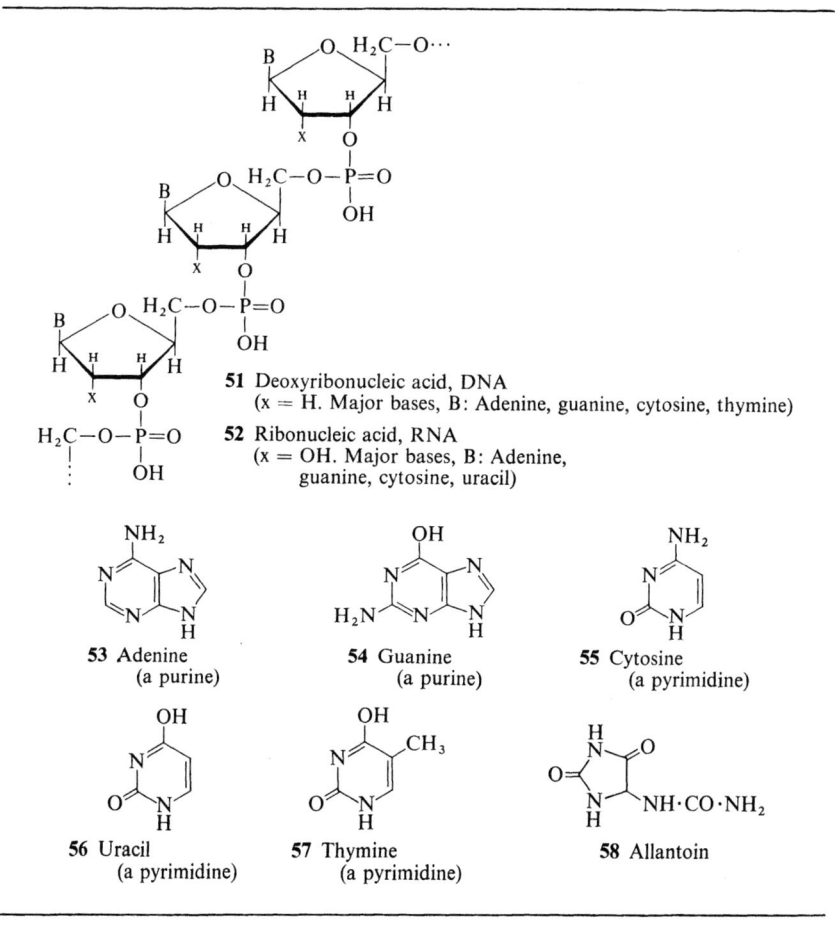

51 Deoxyribonucleic acid, DNA
(x = H. Major bases, B: Adenine, guanine, cytosine, thymine)

52 Ribonucleic acid, RNA
(x = OH. Major bases, B: Adenine, guanine, cytosine, uracil)

53 Adenine (a purine)

54 Guanine (a purine)

55 Cytosine (a pyrimidine)

56 Uracil (a pyrimidine)

57 Thymine (a pyrimidine)

58 Allantoin

The stages of nucleic acid breakdown are usually (*i*) hydrolysis by nucleases, with the release of nucleotides (base–sugar–phosphate), (*ii*) the hydrolysis of nucleotides, by nucleotidases or phosphatases, to give inorganic phosphate and nucleosides (base–sugar), and (*iii*) the hydrolysis of nucleosides, by nucleosidases, to yield free bases (purines and pyrimidines; 53–57) and the sugars ribose (9) or deoxyribose (10). Allantoin (58), a degradation product of purines, also occurs in malt. The effects of kilning on the enzymes involved in the degradation of nucleic acids can be inferred from indirect evidence. Thus at least one ribonuclease is stable to heat and survives kilning. Highly kilned malts yield worts containing a higher proportion of nucleosides than worts from moderately kilned malts, which are richer in free bases. In neither wort type are nucleotides present. Thus it is concluded that nucleases and nucleotidases (phosphatases) are comparatively stable, but nucleosidases may be inactivated during kilning. Some nucleosides have flavour-enhancing characteristics. The nucleosides and free bases can serve as yeast nutrients.

4.3.8 OTHER NITROGENOUS GRAIN COMPONENTS [1–3]

Rootlets contain substantial quantities of tyramine (59), formed by decarboxylation of the amino acid tyrosine (46), and the methylated derivatives

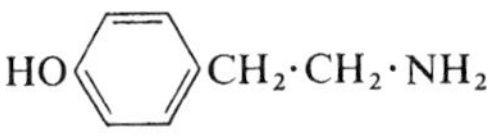

HO–C₆H₄–CH₂·CH₂·NH₂
59 Tyramine

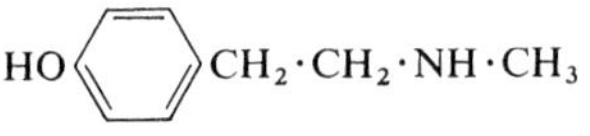

HO–C₆H₄–CH₂·CH₂·NH·CH₃
60 *N*-Methyltyramine

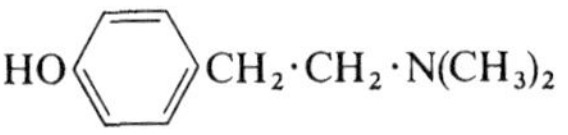

HO–C₆H₄–CH₂·CH₂·N(CH₃)₂
61 Hordenine

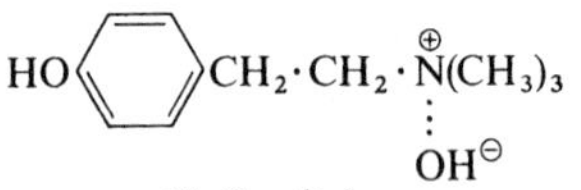

HO–C₆H₄–CH₂·CH₂·N⁺(CH₃)₃ OH⁻
62 Candicine

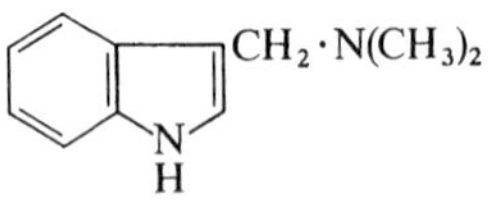

Indole–CH₂·N(CH₃)₂
63 Gramine

HO·CH₂·CH₂·N⁺(CH₃)₃ OH⁻
64 Choline (a B-vitamin)

HO·CH₂·CH₂·NH₂
65 Ethanolamine (2-aminoethanol)

(CH₃)₃·N⁺·CH₂·COO⁻
66 Betaine

NH₃
67 Ammonia

CH₃·NH₂
68 Methylamine

(CH₃)₂·NH
69 Dimethylamine

CH₃·CH₂·NH₂
70 Ethylamine

CH₃·(CH₂)₃·NH₂
71 Butylamine

CH₃·(CH₂)₄·NH₂
72 Amylamine

N-methyltyramine (**60**), hordenine (**61**) and candicine (**62**). Traces of these substances occur in roots. Gramine (**63**) occurs mainly in the leaves of mature plants, but traces may occur in malt. Many other nitrogenous materials occur either free or combined in green malt, including choline (**64**), ethanolamine (**65**), betaine (**66**), ammonia (**67**) and a range of simple volatile amines (**68–72**). *p*-Hydroxybenzylamine and pyrrolidine have also been recognized, and traces of polyamines such as putrescine and cadaverine also occur. The importance of these substances in beer flavour is uncertain. However, in kilning and especially in roasting many highly flavoured heterocyclic nitrogenous compounds can be formed. Some of these amines may be involved in the formation of nitrosamines during kilning.

4.3.9 LIPIDS [1–3, 12]

In most instances the total lipid contents of barleys are about 3·5% of the grain's dry weight. Extraction techniques commonly used in the past failed to extract the more-polar lipids, and so older estimates are lower and mainly represent the less-polar 'petrol-soluble' fraction. Of the petrol-soluble lipids about one-third occur in the embryo, and of the remaining two-thirds more than 90% is located in the aleurone layer. A variable quantity, often 10–12%, of the petrol-soluble lipids, mainly triacylglycerols, are utilized in malting. Probably they are mainly oxidized to support respiration, but some may be converted to sugars via the glyoxylate pathway, which operates at least in the aleurone layer [1]. Starch granules have lipids on their surfaces and within their structures [12]. The 'internal' starch lipids are polar, and may exist as inclusion complexes with the polysaccharides, altering the gelatinization properties of the starch granules. The barley grain lipids contain 65–78% neutral lipids (mainly triacylglycerols, **73**), 7–13% glycolipids and 15–26% phospholipids [12]. Triacylglycerols are the major class of neutral lipids, but 1,2–diacylglycerols and 1,3-diacylglycerols also occur, as do free fatty acids, sterols, sterol esters and sterol glycosides, plus various hydrocarbons (including alkanes and carotenoids). Minor components include 5-*n*-alkyl resorcinols and tocopherols (**83**). Of the total fatty acids of the grain, palmitic acid (16 : 0; **74**), oleic acid (18 : 1; **75**), and linoleic acid (18 : 2; **76**) are the most abundant. Recorded frequencies of total fatty acids are 14 : 0, 0·3–0·5%; 14 : 1, traces; 16 : 0, 18–27%; 16 : 1, 0·1–0·3%; 18 : 0, 0·4–1%; 18 : 1, 9–21%; 18 : 2, 50–59%; 18 : 3, 4–7%.

Phospholipids include phosphatidylethanolamine (**77**), phosphatidylcholine (**78**), phosphatidyl-*myo*-inositol, and lysophosphatidylcholine. The glycolipid fraction probably contains various galactosylglycerides and glycosides of sterols, such as β-sitosterol (**79**).

Most malt lipids are retained in the spent grains during mashing. It has been proposed that these be crudely extracted and used as anti-foaming

$$\begin{array}{l}\text{CH}_2\text{O}\cdot\text{CO}\cdot\text{R}_1\\ \text{CHO}\cdot\text{CO}\cdot\text{R}_2\\ \text{CH}_2\text{O}\cdot\text{CO}\cdot\text{R}_3\end{array}$$

73 Generalized Triglyceride

$$\text{CH}_3\cdot(\text{CH}_2)_{14}\cdot\text{COOH}$$
74 Palmitic acid (16:0)

$$\text{CH}_3\cdot(\text{CH}_2)_7\cdot\text{CH}=\text{CH}\cdot(\text{CH}_2)_7\cdot\text{COOH}$$
75 Oleic acid (18:1)

$$\text{CH}_3\cdot(\text{CH}_2)_4\cdot\text{CH}=\text{CH}\cdot\text{CH}_2\cdot\text{CH}=\text{CH}\cdot(\text{CH}_2)_7\cdot\text{COOH}$$
76 Linoleic acid (18:2)

$$\begin{array}{l}\text{CH}_2\text{O}\cdot\text{CO}\cdot\text{R}_1\\ \text{CHO}\cdot\text{CO}\cdot\text{R}_2\\ \quad\;\;\text{O}\\ \quad\;\;\|\\ \text{CH}_2\text{O}\cdot\text{P}\cdot\text{O}\cdot\text{CH}_2\cdot\text{CH}_2\cdot\text{NH}_2\\ \quad\;\;\text{OH}\end{array}$$

77 Phosphatidyl ethanolamine (cephalin)

$$\begin{array}{l}\text{CH}_2\text{O}\cdot\text{CO}\cdot\text{R}_1\\ \text{CHO}\cdot\text{CO}\cdot\text{R}_2\\ \quad\;\;\text{O}\\ \quad\;\;\|\\ \text{CH}_2\text{O}\cdot\text{P}\cdot\text{O}\cdot\text{CH}_2\cdot\text{CH}_2\cdot\overset{\oplus}{\text{N}}\cdot(\text{CH}_3)_3\\ \quad\;\;\text{OH}\qquad\qquad\qquad\quad\text{OH}^{\ominus}\end{array}$$

78 Phosphatidyl choline (lecithin)

79 β-Sitosterol

agents in fermentation. However, the small quantities dispersed in wort have important effects. Some lipids tend to collapse beer foam while some polar lipids may stabilize it (Chapters 9 and 22). Yeast requires sterols and unsaturated fatty acids for growth and maintenance of membranes under particular conditions. Further, some degradation products of fatty acids may give rise to flavour defects in beers. Two lipoxygenase enzymes of barley convert linoleic acid to the 9- and 13-hydroperoxides which are isomerized and further altered via ketols to produce, e.g. trihydroxy fatty acids, which are found in beer. These, and other products, serve as precursors of various aldehydes such as hexanal, *trans*-2-hexenal and other substances which give beer stale off-flavours.

4.3.10 PHOSPHATE AND PHOSPHATE ESTERS [1–3]

Phosphate is widely utilized in intermediary metabolism and it (as P_2O_5) makes up about 1% of the grain's dry weight. In the grains it occurs combined in a wide range of substances such as phospholipids, nucleic acids and *myo*-inositol hexaphosphate (phytic acid, **80**). Phytic acid, which contains about half of the grain's combined phosphate, occurs as a mixed salt ('phytin') in the aleurone grains of the embryo and aleurone layer. During malting, phosphatase(s) which are able to hydrolyse phytic acid (phytase) increases in amount, and inorganic phosphate is released from the aleurone layer and is accumulated in the embryo. During this process 10–30% of the grain's phytate is hydrolysed. The survival of phytase is greatly influenced by temperature so that during the decoction mashing of lightly kilned lager malts 60–90% of the residual phytate is hydrolysed, while little breakdown of phytate occurs during the infusion mashing of well-kilned ale malts. Phosphate residues are sequentially hydrolysed, from phytate, producing inorganic phosphate, and, eventually, *myo*-inositol (81), which is a 'vitamin' essential for yeast growth. Phytate, even more than inorganic phosphate, has a high affinity for calcium ions. When they interact in the mash and copper boil hydrogen ions are released, and so the pH is reduced (Chapter 9).

80 Phytic acid, *myo*-inositol hexaphosphate

81 *myo*-Inositol (a B-vitamin)

4.3.11 INORGANIC CONSTITUENTS OF BARLEY [1, 32, 33]

After combusion, 2–3% of the barley's original weight remains as ash. The quantity and nature of the ash differs markedly in grain samples grown under different conditions. The husk, and particularly its outer layer, is rich in silica (SiO_2), which gives it an abrasive character. Otherwise minerals are particularly concentrated in the embryo and the aleurone layer. During malting some minerals are leached into the steep liquor. During malting there is a movement of minerals from the aleurone layer to the embryo. A proportion of the embryo's minerals occurs in the roots, and is removed with them. Phosphorus and potassium are among the major components of ash, but a wide range of substances is present. Malt contains all the mineral substances needed to support yeast growth.

4.3.12 VITAMINS [1–3, 34, 35]

Barley and malt are rich in substances that are growth factors for yeasts and/or vitamins for human beings (Table 4.11). Usually these materials are concentrated in the living tissues, the embryo and aleurone layer. They often occur combined in more complex substances that act as cofactors in

TABLE 4.11

Some reported vitamin contents of barleys and malts
BRIGGS (1978) [1]; KNORR (1952) [34]; VOSS and PIENDL (1978) [35]

Vitamin	Barley	Malt
B$_1$; Thiamin, **90** (µg/g)	1·2–16	2·4–8·0
B$_2$; Riboflavin, **84** (µg/g)	0·8–3·7	1·2–5·0
Nicotinic acid **89** (µg/g)	47–147	48–150
Pantothenic acid, **85** (µg/g)	2·9–11	4·3–12·9
B$_6$; Pyridoxin, etc., **86** (µg/g)	2·7–11·5	3·8–7·5
Vitamin E, **83** (µg/g)	2·1–5·2	2·0–2·5
Biotin, **87** (µg/g)	0·11–0·17	0·09–0·22
Folates, **88** (free; µg/g)	0·1	—
(combined; µg/g)	0·2	—
Choline, **64** (mg/g)	1·0–2·2	—
myo-Inositol, **81** (free; mg/g)	0·18	—
(bound; mg/g)	1·4–3·2	—

intermediary metabolism. Some partly move into the rootlets during malting and are lost to the malt. Vitamin C (ascorbic acid, **82**) and dehydroascorbic acid increase during malting but are destroyed during kilning. Vitamin E (e.g. α-tocopherol, **83**), occurs in barley oil, which contains a complete range of tocopherols and tocotrienols [12]. Riboflavin (**84**), pantothenic acid (**85**) and the vitamin B$_6$ group (**86**) increase during malting, while others do not. The vitamin B complex is rich in factors that are important for yeast growth during fermentation, including biotin (**87**), *myo*-inositol (**81**) and pantothenic acid (**85**). Other vitamins present include folic acid (**88**) and related substances, nicotinic acid (**89**) and thiamin (**90**). Carotenoids and sterols occur, and while these are not vitamins themselves, some of them can give rise to vitamins A and D respectively when ingested by animals. As already noted some sterols and unsaturated fatty acids are essential growth factors for yeast. The malt vitamins carried forward into beer may have nutritional significance for consumers.

4.3.13 PHENOLIC AND POLYPHENOLIC MATERIALS [1–3]

Barley contains numerous phenolic materials, the simplest being phenolic acids, such as *p*-hydroxybenzoic acid (**91**), vanillic acid (**92**) and caffeic acid (**93**). Other substances occur which may be involved in the metabolism of lignin, a complex and poorly defined insoluble polymer containing hydroxylated and methoxylated phenylpropane residues. Its major role in grain is to confer strength and rigidity to the cell walls of the husk. It does not alter significantly in malting. Other types of phenolic materials are chlorogenic acid (**94**), coumarin (**95**) and herniarin (**96**). Other substances, 'anthocyanogens' or proanthocyanidins (of which **97** and **98** are conceptually the simplest),

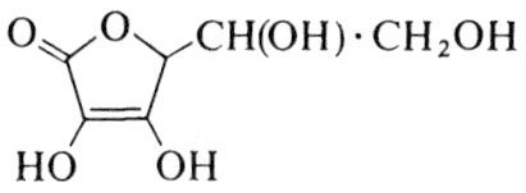

THE BIOCHEMISTRY OF MALTING GRAIN

82 Vitamin C, ascorbic acid

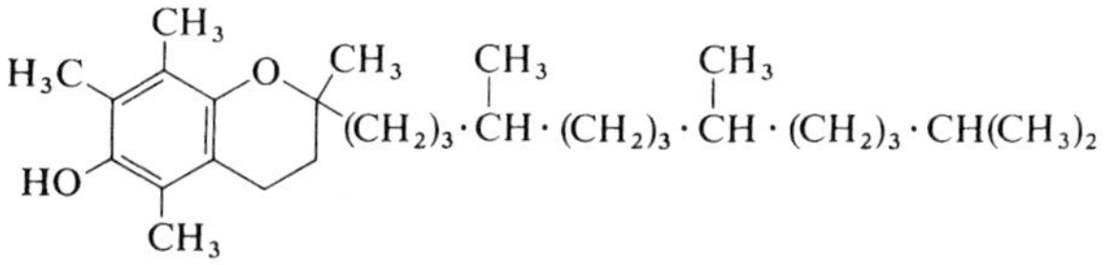

83 α-Tocopherol (one component of vitamin E)

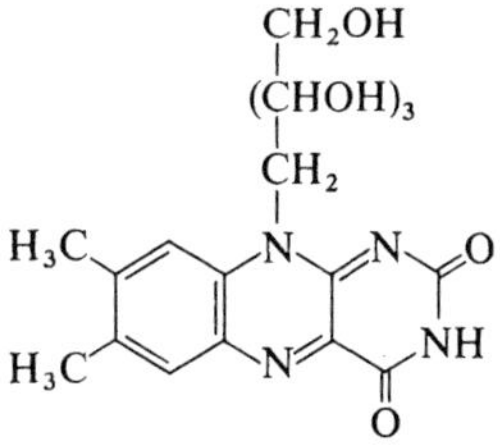

84 Riboflavin, vitamin B₂

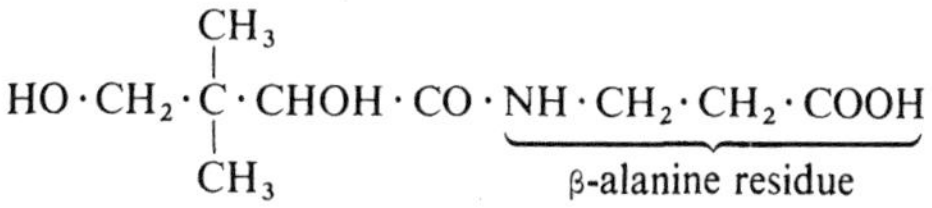

$HO \cdot CH_2 \cdot C(CH_3)_2 \cdot CHOH \cdot CO \cdot NH \cdot CH_2 \cdot CH_2 \cdot COOH$
                                              β-alanine residue

85 Pantothenic acid

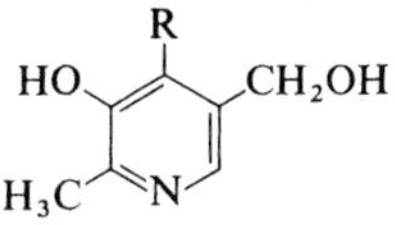

86 Vitamin B₆
Pyridoxin (R = CH₂OH)
Pyridoxal (R = CHO)
Pyridoxamine (R = CH₂·NH₂)

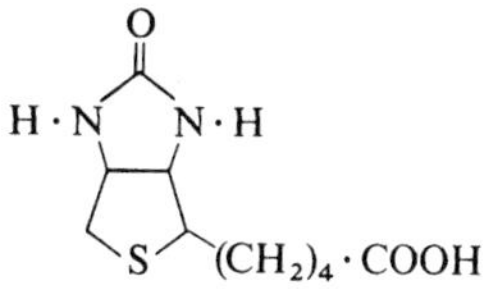

87 Biotin

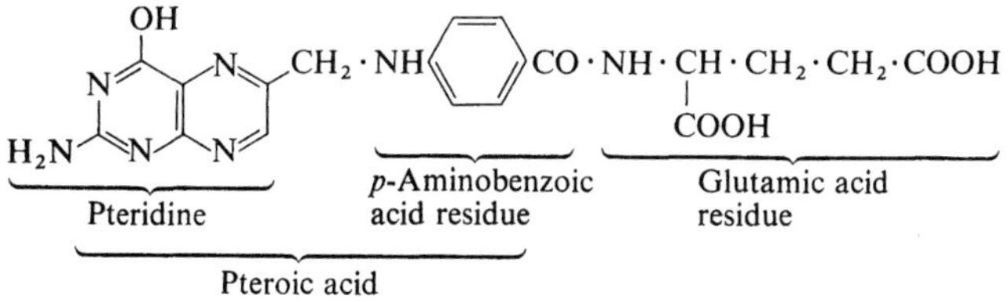

88 Folic acid
Pteridine — p-Aminobenzoic acid residue — Glutamic acid residue
Pteroic acid

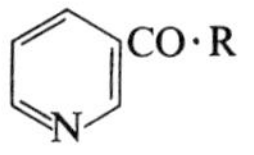

89 Nicotinic acid (R = OH)
Nicotinamide (R = NH₂)

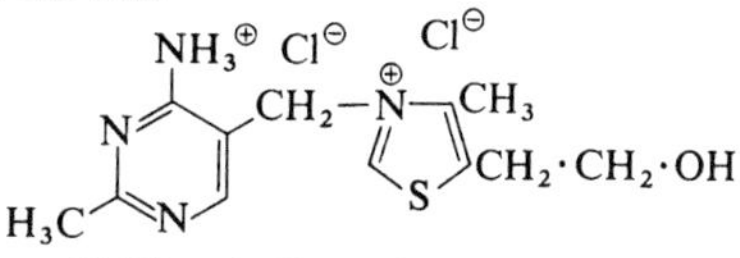

90 Vitamin B₁ (Thiamin, Aneurine)

give rise to bright red anthocyanidin pigments (such as **99** and **100**) when heated with mineral acids. These compounds occur as polymers of various degrees of complexity, dimers (e.g. the biflavonoids **101**, **102**), trimers, tetramers etc., incorporating anthocyanogens and catechin (**103**) which, like epicatechin, also occurs free in barley. The catechins are not anthocyanogens.

Probably biflavanoids are the most important tannin precursors (tanninogens) in barley. On treatment with acids they also give rise to red pigments, so they are anthocyanogens. These compounds are labile and readily polymerize under acidic or oxidizing conditions. Some barleys are coloured by anthocyanidin pigments in the aleurone layer or pericarp. These do not make the grain unfit for malting.

91 *p*-Hydroxybenzoic acid   **92** Vanillic acid   **93** Caffeic acid

94 Chlorogenic acid

95 Coumarin
(Lactone of *o*-coumarinic acid)

96 Herniarin
(7-Methoxy coumarin)

The importance of phenolic compounds to the grain is not clear, although they may help to regulate micro-organisms or serve as endogenous growth regulators. Many simple phenols are removed in steeping and, as many have growth-inhibitory properties, one benefit may be to enhance subsequent germination. Polyphenols are said to add 'astringency' to beer flavour. To the brewer, probably the most important characteristic of malt phenolic materials is the ability of some of them to act as 'tannins' or tannin precursors. The word 'tannin' originally applied to concentrates of vegetable polyphenols used to treat hide, so 'tanning' it and converting it into leather. Tannins interact with proteins, reversibly at first but later irreversibly, eventually forming covalent cross-links between protein molecules which form progressively less-soluble aggregates. Not all polyphenols have equal tanning ability, more polymerized materials being more effective than simple monomers. It seems that, in barley, anthocyanogen polymers are the most important class of tannins. Polyphenol–protein interactions are important in at least two stages in brewing: (*i*) in the formation of the precipitates, 'breaks' or 'trub' during the boiling of the wort with hops in the copper, and subsequent cooling (Chapter 14), and (*ii*) in the formation of non-biological hazes in beers, both chill hazes and permanent hazes (Chapter 22). A mutant *ant-13* of the variety Foma contains only traces of catechin and proanthocyanidins. Beer brewed from malt made with this barley has a remarkably high colloidal stability, i.e. it is unlikely to become hazy [36–38]. Perhaps unexpectedly, the flavour of the flavan-less beer is perfectly normal.

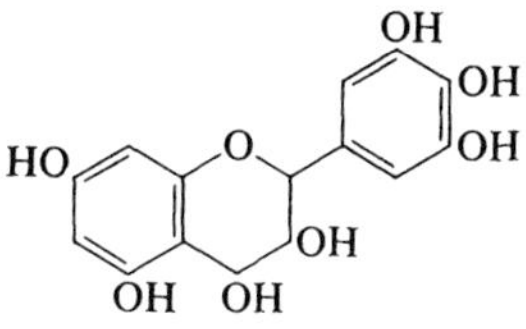

97 Delphindinogen (an anthocyanogen)

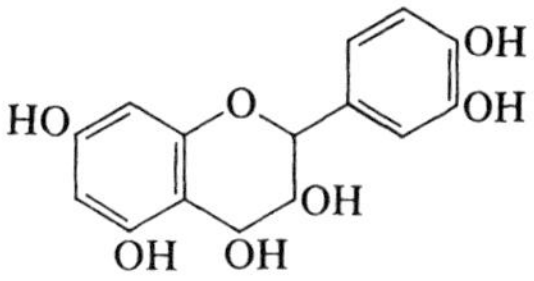

98 Cyanidinogen (an anthocyanogen)

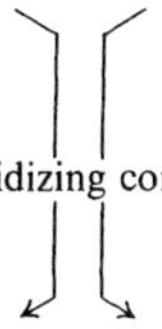

H+, oxidizing conditions

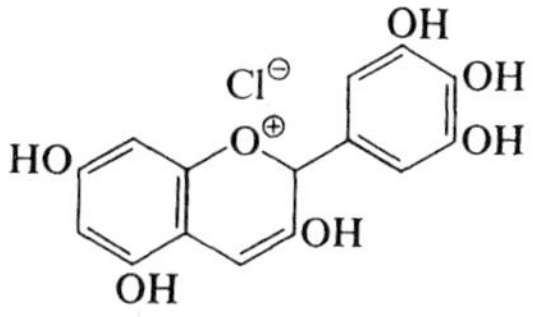

99 Delphinidin chloride

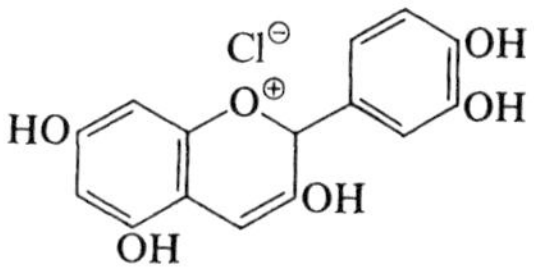

100 Cyanidin chloride

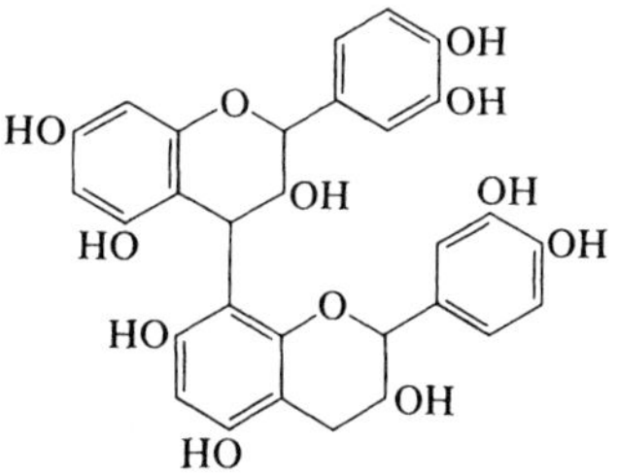

101

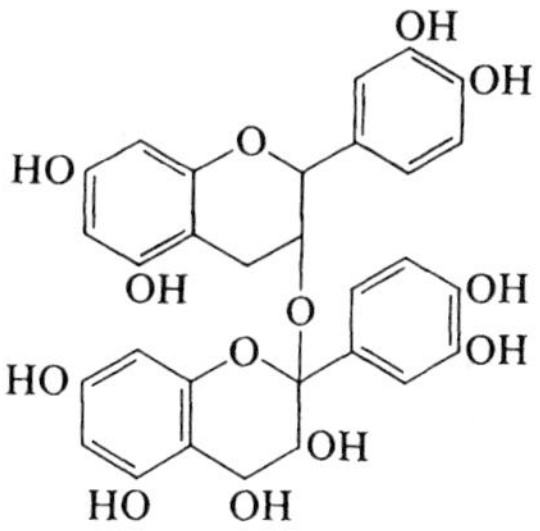

102

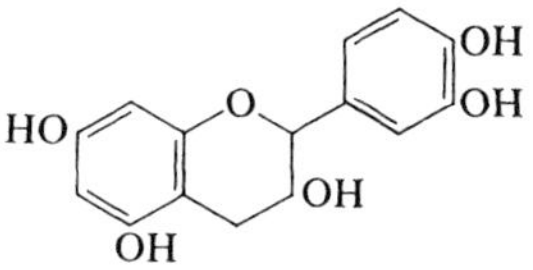

103 D(+).Catechein

The techniques used to analyse the very complex mixtures of barley phenols give results that are difficult to interpret. It can be useful to determine 'oxidized' polyphenols that have tanning ability and give an immediate turbidity or precipitate with cinchonine sulphate and 'oxidizable' polyphenols that may be oxidatively polymerized to yield tannins, and therefore are tannin precursors or 'tanninogens'. The colour given by anthocyanogens when heated with mineral acids is used to estimate them, but the colour yield per flavan residue is greater for the simpler compounds (**97, 98**) than for the more complex materials that have actual tanning ability. A wider range of polyphenol types gives a blue–black colour with ferric ions, which is used for quantitative studies, but again the colour yield given by different phenols varies.

Analyses of whole and decorticated barley samples, the malt 'husk' sieve fraction, and staining tests on grain sections indicate that polyphenols are concentrated in the pericarp and in the aleurone layer. In barleys having blue grains anthocyanin pigments, such as cyanidine-3-arabinoside, also occur in these layers. However, phenols also occur in the 'true' husk and may be partly removed, together with other components of the 'testinic acid' mixture using alkaline steeps, giving rise to less astringent beers [9]. Malts with low polyphenol contents give rise to beers with a reduced tendency to form haze [36–38]. Polyphenol levels can be reduced by steeping or resteeping grain in dilute solutions of formaldehyde, and treated malts give rise to more stable beers [39, 40]. Formaldehyde probably cross-links the polyphenols with proteins, producing insoluble polymers that are not extracted during mashing.

Green malt, and to a lesser extent lightly kilned malts, contain oxidase enzymes that catalyse the oxidative polymerization of polyphenols in well-aerated crushed grains, giving rise to tannins which are retained as insoluble complexes in the mash. Thus aerated mashes made with such malts give worts with reduced polyphenol levels and more stable beers [41, 42]. Similarly green malt peroxidase survives light kilning, and can catalyse the destruction of polyphenols in mashes when small quantities of hydrogen peroxide are added to the mash, leading to more stable beers [40]. As the hydrogen peroxide decomposes to water and oxygen it leaves no undesirable residues.

4.4 *Grain organization, growth conditions and the biochemistry of malting* [1–3]

The intermediary metabolism of the living tissues of barley is complex. Molecules are synthesized, degraded, interconverted and incorporated into complex structures. The energy required to drive these processes is made available mainly by the respiratory oxidation of carbohydrates. Details of many of the metabolic pathways are not important to maltsters, but the consequences of the metabolic fluxes certainly are. There is a need to know how alterations in malting conditions or applications of additives will alter respiration rate, heat output, rootlet growth, increases in malting loss, enzyme accumulation, the rate of modification, changes in hot-water extract and soluble nitrogen, and any other significant malt characteristics (Chapter 5). All these changes are dependent on the biochemical activities of the grains, and some understanding of these processes is needed if malting conditions are to be altered in a rational manner. It has become clear that an understanding of the separate roles of the embryo, the aleurone layer, and the starchy endosperm in particular, and the interactions between these tissues is essential. The development of many of the newer malting methods, such as those designed to hasten malting and make malt with very small attendant malting losses, has been guided by such understanding. Many of the changes in malt characteristics associated with altering germination conditions are considered in Chapter 5, after malt analyses have been discussed. In the following sections the more biochemical aspects of grain germination are considered.

4.4.1 THE NATURE OF THE GRAIN; MOISTURE CONTENT AND MALTING [1–3, 43–45]

Barley varieties differ in their suitability for malting. Some are unsuitable because they make malt with a low yield of extract, others malt with an unduly high malting loss, and the endosperms of others are slow to modify, and yet others germinate slowly. A sample of a genetically good 'malting barley' may be unusable because of poor viability, heavy fungal contamination, small grain size or an unduly high nitrogen content. Barleys with a high crude protein content are converted to malts giving a low extract. They are rich in some enzymes and they respire and grow vigorously (processes which lead to high malting losses), but they are often 'stubborn' in that their endosperms are slow to modify. They may be used in making enzyme-rich highly diastatic malts.

In malting, the water content of barley must be adjusted exactly. At less than about 32% moisture, the grain will not germinate. At about 35% moisture it will germinate vigorously (even if water-sensitive), but is too dry to modify adequately, and must subsequently be wetted further (Chapters 3 and 5). Low-nitrogen (1·4%) mellow barley may be steeped to 42–43% moisture and malted. More steely, stubborn and higher nitrogen grains may be steeped to 46% or even 48% moisture. In special instances, e.g. in re-steeping, even higher moisture contents are reached.

The moisture content is chosen to achieve rapid even modification of the grains, but at the same time to limit embryo growth and respiration (Fig. 4.11), and hence malting losses. As initial moisture contents are increased, the extracts of the final malts tend to increase up to a maximum value, and their hardness declines. In a series of barley samples malted in a similar way α-amylase levels, malting losses, rootlet yields, soluble nitrogen ratios and levels of soluble polyphenols increased with increasing moisture contents, while the β-glucan levels declined [43]. During malting, moisture is lost by evaporation and incorporation into the products of hydrolysis. It is gained as a product of respiration and, on occasion, by 'sprinkling' or spraying water onto the germinating grain. During malting it is withdrawn into the growing embryo from the endosperm, and the progressive drying of the endosperm causes modification to slow or even, in extreme cases, cease. Water sprinkled on to grain to rewet it during the germination period may be retained by the embryo, and so enhance its growth, and so increase malting losses without, however, improving endosperm modification (Chapter 3).

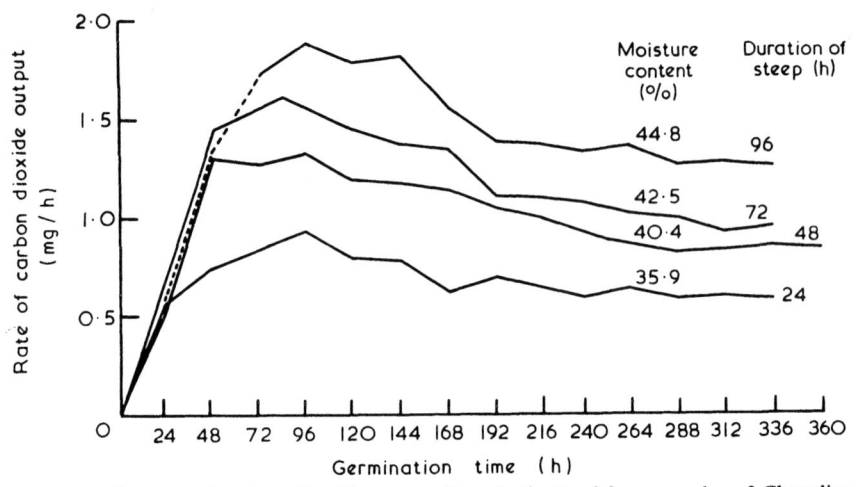

Fig. 4.11 The rates of carbon dioxide output ('respiration') of four samples of Chevalier barley steeped for the times, and to the moisture contents, indicated, during malting at about 14°C (57–58°F). After Day (1896) [46].

4.4.2 TEMPERATURE AND GERMINATION [1–3, 43–45]

At higher temperatures, e.g. 18–20°C (about 64–68°F), dormant grains germinate markedly less well than at lower temperatures (of 12–13°C; about 54°F; Chapter 3). In samples of fully mature grain, steeped uniformly and then malted at different temperatures (12–25°C; about 54–77°F), grain grown at higher temperatures grows faster, produces roots faster and initially produces many enzymes faster. However, the rate of formation of some enzymes, such as α-amylase and protease, rapidly falls off, so that after a while grain growing at a lower temperature, and producing enzyme at a slower but steady rate comes to contain more than the 'high' temperature grain. By beginning germination at a high temperature, e.g. 17°C (62·6°F), to gain rapid initial enzyme production, then after several days dropping the temperature, e.g. to 13°C (55·4°F), to maintain enzyme production it is possible to obtain malt with unusually high enzyme contents in a short time [47, 48] (cf. Chapter 5). In traditional malting, germination temperatures of 12°C (about 54°F) were usual, but in modern practice temperatures of 18-20°C (about 64-68°F) are common, and in special instances even higher temperatures are used, particularly with poor quality protein-rich barleys. Growing barleys at 25-30°C (77-86°F) after an unusual steeping schedule involving warm steeping can limit embryo growth and so reduce malting losses [49]. Because germinating at elevated temperatures does not accelerate all the grains metabolic processes to the same degree, it follows that malts produced at higher temperatures differ from those produced at lower temperatures; they are not merely the same type of malt produced in a short time.

4.4.3 RESPIRATION AND ENZYME DEVELOPMENT IN MALTING GRAINS [1-3]

When grain is steeped, the oxygen dissolved in the water is rapidly utilized and the grain begins to ferment ('respire anaerobically'), the major products being ethanol and carbon dioxide. These are derived from stored sugars via the glycolytic pathway. A considerable proportion of the ethanol is leached into the steep liquor. Steep aeration reduces ethanol production and subsequently leads to more vigorous germination. Even so, at steep out, grain ethanol levels continue to rise until chitting occurs, when the surface layers of the grain are broken and oxygen reaches the living tissues more easily (Chapter 3). Oxygen uptake is limited by the surface moisture film of the grain, the husk and the pericarp. Grain respires faster when the surface film of moisture has been taken up, and decorticated grain respires more rapidly than intact grain. Also the microbes present in the surface layers actively compete for oxygen with the grain tissues. When the grain has chitted, ethanol formation has ceased and residual ethanol has been oxidized, then oxygen uptake and carbon dioxide production occur at approximately equal rates. The main respiratory substrate is carbohydrate, but a little lipid is utilized. Respiration provides the energy which drives grain metabolism. Heat is also produced, and during malting this must be dissipated. As oxygen partial pressures are reduced, grain respiration is suppressed, so that in 5% oxygen respiration occurs at about 60% of the rate found in air [1]. A lack of oxygen at any stage will stop growth and enzyme production, and prolonged anaerobiosis is progressively damaging and finally lethal to the grain. Within limits, elevated oxygen levels increase grain respiration and accelerate enzyme production. High concentrations of carbon dioxide suppress respiration and, to a lesser extent, fermentative processes. Various special malting processes have used anaerobic conditions or enhanced levels of carbon dioxide to check grain growth and the accumulation of malting losses (Chapter 5).

The respiration of the grain at steep out is about equally due to the embryo and the aleurone layer, with an irregular (but probably substantial) contribution from the microflora. As germination continues the respiration rate of the aleurone increases, but the main rise is ascribable to the growing embryo. Thus the embryo is responsible for rootlet production and most of the respiratory losses and is chiefly responsible for the decline in grain dry matter that occurs during malting. The intensity of grain respiration is greatly influenced by the conditions used in malting. Thus steep aeration or additions of gibberellic acid result in vigorous germination and enhanced respiration rates. Also, within limits, grain respires faster at higher temperatures, but while 'malting' at around 4–5°C (40°F), respiration increases linearly with advancing germination time at 18–20°C (64–68°F), the respiration rate peaks and declines, while at intermediate temperatures of 12–13°C (about 54°F) the respiration rate eventually levels off and changes little in the later stages of germination. Similarly, although extended steeping often checks grain respiration in the initial stages of germination, the wetter grains subsequently grow more strongly and respire more vigorously (Fig. 4.11). In 'traditional' malting the maximal respiration rate occurs between day 3 and day 5, and the rate during this period correlates well with the ultimate malting loss [51].

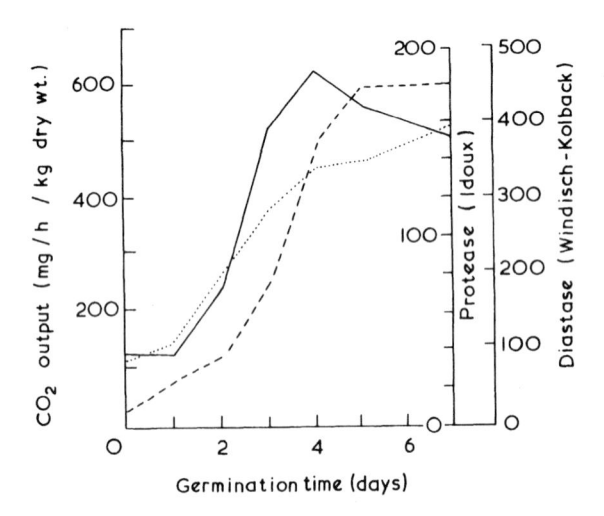

Fig. 4.12 Variations in the rate of carbon dioxide output, and in protease and diastase activities in malting barley. After De Clerck and Cloetens (1940) [50]. ———, Respiration (CO_2 output), – – – –, Diastatic activity, ..., Proteolytic activity.

Embryo growth, respiration and the development of various enzymes are not tightly linked in the sense that they do not all change in parallel in a single sample of malting grain (Fig. 4.12), and furthermore they may often be varied independently by changing malting conditions. However, using one malting system, processing one type of barley, the experienced maltster may be able to use acrospire growth, rootlet production or (in principle) carbon dioxide output as indirect guides to the progress of modification.

104 Gibberellin A$_1$, GA$_1$

105 Gibberellic acid, GA$_3$

The enzymes that catalyse modification are derived mainly from the scutellum and aleurone layer. In the initial stages of germination the scutellum releases many hydrolytic enzymes which begin to degrade the cell walls of the crushed cell layer, and the walls, protein and starch granules of the starchy endosperm. The continued production of α-amylase, and probably the other enzymes is dependent on respiration, a supply of nutrients and, probably in most instances, gibberellins. The rather similar gibberellins GA$_1$ (**104**), GA$_3$ (**105**; gibberellic acid), GA$_4$ and GA$_7$ have been characterized in barley, and, of these, gibberellic acid seems to play the major role in controlling endosperm modification. However, others probably remain to be recognized. When a solution of gibberellic acid, (which is prepared commercially from fungal cultures) is applied to grain it gains entry at the embryo, augments the endogenous supply and greatly accelerates the malting process (Chapter 5). The endogenous gibberellin responsible for controlling endosperm modification is formed from a stored precursor, probably the hydrocarbon *ent*-kaurene [1, 14]. Gibberellin production begins at about the time when respiration has reduced the soluble sugar reserves of the embryo, and may be triggered by this depletion. In addition to enzymes, the scutellum releases gibberellins (probably mainly GA$_3$) which diffuse down the grain and stimulate and alter the metabolism of the aleurone layer. As a consequence its respiration rate increases, phytate is hydrolysed and inorganic phosphate is released with other mineral substances, lipids are hydrolysed and the fatty acids are oxidized or converted to sugars via the glyoxylate pathway. In addition, simple molecules are taken up, metabolized and the metabolic products are released so that glucose is converted to sucrose and several amino acids are turned into glutamine. However, the most striking occurrence is the gibberellin-triggered production and release of numerous hydrolytic enzymes by the aleurone layer at the expense of its own reserves. In contrast with the scutellum, it does not require an external supply of nutrients to support its metabolic activities [1, 14]. All these changes are dependent on respiration and are enhanced by improving the access of oxygen to the aleurone, e.g. by decorticating grain. Various hydrolytic enzymes in the aleurone layer differ in their responses to the arrival of gibberellin. Thus some enzymes, such as phosphatase and laminarinase, increase when the aleurone is hydrated, but their release into the starchy endosperm is gibberellin-dependent. Other enzymes, e.g. α-glucosidase, occur in the resting tissues, but their levels increase in response to gibberellin. α-Amylase is absent from hydrated aleurone, and its synthesis *de novo* and release are independently triggered by the arrival of the hormone. Enzymes are released in sequence, and not simultaneously, when gibberellins reach the aleurone layer. The observed pattern of modification (Fig. 4.1) is fully explained by this sequence of events. α-Amylase is the enzyme most thoroughly studied. It finally occurs in a concentration gradient along the grain, most being at the embryo end, and in roughly equal amounts on the dorsal and ventral sides of the grain. It is estimated that in malt about 85% of this enzyme originates from the aleurone layer and 15% from the scutellum. Of this latter, about half is retained in the scutellar tissue and half is released into the starchy endosperm. The levels of α-amylase observed in grain germinated on wet filter paper under non-malting conditions are the result of a balance between the rate at which enzyme is being formed and the rate at which it is being destroyed. It seems likely that, at least, under some malting germination conditions, some limited destruction of enzymes, e.g. α-amylase and β-glucanase, also occurs. In addition to enhanced levels of hydrolytic enzymes the embryo ends of grains modify sooner and yield higher hot-water extracts, cold-water extracts, and levels of soluble nitrogen than the distal ends of the grains so in a sense the modification of grain apices is always behind that of the embryo ends [52, 53].

As a consequence of the activity of hydrolytic enzymes the 'latent' β-amylase of the starchy endosperm is freed, and the degradation of the components of the starchy endosperm proceeds. Consequently there is an accumulation of hydrolysis products, including minerals, amino acids and sugars, which diffuse to the embryo and are taken up by it and used to support its growth. The sugars taken up by the scutellum restore its internal level of soluble carbohydrates, and 'switch off' gibberellin supplies. Thus a feedback loop exists which normally controls the rate at which endosperm modification occurs (Fig. 4.13). However, when gibberellic acid is added to grain, this control mechanism is by-passed and the modification rate is greatly accelerated. Added gibberellic acid also accelerates embryo growth, but proportionately less than the rate of modification, as well as enzyme production, respiration rate, and heat output [1, 14] (Chapter 5).

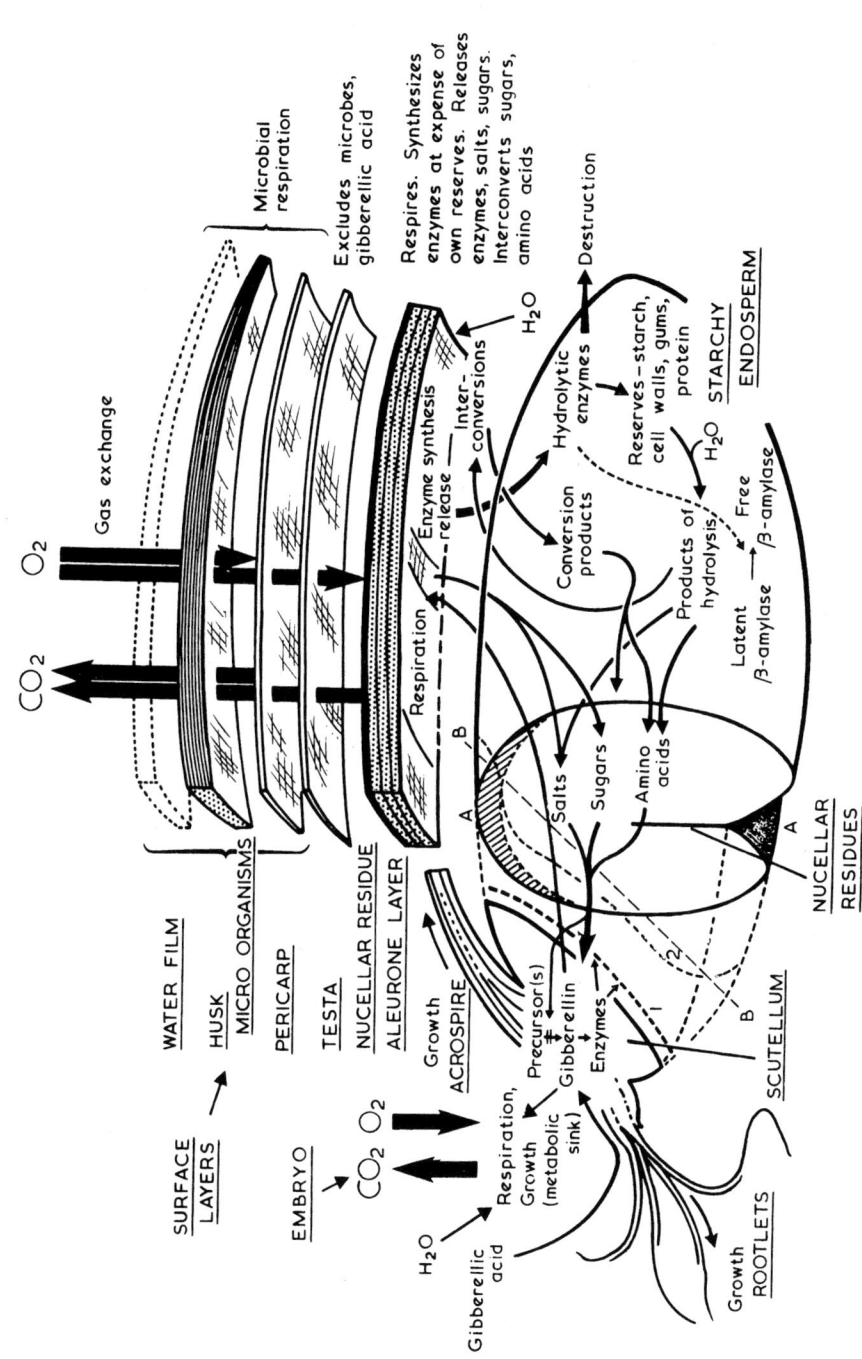

Fig. 4.13 Diagram indicating some of the metabolic interrelationships occurring in malting barley.

If barley is malted under 'traditional' conditions, and at intervals some grains are degermed, then is held for the full malting period before being analyzed, then it is found that, in samples where embryo removal is before about day 3, the 'malt extract' and α-amylase levels are well below normal, but in samples in which embryos are removed after day 3 analyses are more nearly normal. The explanation is that a 'pulse' of gibberellin, and a small dose of enzymes, moves from the scuttellum on about day 3 and triggers enzyme production in the aleurone layer and hence endosperm modification [1, 14]. On the industrial scale it has been possible to take advantage of this finding and damage the embryo on or after germination day 3, thereby reducing malting losses but obtaining comparatively 'normal' modification (Chapter 5). Techniques involving deliberate damage to embryos are more reliable if the grain is also dosed with gibberellic acid. However, the embryo is not merely a source of endogenous gibberellin. It is also a 'sink', in that it consumes the products of endosperm modification, so that, by checking its growth, the level of soluble materials in the grain (the cold-water extract) is elevated and the final malt, prepared using an 'embryo-damaging' technique, has a slightly different composition from that of a conventional malt, being comparatively rich in soluble sugars and nitrogenous substances. These readily interact on kilning to give a dark colour to the malt.

4.5 *The chemistry of kilning* [1-3, 7, 48, 54, 55]

Some kilning schedules are given in Chapter 5, and the physical principles involved in water removal during kilning are indicated in Chapter 6. Some of the chemical changes occurring during kilning are discussed here. The chemistry of melanoidin formation is surveyed in Volume 2.

Kilning consists of passing a flow of hot dry air through a bed of malt at various rates and at increasing temperatures as the product dries. Initially malt is normally dried in a rapid air flow at comparatively low temperatures and, when its moisture content has been reduced sufficiently, it is 'cooked' at higher temperatures. In the process the malt is dried, and its composition is altered. The dry brittle roots are easily broken and separated from the friable product. Finished malt is stable on storage, provided that it is kept dry. There are many types of malt, and their characteristic flavours and aromas depend on the state of the green malt when kilning begins and on the nature of the kilning cycle. To some extent malt character can be altered by the fuel used in direct kilning, as can the risk of accumulation of nitrosamines.

Enzymes, including those important in mashing, are destroyed to a variable extent during kilning. Some enzymes, e.g. α-glucosidase, are appreciably inactivated even at 45°C (113°F), which is well below normal kilning temperatures. At 80°C (176°F) activities of enzymes such as β-glucanase and β-amylase are reduced, and inactivation is even greater at 90°C (194°F) when more stable enzymes such as α-amylase and endopeptidases are inactivated to significant extents. The degree of enzyme destruction at any temperature is greater when the malt is wet. Thus in traditional British ale malts that are very dry, e.g. 2–3% moisture, before the onset of curing at 105°C (221°F), appreciable quantities of enzymes survive, although others, such as peptidases and phytase are almost totally inactivated. However, in roasted malts (or barley), enzyme destruction is complete. Highly enzymic malts are kilned at a low temperature and in a rapid airflow to ensure that the grain is cooled by evaporation. Thus enzyme survival is greatly influenced by the exact details of the kilning cycle.

The composition of malt changes during kilning, so that normally its enzyme complement declines, its colour, aroma and flavour increase, the pH of its extract falls, its hot-water extract, cold-water extract and soluble nitrogen yield all fall, while its content of extractable polyphenols increases. However, in the initial stages of kilning, and particularly if a warm moist air flow is passed through the grain, so that it dries slowly and may even continue to grow and modification of the grain may continue. Gently 'stewing' the grain may cause the accumulation of high levels of soluble sugars and amino acids, which afterwards can give rise to high colours during curing. On the other hand enzyme destruction proceeds rapidly in grain stewed intensively, at higher temperatures. Intensive stewing is used in the production of crystal and caramel malts (Chapter 5). It is not unusual to finish modification on the top floor of European-type two- and three-floor kilns, where the grain is warm and damp.

During kilning many reducing sugars decrease in quantity, but sucrose levels often increase, possibly owing to hydrolytic enzymes acting in reverse under the strongly drying conditions, but more probably via some other route. Levels of amino compounds decline, and melanoidin precursors (adducts between sugars and amino acids) and melanoidins increase in quantity. Many changes that occur during kilning affect malt colour and flavour. Melanoidin formation involves complex Maillard reactions and Amadori rearrangements involving the condensation of amino acids and reducing sugars, followed by polymerization and a series of reactions that give rise to coloured polymeric substances resembling caramels, flavoured compounds, aroma compounds and reductones. Condensations *in vitro* between mixtures of different sugars and amino acids give rise to differently flavoured products (Volume 2). Conditions favouring melanoidin formation include high temperatures, high moisture levels and high concentrations of amino acids and sugars. Kilning conditions chosen to favour melanoidin formation (overmodification, 'stewing', and raising the temperature while the grain is still comparatively moist) in the preparation of dark malts also favour enzyme destruction and the reduction of hot-water extract and wort fermentability.

The low-temperature kilning of lager malts favours the production of very pale malts in which enzyme activities are well preserved.

$$\begin{array}{c}H_3C\\H_3C\end{array}\!\!\overset{\oplus}{S}\cdot CH_2\cdot CH_2\cdot \underset{NH_2}{CH}\cdot COOH + O\overset{\ominus}{H} \rightarrow \begin{array}{c}H_3C\\H_3C\end{array}\!\!S + HO\cdot CH_2\cdot CH_2\cdot \underset{NH_2}{CH}\cdot COOH$$

107 S-Methyl methionine, SMM   106 Dimethyl sulphide, DMS   109 Homoserine

$$\text{reduction} \updownarrow \text{oxidation}$$

$$\begin{array}{c}H_3C\\H_3C\end{array}\!\!S=O$$

108 Dimethyl sulphoxide, DMSO

In addition to the formation of desirable products, kilning also removes unwanted 'green-grain' flavours from malt (experimental beers prepared from unkilned malt can taste appalling). If they are not removed during kilning, these 'off-flavour' substances can be removed by the unusual step of steam-stripping the wort. To a limited extent polyphenols may be converted to coloured 'phlobaphenes' during kilning, and probably other changes occur which are significant in the development of colour or flavour. Currently there is considerable interest in dimethyl sulphide precursors in malt [56]. Dimethyl sulphide, DMS (106), adds a characteristic flavour to lager beer. In part, at least, it arises from a precursor (or two precursors) which is formed during grain germination, and which can be destroyed by strong kilning (as in making ale malts) or during the hop boil. One DMS precursor is probably the sulphonium compound S-methylmethionine, or a peptide containing it (107). Another precursor may be dimethyl sulphoxide (DMSO, 108). Probably during kilning some S-methylmethionine is broken down to yield DMS. Some of this is volatilized and lost, but at least part of the remainder may be oxidized to DMSO (108). This may be reduced to DMS by yeast.

Sulphur dioxide is sometimes fed into kilns, particularly in N. America, and is often present in the combustion gases of fuels. It tends to bleach malt, reducing its colour and increasing its 'brightness', as well as reducing the pH of worts made from it and increasing the levels of soluble nitrogen, both by virtue of its acidity and reducing properties, both of which favour proteolysis. Rarely poor-quality fuels give off-flavours to malts, and care must be taken to avoid this and to ensure that no arsenic, for example, is taken up from the gases generated by burning, particularly samples of anthracite. There is currently concern that in some conditions, and particularly where the fuel is natural gas, oxides of nitrogen (NOX) may occur in the furnace gases, and in turn may give rise to carcinogenic nitrosamines, such as dimethyl nitrosamine, by interacting with amino compounds in malts. It seems that drying malt at low temperatures for short periods results in less nitrosamine formation. In addition, the use of sulphur-containing fuels or burning sulphur in the kiln furnace creates sulphur dioxide which reduces nitrosamine formation. However, the oxides of sulphur may create a pollution problem and cause kiln deterioration. In addition, fuels have been used to flavour malts, e.g. by using the smoke from burning hardwood. Smoke from burning peat is still used to flavour Scottish distillers malts which are used to make whisky. Probably flavour derived from the peat 'reek' is due to a mixture of volatile phenols.

Many other flavour compounds have been recognized in dark malts and particularly in roasted or black malts. For example, hydroxymethylfurfural, maltol and many heterocyclic compounds, such as pyridines, pyrazines, thiazoles and bitter cyclic dipeptides (β-diketopiperazines), have been recognized. Roasting also destroys enzymes, partially degrades β-glucans and alters the structure of the dry granular starch so that it undergoes partial pyrodextrinization, becomes pale yellow, is less readily converted during mashing, and some of its glucose residues are converted to 1,6-anhydro-β-D-glucose.

A number of independent variables are involved in colour and flavour development in malts. Aroma and flavour are not easily estimated, but fortunately their intensity is related to the coloration achieved during kilning. Factors involved in colour development are: (*i*) the degree of modification and the analysis of the green malt loaded on to the kiln; (*ii*) the durations and levels of the temperature time sequences of the kilning cycle, (*iii*) the moisture content of the grain at different stages of the cycle. Skill and experience are needed to make good coloured malts. Highly kilned malts may give rise to turbid worts on mashing, but the reasons for this are unknown. However, even pale lager malts are usually cured at about 85°C (185°F) as are pale ale malts [traditionally at about 105°C; 221°F].

Curing ensures that the malt confers a particular character to beer, and is said to reduce the likelihood of the formation of chill or permanent non-biological hazes.

REFERENCES

[1] BRIGGS, D. E. (1978). *Barley*, Chapman & Hall, London, 612 pp.
[2] POLLOCK, J. R. A. (1962). *Barley and Malt* (ed. COOK, A. H.), Academic Press, London, p. 363.
[3] HARRIS, G. (1962). *Barley and Malt* (ed. COOK, A. H.), Academic Press, London, p. 431; p. 583.
[4] BROWN, H. T. and MORRIS, G. H. (1890). *J. Chem. Soc.*, 57, 458.
[5] BRIGGS, D. E. (1972). *Planta*, 108, 351.

[6] HUDSON, J. R. (1960). *Development of Brewing Analysis*, The Institute of Brewing, London, p. 21.
[7] HOPKINS, R. H. and KRAUSE, B. (1947). *Biochemistry Applied to Malting and Brewing*, George Allen & Unwin, London, p. 168.
[8] POLLOCK, J. R. A., ESSERY, R. E. and KIRSOP, B. H. (1955). *J. Inst. Brewing*, **61**, 295.
[9] STEVENS, R. (1958). *J. Inst. Brewing*, **64**, 470.
[10] COLLINS, E. J. (1918). *Ann. Bot.* **32**, 381.
[11] MORRALL, P. C. and BRIGGS, D. E. (1978). *Phytochemistry*, **17**, 1495.
[12] MORRISON, W. R. (1978). *Adv. Cereal Chem.*, **2**, 221 (ed. POMERANZ, Y.), American Association of Cereal Chemists, St. Paul., Minn., U.S.A.
[13] HALL, R. D., HARRIS, G. and MACWILLIAM, I. C. (1956). *J. Inst. Brewing*, **62**, 232.
[14] BRIGGS, D. E. (1973). *Biosynthesis and its Control in Plants* (ed. MILBORROW, B. V.), Academic Press, London, p. 219.
[15] MAEDA, I., KIRIBUCHI, S. and NAKAMURA, M. (1978). *Agric. Biol. Chem.*, **42**, 259.
[16] THOMPSON, R. G. and LA BERGE, D. E. (1977). *MBAA Tech. Q.*, **14**, 238.
[17] BATHGATE, G. N. and DALGLIESH, C. E. (1975). *Proc. Am. Soc. Brew. Chem.*, **33**, 32.
[18] FULCHER, R. G., SETTERFIELD, G., MCCULLY, M. E. and WOOD, P. J. (1977). *Aust. J. Plant Physiol.*, **4**, 917.
[19] BAXTER, E. D. (1978). *J. Inst. Brewing*, **84**, 271.
[20] MACLEOD, A. M., TRAVIS, D. C. and WREAY, D. G. (1953). *J. Inst. Brewing*, **59**, 159.
[21] BISHOP, L. R. and MARX, D. (1934). *J. Inst. Brewing*, **40**, 62.
[22] BISHOP, L. R. (1928). *J. Inst. Brewing*, **34**, 101.
[23] BISHOP, L. R. (1929). *J. Inst. Brewing*, **35**, 316.
[24] BISHOP, L. R. (1929). *J. Inst. Brewing*, **35**, 323.
[25] BISHOP, L. R. (1930). *J. Inst. Brewing*, **36**, 336.
[26] BROWN, H. T. (1909). *J. Inst. Brewing*, **15**, 184
[27] MACEY, A. (1957). *J. Inst. Brewing*, **63**, 306.
[28] JONES, M. and PIERCE, J. S. (1963). *Proc. Eur. Brewery Conv. Congr. Brussels* Elsevier, Amsterdam, p. 107.
[29] ENARI, T.-M. and MIKOLA, J. (1977). *Ciba Found. Symp.*, **50**, 335.
[30] BURGER, W. C. and SCHROEDER, R. L. (1976). *J. Am. Soc. Brew. Chem.*, **34**, 133; 138.
[31] MIKOLA, J. and ENARI, T.-M. (1976). *J. Inst. Brewing*, **76**, 182.
[32] POMERANZ, Y. and DIKEMAN, E. (1976-July). *Brewer's Digest*, p. 30.
[33] HOPULELE, T. and PIENDL, A. (1973). *Proc. Annu. Meet. Am. Soc. Brew. Chem.*, p. 75.
[34] KNORR, F. (1952). *Brauwissenschaft*, p. 70.
[35] VOSS, H. and PIENDL, A. (1978). *MBAA Tech., Q.* **15**, 215
[36] VON WETTSTEIN, D., JENDE-STRID, B., AHRENST-LARSEN, B. and SORENSEN, J. A. (1977). *Carlsberg Res. Commun.*, **42**, 341.
[37] VON WETTSTEIN, D. (1979). *Proc. Eur. Brew. Conv. Congr. 17th, Berlin (West)*, p. 587.
[38] AHRENST-LARSEN, B. and ERDAL, K. (1979). *Proc. Eur. Brew. Conv. Congr. 17th, Berlin (West)*, p. 631.
[39] WITHEY, J. S. and BRIGGS, D. E. (1966). *J. Inst. Brewing*, **72**, 474.

[40] WHATLING, A. J., PASFIELD, J. and BRIGGS, D. E. (1968). *J. Inst. Brewing*, **74**, 525.
[41] MACWILLIAM, I. C., HUDSON, J. R. and WHITEAR, A. L. (1963). *J. Inst. Brewing*, **69**, 303.
[42] CHAPON, L. and CHEMARDIN, M. (1964). *Proc. Annu. Meet. Am. Soc. Brew. Chem.*, p. 244.
[43] BALLESTEROS, G. and PIENDL, A. (1977-Oct.) *Brewer's Digest*, p. 36; 43.
[44] SHANDS, H. L., DICKSON, A. D., DICKSON, J. G. and BURKHARD, B. A. (1941). *Cereal Chem.*, **18**, 370.
[45] SHANDS, H. L., DICKSON, A. D. and DICKSON, J. G. (1942). *Cereal Chem.*, **19**, 471.
[46] DAY, T. C. (1896). *Trans. Proc. Bot. Soc. Edinburgh*, **20**, 492.
[47] NARZISS, L. (1975). *EBC Barley and Malt Symp.*, Zeist. (*Monogr. II*), p. 62.
[48] NARZISS, L. (1976). *Die Bierbrauerei*, 1, Die Technologie der Malzbereitung, Ferdinand Enke, Stuttgart, 382 pp.
[49] MACEY, A. (1977-Sept.). *Brewer's Guard.* **106** (9), 81.
[50] DE CLERCK, J. and CLOETENS, J. (1940). *Bull. Assoc. Anc. Étud. École Sup. Brass.*, Louvain, **40**, 41.
[51] MEREDITH, W. O. S., ANDERSON, J. A. and HUDSON, L. E. (1962). *Barley and Malt* (ed. COOK, A. H.), Academic Press, London, p. 207.
[52] DICKSON, A. D. and BURKHARD, B. A. (1942). *Cereal Chem.*, **19**, 251.
[53] KIRSOP, B. H. and POLLOCK, J. R. A. (1958). *J. Inst. Brewing*, **64**, 227.
[54] HIND, H. L. (1943). *Brewing Sci. Pract.* **1**, p. 8.
[55] BATHGATE, G. N. and BRENNAN, H. (1971), *Proc. Annu. Meet. Am. Soc. Brew. Chem.*, p. 22.
[56] WHITE, F. H. (1977-May). *Brewer's Digest*, p. 38.

Chapter 5

MALTING CONDITIONS AND MALT TYPES

Malting processes were originally developed by trial and error. More recently systematic investigations have greatly improved our understanding of the process, and allowed accelerated malting and the more regular production of high-quality malts. It is convenient to follow some of these developments in this chapter. A wide range of malting techniques are in use, varying greatly in sophistication. In general, advances in malting have led to: (*i*) more even modification due to more uniform germination; (*ii*) faster throughputs achieved by reducing processing times; (*iii*) more malt made from a given weight of barley, i.e. malting losses have been reduced; (*iv*) better quality malt having higher hot-water extracts; (*v*) malts having other more exactly specified characters, e.g. of colour, levels of soluble nitrogen or wort fermentability. (*vi*) In addition there is a greater ability to malt barleys of less good malting quality, e.g. with inconveniently high nitrogen contents or 'feed' varieties, or showing some degrees of dormancy.

Malt analyses often include estimates of hot-water extract (HWE), cold-water extract (CWE), total nitrogen (TN), permanently soluble nitrogen (PSN), or total soluble nitrogen (TSN), diastatic power (DP) and α-amylase. Sometimes other analyses, such as estimates β-glucanase, are also made. In addition, analytical 'worts' may be evaluated for their colour, fermentability, content of α-amino nitrogen, various polyphenol fractions, viscosity, and so on (Chapter 8). Also note is taken of the weight of rootlets produced, and the total loss in dry matter in turning barley into screened malt.

5.1 *Floor malting* [1, 2]

In the simplest form of traditional floor malting cleaned barley is steeped, preferably in a shallow bed of about 60 cm (2 ft) at the well water temperature, usually 10–15°C (50–59°F). The water is changed two to four times during the steeping period without the use of extended air rests. Immersion may be continued for 60 hr or more, depending on the barley sample and the temperature of the grain-water mix. When the grain has swollen to about 1·3–1·4 times its original volume and has reached a chosen moisture content, e.g. of 43% (fresh wt.) in the UK or 41% in Europe, the water is drained away, and after 6–10 hr the grain is moved from the steep and made up into a heap (ruck or couch) of 23–76 cm (9 in–2 ft 6 in) deep. The build-up of heat accelerates the uptake of the film of surface moisture and the onset of germination. After a period, the grain is spread on an impervious floor in a layer 10–15 cm (4–6 in) deep, in a room with cool [less than 12·5°C (56·3°F)] moist air. The grain 'piece' is turned and mixed once or twice a day to equalize the temperature and to prevent the growing roots matting together. To allow greater heat losses and so reduce the temperature, the piece is spread more thinly over a wider area of the floor. Conversely, to allow a rise in temperature, it is thickened. Sometimes the piece 'sweats' on the floor, i.e. becomes covered with drops of moisture, demonstrating the saturation of the air with water vapour. Probably a build-up of carbon dioxide also occurs within the bed of grain. Normally a flooring temperature of 13–16·5°C (55·4–62°F) is aimed for, although at the end of the germination period it may reach 21°C (70°F). For a piece 7·6 cm (3 in) deep, and an air temperature of 9–11°C (48–52°F), the temperature in the grain bed may be 14–16°C (58–61°F) with a temperature differential of around 0·5°C (1°F) between the top and bottom. Similarly, for a piece 15 cm (6 in) deep, with an ambient temperature of 13°C (approx. 56°F) the mean temperature of the piece will be about 20°C (68°F), with a temperature differential of about 1·5°C (2·5°F). These values are not constant, since the heat output of the grain changes with the stage of germination. Turning the piece to ventilate it reduces the temperature by about 1°C (1·8°F). In many floor maltings, the processing is more complex and takes advantage of various modern techniques, such as more complex steeping schedules or the use of additives (see Chapter 4, and below). The germination conditions of the grain in floor maltings, with its still, moisture-saturated, carbon dioxide-enriched air, are significantly different from those encountered in modern pneumatic maltings. When the processes of 'modification' have proceeded far enough and the rootlets are beginning to wither, which may take 7–10 days, the grain is moved to a kiln where it is dried and 'cooked' to various degrees, depending on the type of product needed. The temperature and ventilation schedules used during kilning are very variable, and are considered later. Roots are removed from the dry malt and are sold.

During malting losses in dry matter increase, but at very varying rates (Fig. 5.1). The malting losses are made up of dry matter leached from the grain during steeping, 'losses' as roots (which have a cash value), and losses due to the formation of carbon dioxide and water. The respiration rate, respiratory losses and heat output are greatest at about the middle of the germination period. Rootlet growth occurs most rapidly at an earlier stage. Nitrogen is not lost from the whole grain plus root system during the germination period, but it does move from the main part of the grain into the rootlets, consequently the total nitrogen content (TN, as a percentage) found in the derooted malt at any time (Fig. 5.1) depends on the amount initially present, the quantity that has migrated into the roots and the decline in weight of the

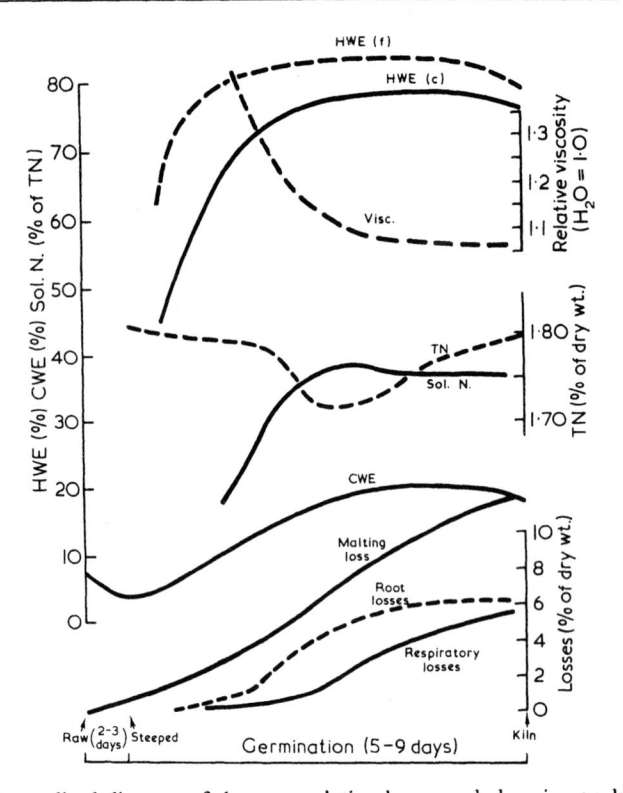

Fig. 5.1 Generalized diagram of the accumulating losses and changing analyses of barley under traditional floor malting conditions. Usually the process is terminated by kilning before the hot-water extract (HWE) begins to decline.
Key: HWE, hot-water extract (% dry matter); (f), determined using a fine grind; (c), determined using a coarse grind; Visc., viscosity of wort, relative to water; TN, total nitrogen of derooted malt, as % dry weight; Sol. N., soluble nitrogen found in an analytical wort as a % of TN; CWE, cold-water extract of malt, expressed as % dry weight; Malting loss, loss in dry matter, as % of initial dry weight of the barley; made up of weight of roots, losses due to the conversion of grain solids to carbon dioxide and water (curves shown) and losses due to dust washing away and grain substances being dissolved during steeping.

root-free grain. Thus TN varies, but under most conditions of germination is nearly the same in the finished malt as in the original barley [3]. As germination progresses, hydrolytic enzymes accumulate in the starchy endosperm, and catalyse the degradation of its components. As a result it is 'modified', and is progressively more easily 'rubbed out' while moist and becomes more friable when dried. Products of hydrolysis, especially sugars, accumulate in the grain and comprise the cold-water extract (CWE or preformed soluble substances) of the grain (Fig. 5.1). Progressive physical modification and the increased levels of hydrolytic enzymes make it easier to recover a good yield of hot-water extract when the malt is milled and mashed, even when it is coarsely ground, but more extract is always recovered when finely ground samples are mashed (Fig. 5.1). The level of soluble nitrogen, as recovered by mashing, rises to a roughly level plateau during germination. The quality of the extract obtained by mashing malts of different ages differs in various ways. Thus the proportions of different nitrogenous substances and the range of carbohydrates (and hence wort fermentabilities) also change. The complexity of high-molecular-weight substances, and particularly β-glucans, is generally less in worts prepared from well-grown malts, and so their viscosities are less (Fig. 5.1). Malt analyses are changed by using different kilning conditions to finish a particular green malt. Clearly malt grown for the shortest acceptable time is obtained in highest yield, and the faster germination can be carried out, the more malt can be made in a particular germination space.

5.2 *Extraction prediction* [3–10]

It would be advantageous to know exactly what quality of malt may be obtained when a given sample of barley is processed but, in fact, a knowledge of a barley's composition only has a limited predictive value. Such knowledge is used in selecting new barleys for malting quality during breeding programmes, when purchasing barleys and in checking whether a malt made from a particular barley is of as good a quality as it should be. Different types of malt may be made from one barley, so 'predictive' approaches, which are based on previous experience, presuppose that a standard form of malting is being used. The regularity principle of barley composition has been noted (Chapter 4). It is further found that for a given variety, grain samples of a known range of compositions, when processed in a standard way give rise to malts in which extract yields also vary in a regular way. Thus it has long been known that malt extract is less in samples which are rich in nitrogen prepared from barleys containing higher nitrogen contents. For example, for a range of malts with crude protein contents of 9, 11, 13 and 15% the average extracts were 81·8, 79·3, 77·3 and 75·6% dry matter respectively [11]. Also higher extracts tend to be given by malts prepared from samples with higher average grain weights or (less certainly) weights per unit packed volume (e.g. kg/hl or lb/bushel). Equations, applicable to British barleys, revised to take account of the 1971 IoB methods of analysis, are:

$$TE = A - 11 \cdot 0N + 0 \cdot 22G \tag{5.1}$$

$$TE = A_1 - 10 \cdot 0N - 0 \cdot 3Z - 0 \cdot 2X \tag{5.2}$$

$$TE = 138 \cdot 2 - 9 \cdot 5N - 3 \cdot 0I \tag{5.3}$$

TE = hot-water extract (dry wt. basis lb/Qr., where Qr. is a quarter of 336 lb) as determined by the Recommended Methods of Analysis of the Institute of Brewing (1971 revision).

A, A_1 = varietal 'constants' for each type of barley. 'A' for 'Proctor' = 112·5. Range of values for two-rowed varieties = 109·5–114·5.

G = dry weight of 1000 corns, TCW (g).
N = nitrogen content (%) of dry barley.
Z = 1000/G, that is the recipocal of the dry weight (g), of one corn.
X = percentage of dead and dormant grain in the sample as found by a specified method.
I = percentage of 'insoluble carbohydrate' remaining after a specified sequence of hydrolyses and extractions.

Corresponding equations for predicting the results of Congress extract determinations are given in Analytica-EBC (2nd edition).

Equation 5.1 uses readily available figures for the prediction, but takes no account of dead or dormant corns, which are allowed for in equation 5.2. Equation 5.3 does not suppose a knowledge of the barley variety and therefore the varietal constant, but is based on a chemical determination of 'insoluble carbohydrate'. This material probably parallels the quantity of carbohydrate that does not contribute to extract when the malt is eventually mashed. Such equations only accurately predict malt extract for a given type of malting procedure that has been used to determine the 'constants', but they indicate the trends in the relationships between barley composition and malt extract when any malting process is used. The varietal 'constant' alters a little between seasons. The thousand corn weight, G, always a tedious and often an imprecise determination, detracts from the precision of the prediction [8, 9]. With 'Proctor' it was shown that an increase of 0·1% in the total nitrogen content of the barley was equivalent to a loss of 1·2 lb/Qr. of extract (coarse grind) in the finished malt. The revised prediction equation (5·4) suggested for 'Proctor', using a particular set of malting conditions on barleys of the 1960 and 1961 crops [8, 9] is:

$$TE = 119 \cdot 9 - 11 \cdot 5N \pm 1 \cdot 1 \text{ (lb/Qr.)} \qquad (5.4).$$

Different units of extract were introduced in 1977 and were revised in 1979. Interconversions between these units are explained in Chapter 9.

When these equations were derived it appeared that the same linear inverse relationship between barley nitrogen (TN) and malt hot-water extract (HWE) applied to all the British barleys in cultivation. This is no longer the case, and studies on wide ranges of barleys indicate that the inverse relationship between TN and HWE is not always strictly linear, so entirely different equations now need to be derived for each barley variety. This seems to be excessively laborious and so either a range of samples must be micromalted and analysed or some other basis for extract prediction must be used. 'Barley extracts' can be determined by mashing very finely ground ungerminated grain with a carefully standardized mixture of hydrolytic enzymes, which includes α-amylase. In many instances linear correlations between barley extracts and the extracts of malts prepared from them have been demonstrated [3,10]. Thus barley extracts, determined by a standardized method, may prove a useful guide for predicting malt extracts.

5.3 Small-scale malting [10]

Increasingly barley evaluation is carried out by malting small subsamples and analysing the 'micromalts' produced. Such small-scale malting techniques are flexible and reliable, and so are also widely used in developmental trials to test the effects of altering the steeping and germination conditions, the effects of additives, the relative 'maltabilities' of different grain samples, and so on. In addition to hot-water extracts, the micromalts can be analysed for other characters, such as total nitrogen content, soluble nitrogen, wort fermentability, total malting losses, and rootlet production. Bag or 'stocking' small-scale malting, with 0·45–1 kg (approx. 1–2·2 lb) samples of barley, has been widely used especially in floor maltings. The weighed barley is held in a bag of muslin or nylon mesh and is steeped, germinated and kilned embedded in a bulk of another barley, which is being processed on the production scale, and which keeps the bag of grain 'conditioned'. Such samples are a good guide to the malting value of a barley, although the hot-water extracts are often about 1% above those obtained when a bulk sample is processed.

'Micromalting' systems handling 5 g–2 kg samples of barley, in which the processing conditions are very exactly controlled, are now very numerous. Systems handling the smaller samples (e.g. 5 g; contained in boiling tubes, for example) are needed by plant breeders who can only spare small quantities of grain, but they are inherently less valuable because fewer analyses can be carried out, and greater sampling errors are involved in selecting such tiny amounts of grain. With larger samples, of say 75 g, drawn from a well-mixed bulk of grain that has been graded to a uniform range of grain sizes, agreement between analyses of replicate samples is usually excellent. In small-scale maltings grain can be steeped at a chosen temperature(s) for selected periods of time, with or without air rests, forced ventilation and changes in steep liquor. Sometimes micromalting equipment is so constructed that the alterations in processing conditions are controlled automatically. In the simplest systems grain samples are put to germinate arranged in heaps on glass shelves, or retained in glass bottles, but in each case in a temperature-controlled high-humidity chamber. In more complex equipment a flow of water-saturated attemperated air is pumped through beds of grain held in containers with perforated bases. In such equipment it is possible to estimate the rate of carbon dioxide generation by the grain by estimating the quantities in the emerging air, and so continuously monitor the intensity of grain metabolism. Germinating grain must be turned regularly, either by hand or automatically.

Kilning small samples of malt presents difficulties in that it is doubtful if the conditions used in small kilns ever truly reproduce the conditions found in a commercial kiln. In 'microkilns' hot air may be allowed to rise through a bed of green malt, or a forced air flow may be used. Drying green malt in ovens in stationary air that is not compelled to flow through the grain mass is usually unsatisfactory. For studying the effects of varying the conditions used during the earlier malting steps it is often convenient to kiln in a comparatively cool rapid air flow at a fixed temperature (e.g. in the range 45–60°C; 113–140°F) to fix the composition of the malt quickly, with a minimum of enzyme destruction, and avoid the complexities caused by prematurely beginning high-temperature 'curing'. This will reduce the moisture level to about 6% in 24 hr. Subsequently these samples may be subjected to a curing phase. However, kilning schedules in which the air temperature rises, and the air flow is reduced, and/or recirculation is introduced, are being used in some 'microkilns' that are built with adequate degrees of instrumentation and automatic control. Different micromalting techniques produce different types of malt differing in analysis and in yield from a single batch of barley. It is of great interest therefore that, in general, they *do* rank different barleys in essentially the same order of ease of malting and ability to make good-quality malts. Micromalting trials, analyses of the malts and small-scale brewing trials with the malts can give exact comparative data, for example on: (*i*) malt yield, (*ii*) root yield, (*iii*) hot-water extract, (*iv*) cold-water extract, (*v*) total malt nitrogen, (*vi*) total soluble nitrogen, (*vii*) wort pH, (*viii*) wort fermentability and (*ix*) levels of polyphenols. However, difficulties are sometimes encountered when trying to scale up a malting programme from a micromalting to a production scale. In part, this is because it takes an appreciable time to move a large bulk of grain, while small bodies of grain can be moved almost instantaneously. Further, a large mass of grain has a large thermal inertia, and so its temperature is more difficult to adjust. Also on the small scale the control of grain quality, and all the processing parameters, is much more exact. Consequently to help in overcoming 'scaling up' problems pilot-scale maltings of various types, capable of handling say 50–1000 kg of barley (approx. 1 cwt–1 ton) are in use. These produce enough malt to allow the production of sufficient beer with pilot brews to be used in further chemical and taste-evaluation tests. These are essential since conventional analyses may fail to detect defects that can give rise to processing difficulties or off-flavours. However, experience shows that small model breweries nearly always produce beers with flavours that do not match those of larger breweries.

5.4 *Malting losses*
Malting losses may be accurately determined in small-scale trials in which all the barley, malt and rootlets can be weighed with precision and without

sampling errors. In production malting weighing is probably rarely better than to 0·5%, and often 'within 1%' is considered realistic. Barley that is ready for malting when weighed 'as is' (i.e. fresh weight basis) usually contains 10–18% moisture. Until recently the weight was expressed in barley quarters (Qr.) of 203·21 kg (448 lb). After malting and the removal of rootlets, the weight of malt, 'as is', at 1–6% moisture, was also expressed in quarters (Qr.) of 152·91 kg (336 lb). It used to be expected that 1 Qr. of barley would yield about 1 Qr. of malt. Consequent upon the different moisture contents and the change in units estimates of malting losses (and gains) expressed in this way meant little. Units of volume (e.g. bushels) were also employed. The valid measure of malting loss, that should always be used in serious work, is the percentage reduction in dry weight occurring in the conversion of barley to the finished malt.

During steeping 'skimmings' (floating corns, awns and trash), dust and soluble materials are removed from the barley, usually giving a 'steeping loss' of 0·5–1·5%, of the initial dry weight. During germination and kilning, further losses accumulate which are due to the production of roots and the loss of dry matter consumed in respiration (Table 5.1; Fig. 5.1). Practically speaking, respiring barley consumes carbohydrates, converting them to carbon dioxide and water (Chapter 4). In a floor malting, barley may be cast at 43–45% moisture and loaded on to the kiln at about 42% moisture, yet there has been a loss of 6–10% dry matter, so that during germination there has been a net loss, due to evaporation, of 'steep water' and 'water of respiration.'

TABLE 5.1

Losses in dry weight (%) incurred in the conversion of barley to malt

Type of malt	German data					British data
	Pale*	Dark*	Floor†	Pneumatic drum†	Kropff†	Current range‡
Steep losses	1·0	1·0	1·5	1·5	1·5	0·5–1·5
Rootlets	3·7	4·5	4·5	3·8	2·4	2·5–4·0
Respiration losses	5·8	7·5	5·4	4·9	3·1	3·5–5·0
Total malting loss	10·5	13·0	11·4	10·2	7·0	6·5–10·5

* Data of KOLBACH cited by DE CLERCK (1957) [12].
† Data of LÜERS (1922) [13].
‡ Excluding malts made with resteeping or applications of bromate.

The bulk of respiration losses (Chapter 4) and all the rootlets originate from the metabolic activity of the embryo. Clearly malting with embryos having reduced viability, using gibberellic acid to trigger modification, offers the chance of dramatically reducing malting losses, and saving large quantities of raw materials. A possible difference of approach between sales maltsters and brewer maltsters is apparent. A sales maltster must produce malt of a specified analysis before he can sell it, whatever malting losses are incurred. A brewer maltster, however, may aim to produce as much acceptable-quality extract from a given weight of barley as is feasible, and the conditions that achieve this may not yield maximal extract per unit weight of malt. In modern malting, losses are less than formerly (Table 5.1). This is partly due to the use of better-quality, more uniform and more easily modified barley, and to the acceptance of malt having a different set of characteristics and appearance. However, the major savings come from the careful application of newer steeping and other techniques in the malting process and the use of gibberellic acid and other additives, where these are permitted.

5.5 *Variations in steeping conditions* [14–19]

An outline of some steeping schedules is given in Chapter 3. In traditional British malting mellow, two-rowed barley was discharged to the germination flour with a moisture content of about 43%. Increasingly higher 'steep-out' moisture levels have come into use, so that 45% is a common value, and very poor, steely barleys may be discharged at 46–48% moisture. Putting barleys to germinate at elevated moisture levels alters the pattern of changes which occur during the subsequent germination phase, and accelerates the overall process. The results vary with variety and germination temperature (Fig. 5.2), but in general at higher moisture levels the maximum extract attained is less than in drier conditions. Malting losses, rootlet production, respiration with its associated heat output, α-amylase levels, diastatic power, level of soluble nitrogen and nitrogen index of modification are all enhanced. On the other hand, it has recently been suggested that by casting at 35% moisture, and by applying heavy doses of gibberellic acid, it may be possible to make adequate malt that is cheaper to kiln, since it contains much less moisture needing to be evaporated [20]. Traditional steeps were largely anaerobic. By aerating steeps by pumping over grain–water mixtures between steep tanks or by recirculating around one tank, by sparging air into the steep or by air resting the grain, more uniform and rapid grain germination is obtained, allowing better malt to be produced in a shorter time. Often newer air-rests, multiple-steeping and spray-steeping procedures ensure that germination has started before the end of the 'steeping' period, and indeed grain may be 'sprinkled' during the germination process to counteract evaporative losses. So the clear distinction between the steeping and germination phases, characteristic of traditional malting, has become blurred. In resteeping procedures (see below) a reimmersion is used to control malting losses.

Increasingly arrangements are being made for adjusting steep water temperatures. Slightly alkaline waters are preferred for steeping, but, provided that they are not grossly contaminated, soft waters are adequate (Chapter 3).

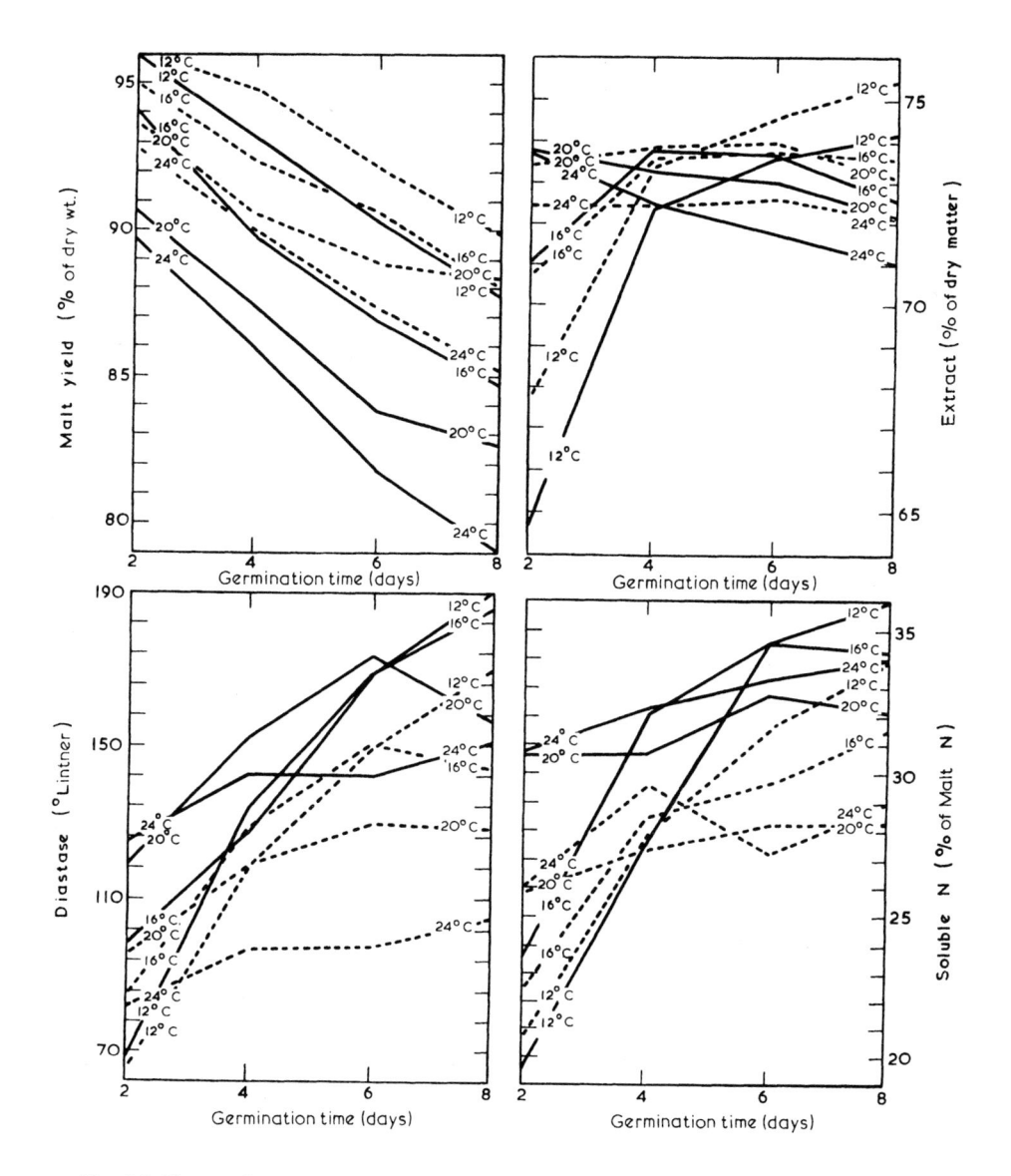

Fig. 5.2 Changes in the average analyses of four barley varieties malted at four different temperatures (12, 16, 20, 24°C; 53·6, 60·8, 68·0, 75·2°F) and at two different moisture contents 43% (- - - -), and 49% (———). {Shands *et al.* (1941) [15]}.

Dilute formaldehyde has been used as a disinfectant in steep liquors [21]. This is so effective in controlling microbes that treated steep liquors can be reused, so reducing water use and effluent production. Formaldehyde also reduces some types of dormancy. Treated grain grows slightly less readily, and so may turn into malt with a reduced malting loss and less rootlet production. In addition formaldehyde applied in the last steep (500–1000 mg/litre) reacts with, and presumably cross-links, barley polyphenols, so that beer made from malt prepared in this way is less prone to form haze [22, 23]. Formaldehyde is potentially dangerous. Hydrogen peroxide (e.g. 0·1%) is sometimes added to steeps to help break barley dormancy. However, this is an expensive way of using this additive, which, like gibberellic acid, bromates and some other additives (see below) may be effectively applied by spraying solutions on to grain. Another proposal is to spray-steep with ozonized water. The positive benefits of this treatment have not been described, but may resemble those given by hydrogen peroxide.

5.6 Pneumatic malting

Economic considerations demand that larger batches of grain should be malted, using less manpower. Thus maltings are increasingly mechanized and built to process larger quantities of grain. As a consequence, floor maltings, which are not readily automated, are being displaced by 'pneumatic maltings' in which grain germinates in a bed 1 metre (39 in) or more in depth; it is turned mechanically, and is attemperated and ventilated with a forced flow of 'conditioned' air, i.e. air at a controlled temperature, as nearly saturated with water vapour as possible. In general the rate of air flow is constant, but the temperature of the 'air on', i.e. that entering the bed of grain, may be reduced or allowed to rise to regulate the average temperature of the grain bed. The temperature differential across the bed is typically 1–2°C (approx. 2–4°F), but may be 3°C (5·4°F) or even more. The grain is turned (i) to prevent rootlets matting, (ii) to ensure that on average all grains are treated as nearly evenly as possible, (iii) to break up hot spots and (iv) to 'lighten the piece' i.e. reduce its bulk density, and so allow a less impeded passage to the attemperating air stream. Under ideal conditions the air enters and leaves the grain bed saturated with water vapour, i.e. with a relative humidity (RH) of 100%. However, the air leaving the grain bed is warmer and carries more water than the air entering, so inevitably water is carried away, and the grain dries. Grain may be sprinkled with water during turning in an attempt to make good these losses, but this is not wholly satisfactory, and grain set to germinate in a pneumatic maltings may, therefore, have a higher initial moisture content than would be used in a floor maltings. In one instance the air entering a grain bed was saturated with moisture, and on average the grain lost water at the rate of 0·5% per day. However, when the RH of the 'air on' fell to 95% the grain moisture declined by 2·5% per day, and in 2 days the grain was so dry that modification ceased.

5.7 Germination temperatures and malting [3, 14-19]

In traditional malting, after an initial warm period 'in the couch', germination was carried out for example at 13-16°C (55-61°F) for 8-10 days for a well-modified ale malt, or 6-8 days for a less-well-modified lager malt. In pneumatic maltings temperature control can be more exact, and indeed can be programmed, and germination temperatures of 16-19°C (approx. 61-66°F) are commonly used, and at the end of the germination period may be allowed to rise to 25°C (77°F) or even higher. In unusual cases germination temperatures of 25-30°C (77-86°F) are used (Chapter 4). In general, attempts to carry out the whole malting process at 25°C (77°F) or more have been unsuccessful, at least with British barleys, so there are upper limits to the temperatures that can be used. The use of higher germination temperatures, together with the use of more sophisticated steeping schedules and additives which can offset some of the undesirable effects of elevated temperatures, have resulted in germination periods of as little as 3·5-5 days. Higher temperatures reduce the germination percentage of dormant grains and so can create 'unevenness' in malting barley. Also there is a greater danger of fungal growth on grain germinating warm. When non-dormant grain is germinated at higher temperatures many metabolic processes are accelerated, but not all to the same extent, and the pattern of changes observed alters with time (Fig. 5.2) [3, 15, 17]. Consequently at elevated temperatures green malt may be ready to kiln in a shorter time, but its composition will differ significantly from that of a green malt prepared using a long cool germination. Thus malting losses accumulate faster at higher temperatures, but rootlets do not grow much faster as the temperature is increased over about 22°C (71·6°F). The rate of modification is greater at elevated temperatures, but while extracts increase faster initially, they do not reach such high maximal levels in warm-grown grains, nor ultimately are the levels of diastase, α-amylase, soluble nitrogen, or the soluble nitrogen ratio (TSN/TN) as great in warm-grown grains. Consequently the extract difference between fine and coarse grinds, saccharification times, and often wort viscosities are increased by 'warm germination'. Such results can be countered by the use of additives, or by altering the moisture content of the grain. The temperature of the grain may usefully be altered during germination. For example, by starting germination warm, so that initially enzyme formation is rapid, then by reducing the temperature at a chosen point in the germination period, the enzyme formation rate is maintained, and good-quality malt can be produced in a reduced time, and with a reduced malting loss [16-18] (Table 5.2). Such techniques are particularly valuable in countries where the use of additives is forbidden. In other processes involving ventilated multiple steeping and the use of additives, germination may be carried out at temperatures above 20°C (68°F), and grain may even receive its final steep at an elevated temperature [24] (Chapter 4).

TABLE 5.2

Comparison of malts produced by different germination schemes (Barley-Carina, 1974)
{Data of NARZISS (1975) [17]}

Temperatures† (°C)	12/15/17	17/13	17/13	18/12	18/12*
Moisture‡ (%)	44	38/43/47	38/43/48·5	38/43/48·5	38/50
					RESTEEP
Steep + germination (days)	9	7	6	6	6
Extract (% dry matter)	82·5	82·8	82·4	82·7	82·8
Fine–coarse difference (%)	1·5	1·3	1·5	1·5	1·0
Viscosity (cP)	1·55	1·53	1·57	1·57	1·60
Protein (% dry matter)	10·2	10·2	10·3	10·3	10·4
Kolbach index (%)	40·2	44·7	42·4	44·0	44·2
α-Amylase** (ASBC)	55	64·2	57·1	60·9	60·3
Colour (EBC units)	3·0	3·0	2·8	2·8	2·5
Colour of boiled wort (EBC units)	5·0	5·5	4·8	5·2	5·0
Malting loss (%)	9·2	8·5	7·7	7·7	6·3

* Resteeping process.
† Temperatures at successive stages of germination.
‡ Moisture levels at successive stages of germination.
**Method of the American Society of Brewing Chemists.

5.8 Additives in malting and techniques for reducing malting losses

Additives and various 'special' techniques are, or have been, used in malting: (*i*) to reduce the microbial populations on grain (hypochlorites, alkalis, formaldehyde); (*ii*) to reduce the soluble polyphenol contents of malt (formaldehyde); (*iii*) to secure rapid and even germination of grain (hydrogen peroxide, steep aeration, air rests, ozone, gibberellic acid); (*iv*) to accelerate all aspects of grain modification, and break dormancy (gibberellic acid); (*v*) to augment the extract, and possibly alter the nature of the final product (glucose); (*vi*) to control the development of soluble nitrogen (bromates; salts of octanoic acid); (*vii*) to check malting losses and reduce rootlet growth (many chemical additives and physical treatments have been proposed). Some of these additives have been discussed in Chapters 3 and 4.

Where permitted by customers and local legislation gibberellic acid (GA_3) is the most widely used additive [25]. This hormone occurs naturally in barley (Chapter 4). It is prepared on an industrial scale from culture filtrates of the fungus *Gibberella fujikuroi* (syn. *Fusarium moniliforme*). Added GA_3 augments the grain's own supplies, and enhances embryo metabolism and growth, but to a relatively greater extent induces an increase in the hydrolytic enzymes which catalyse the processes of modification (Chapter 4). Additions of GA_3 to barley: (*i*) break dormancy; (*ii*) accelerate the whole malting process, increasing the degree of modification achieved in a short time and shortening the germination period; (*iii*) increase the respiration rate and heat output; (*iv*) enhance rootlet growth and, more markedly, acrospire growth; (*v*) stimulate the production of some enzymes, including some proteases, peptidases, α-amylase and hence diastatic power, some other carbohydrases and phosphatases. However, not all enzymes increase by the same proportional amount in response to an application of GA_3. Because most of the enzymes involved in modification are produced in the aleurone layer and are formed, or at least released, in response to gibberellic acid, and most malting losses are associated with embryo growth and metabolism, many treatments have been proposed in which the embryo is damaged or killed. In these cases modification is induced by GA_3, resulting in the production of malt with minimal malting losses. Response to added GA_3 is highly dose-dependent, and differs between varieties. Within limits the major responses are often proportional to the logarithm of the dose applied [25] (Fig. 5.3).

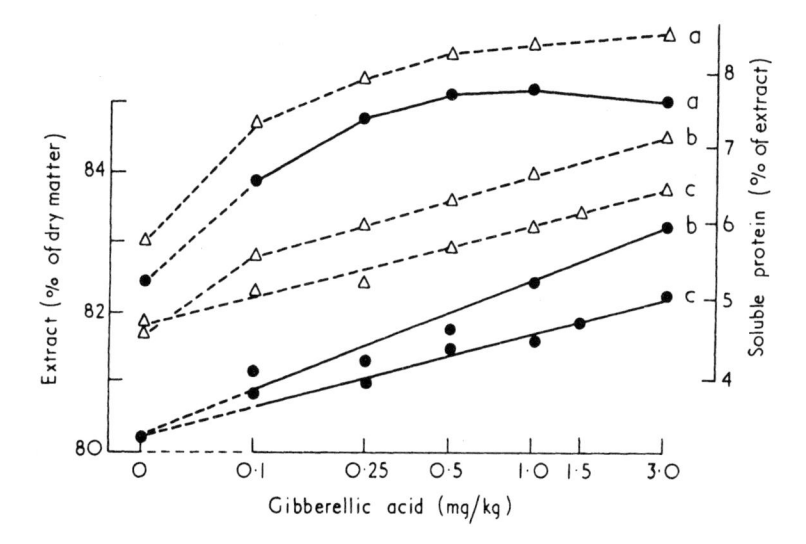

Fig. 5.3 Increased values of extract (●) and soluble nitrogen (△) found after malting three varieties of barley for a fixed period after dosing with various amounts of gibberellic acid. Varieties were (a) Beka, (b) Carlsberg and (c) Aurore. The gibberellin-dosage axis is on a logarithmic scale. {Data of Secobrah cited by Briggs (1963) [25]}.

With suitable applications of GA_3, maximal extract is obtained 1–3 days sooner than would otherwise be the case, and in up to 3% greater yield than is obtained from an untreated control. If germination is terminated in a shorter time, at the right stage, malting losses are not enhanced, and may even be reduced by 4% by a dose of GA_3, apparently because the rate of modification is accelerated relatively more than the rate of increase of malting losses. With a suitably chosen dose (e.g. 0·025–0·25 mg of GA_3/kg of barley), all the advantages mentioned may be achieved, with the production of a malt that is at least equal to many traditional samples. This is achieved more simply by the additional use of bromates (see below). If excessive doses of GA_3 (e.g. 3 mg/kg of grain) are used, then: (*i*) malt is over-modified; (*ii*) its cold-water extract and content of preformed soluble nitrogenous substances are unusually great; so (*iii*) it becomes dark during kilning, and the derived wort is dark; (*iv*) the soluble nitrogen level of the wort is increased, and so is its soluble nitrogen ratio (SNR); (*v*) the ratio of simpler to complex nitrogenous substances in the wort, e.g. the amino acid/protein ratio, is increased; (*vi*) the wort contains a greater proportion of simpler carbohydrates, so the sugar/dextrin ratio and fermentabilities are unusually high. The manner of application of a dose of GA_3, which is always used in aqueous solution, is also critical [26]. The hormone may be applied in the steeps, in which case the last steep is most efficient, but still much dissolved GA_3 runs to waste. If a solution of gibberellic acid is sprayed evenly on to the grain as it is conveyed at 'steep-out', only about 20–25% of the dose used in the last steep is needed to obtain a comparable response. In recent multiple-steeping schedules, GA_3, and other additives, may be sprayed on to grain at the onset of an air-rest period between immersion or spray-steeps. Hydrogen peroxide may be incorporated into the gibberellic acid solution sprayed on to grain, often with a useful increase in the rate of chitting.

The selectively permeable surface structures, and specifically the testa, exclude water and gibberellic acid, except at the micropyle [27]. Both gain entry faster when grain chits, when a film of surface moisture including some of the GA_3 dissolved in it is rapidly taken up. By lightly crushing the grain so that small splits occur in the pericarp and testa the uptake of a GA_3 solution is greatly accelerated [28]. The time course of enzyme development in such crushed GA_3-treated grain is different from that occurring in whole grain, but eventually the modification is unusually rapid, and is completed in about one-third less time than is taken by entire grain receiving an equal dose of GA_3. Split grains cannot be used on the commercial scale because of handling problems and the frequent occurrence of heavy microbial infestations. Barley that was battered by an impeller during steeping subsequently malted at an enhanced rate when treated with gibberellic acid [29]. Malting barley that had been 'dehusked' by mechanical treatments that removed the surface structures (7–9% dry wt.) malted rapidly when dosed with GA_3, but was not used commercially because of the difficulties mentioned in connection with crushed grain [30]. In attempts to overcome these problems the process of 'abrasion' was introduced [31, 32] (Chapter 6). Initially grain abrasion was achieved by wire brushes rotating in a metal housing through which the grain passed. It was believed that abrasion of the apex of the grain permitted the entry of externally applied GA_3 and water at both the apex and basal embryo ends causing 'two-way modification'. However, this is clearly not the mechanism involved. Instead it seems probable that the abrasive treatments damage the structure of the grain, causing faults in the surface layers which permit more oxygen to reach the living tissues within [27]. This permits the enzyme-forming tissues to respond better to the GA_3, which still gains entry at the embryo end, by making and releasing more enzymes more rapidly. The pattern of modification encountered in abraded grains is generally similar to that encountered in normal grains, but the endosperm degradation advances more rapidly [27], maximal hot-water extracts are reached in a shorter time, and soluble nitrogen levels are elevated. On occasion but not invariably, abrasion and similar treatments enhance the germination of dormant grains and marginally accelerate the rate of water uptake during steeping. Abrasive treatments vary in effectiveness when used on different grain samples. Among the reasons for this may be that many grain samples are already 'abraded' by rough threshing and passages through handling machinery, and that the surface layers of some grains may be tougher in some samples than others.

In a particular process glucose syrup is applied to germinating grain [33]. Claims are made that the malt produced has a higher extract, a greater yield of roots and a higher degree of modification after a fixed period of germination compared with untreated controls. Residual sugar on the malt will increase its extract simply by redissolving during mashing, and acid produced from the sugar by micro-organisms on the grains may reduce the pH of the mashed malt to a useful extent [33]. Thus the treatment may result in a more desirable malt, but whether it alters the metabolism of the barley itself is uncertain.

After gibberellic acid, and generally used in conjunction with it, the most widely used additives are sodium and potassium bromates. These are apparently interchangeable. These soluble bromates, which are used as 'flour improvers' in baking, reduce malting losses and by checking respiration usefully reduce heat generation in the grain bed and hence the need for refrigeration. They cause a reduction in rootlet growth. Rootlets of treated grain are characteristically shortened, twisted and swollen in appearance. Because of the reduction in volume of treated germinating green malt, owing to reduced rootlet production, a given germination compartment will contain green malt from a larger quantity of barley than would otherwise be the case, so the capacity of the germination area is effectively increased. In addition, at some stages of germination, malts prepared from treated grain give rise to lower levels of soluble nitrogen on mashing because bromate ions inhibit many proteolytic enzymes. Bromate penetrates the grain over the whole surface at least slowly, but probably mainly gains entry at the embryo end [34, 35]. During germination bromate is progressively reduced to bromide.

No bromate survives unreduced into the kilned malt [35]. Applications of potassium bromate in the last steep need about five times the dose (e.g. 500 mg of $KBrO_3$/kg of grain) to achieve the result given by an application sprayed on to the grain at 'steep out' (e.g. 100 mg/kg). Bromate is most effective when applied at the start of the germination period [36]. Levels of soluble nitrogen reached in malts from brain dosed with bromate change with germination time and are dependent on the dose applied, the germination conditions and whether or not GA_3 has also been applied. Thus in a trial using a fixed germination period, increasing doses of bromate, but no other additives, are seen to depress malting losses, but cause an increase and then a decline in the hot-water extract, the cold-water extract and the level of soluble nitrogen (Table 5.3). The different levels of cold-water extract represent a balance

TABLE 5.3

Analysis of malts prepared with additions of potassium bromate
{MACEY and STOWELL (1961) [34]}

$KBrO_3$ (mg/kg of barley, 2nd steep)	0	500	1000	2000
Hot-water extract (litre°/kg at 20°C)	286·1	287·1	285·6	283·6
(lb/Qr.)	96·2	96·9	96·4	95·6
Cold-water extract (%)	17·3	17·8	16·8	16·6
Permanently soluble nitrogen (PSN, %)	0·48	0·52	0·46	0·40
Malting loss (%)	9·1	6·6	5·0	4·1

between preformed sugars 'saved' by the depression of malting losses, and a slowing down in their rate of formation due to a reduced rate of endosperm degradation. Bromate-treated malts release reduced quantities of proline into wort during mashing, and, as yeasts do not use proline under anaerobic conditions, less soluble nitrogen remains in finished beer [37, 38]. Normally bromate is used in conjunction with GA_3. It offsets some of the undesirable effects of the latter by reducing heat output and by reducing the level of soluble nitrogen obtained on mashing (Table 5.4). In commercial malting, but not in micromaltings in which grain temperature is completely controlled (and hence the respiration rate is limited), bromate enhances the hot-water extract (Table 5.4; Fig. 5.4a, b). Increasing doses of bromate applied to grain together with a fixed dose of GA_3 decrease the rates of increase in malting loss and rootlet production to relatively different extents, and delay the attainment of the maximum level of soluble nitrogen (Fig. 5.4). By altering the dose rates of GA_3 and bromate, malts can be obtained at the expense of a wide range of malting losses, hot-water extracts and very varied soluble nitrogen ratios. Thus there are no immutable relationships between these analyses and other malt characteristics, and different 'indices of modification',

TABLE 5.4
Analyses of malts prepared with various additions of gibberellic acid and potassium bromate sprinkled on to the grain, in solution, at 'steep out'
{MACEY and STOWELL (1961) [34]}

Gibberellic acid (mg/kg of barley)	0	0·25	0·25	0·25	0·25
Potassium bromate (mg/kg of barley)	0	125	188	250	375
Hot-water extract {(litre°/kg at 20°C)	302·9	307·1	307·4	306·1	308·1
(lb/Qr.)	102·4	103·9	104·0	103·5	104·2
Cold-water extract (%)	17·0	19·9	19·1	18·7	18·1
Diastatic power (°Lintner)	58	58	58	52	54
Total nitrogen, TN (%)	1·55	1·54	1·52	1·50	1·51
Permanently soluble nitrogen, (PSN, %)	0·56	0·61	0·58	0·53	0·50
Index of modification [(PSN/TN) × 100]	36	40	38	35	33
Malting loss (% dry matter)	7·2	5·3	5·3	4·6	4·5

established using traditional malting, frequently disagree when used to assess the value of a newer type of malt. Bromate is avoided when lager malts are being made, since it suppresses the formation of precursors of dimethyl sulphide. Bromate also reduces the formation of unwanted nitrosamines during kilning.

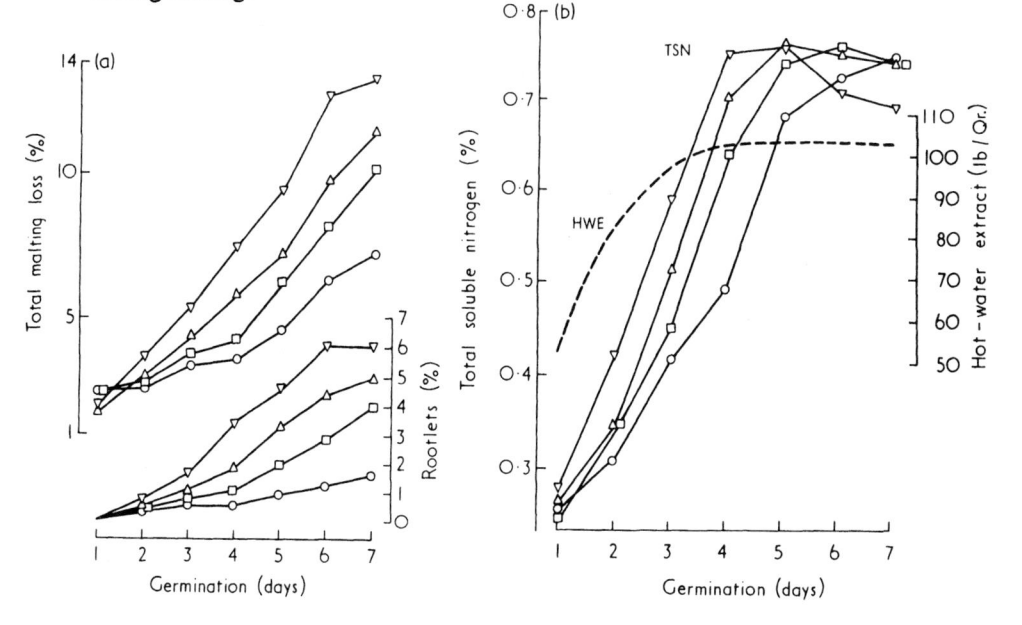

Fig. 5.4 The effects of four different applications of potassium bromate (▽, none; △, 75 mg/kg of barley; □, 125 mg/kg; ○, 250 mg/kg) applied at 'steep out' with gibberellic acid (0.25 mg/kg of fresh grain) during subsequent micromalting. (a) Rootlet production and total malting loss. (b) Hot-water extract and total soluble nitrogen. {Dudley and Briggs (1977) [36]}.

Recently the experimental use of sodium salts of certain fatty acids (C_6–C_{10}) to offset some effects of GA_3 was described [39]. Using applications of sodium octanoate at steep out, applications of 50–100 mg/kg barley began to reduce the levels of soluble nitrogen and free amino-nitrogen encountered in the finished malt. Doses of 150–200 mg/kg started to reduce malting losses, while doses of 400–500 mg/kg were needed to reduce hot water extracts [39]. Octanoate had less effect than bromate on rootlet growth. On the other hand it did not reduce the production of dimethyl sulphide precursors.

Other means have been sought for reducing malting losses by damaging grain embryos, whether or not modification has been accelerated or induced by external applications of gibberellic acid (GA_3). If GA_3 is supplied, then the embryo-damaging loss-checking treatments may be applied in the steep or at 'steep out'. However, in the absence of added GA_3, treatments can only be applied after the embryo has 'dosed' the enzyme-forming tissues with endogenously produced gibberellins, generally after 3 to 4 days germination under most conditions (Chapter 4). Germinating grain has been variously treated on the experimental scale, but only rarely, or never on the production scale, with (*i*) concentrated steep liquors (*ii*) benzoxazolone, (*iii*) naphthalene acetic acid (NAA), (*iv*) lactic acid, (*v*) 2,4-dichlorophenoxyacetic acid (2, 4-D), (*vi*) acetic acid, (*vii*) nitric acid, (*viii*) sulphuric acid, (*ix*) phosphoric acid and acid phosphate salts, (*x*) coumarin, (*xi*) copper sulphate, (*xii*) urea nitrate, (*xiii*) many organic solvents such as methanol and acetone, (*xiv*) ammonia gas or solution, (*xv*) sulphur dioxide as a gas or in solution, and (*xvi*) ethylene gas [22, 40–45]. Ethylene restricts root growth under a range of conditions, but barleys vary in their susceptibility to this gas, and it has not come into commercial use [45]. It is inflammable and can form explosive mixtures with air. Lactic acid has been used for another purpose, to make 'acid' malts and 'enzymic' malts.

The use of some of these compounds requires comment. For example, copper sulphate is poisonous, hastens haze formation in beer, and weakens yeast. It is stated that little copper reached the final beer from treated malt, presumably because it is retained by the spent grains, the trub and the yeast [42]. Treatments with mineral acids often reduce malting losses and give malts with enhanced extracts, but the extract frequently has a reduced fermentability [43, 44]. It is claimed that 'acidulation' with sulphuric acid can be used to check malting losses in a malt made with GA_3 and abrasion, and that wort from the malt has a normal fermentability [32]. Bromates are used to counteract the elevated levels of soluble nitrogen when malting with abrasion and GA_3. In small-scale trials (*i*) acetic acid applied during steeping followed by GA_3 at 'steep out' or (*ii*) applications of sulphur dioxide gas or solution during germination both reduced malting losses, and enhanced yields of extracts which had normal fermentabilities. Such malts have elevated soluble nitrogen ratios [40, 41]. Both ammonia and sulphur dioxide gases are toxic and corrosive and so should be handled with care. Both are so soluble that they are rapidly absorbed from a closed container by moist grain. However, British breweries would generally avoid malts with enhanced contents of nitrogen or sulphur, although in many areas malt is partly bleached and acidified by exposure to sulphur dioxide during kilning (Chapter 6). This may change if it is found that kilning with natural gas (sulphur-free) alters the characters of malts. Sulphur dioxide in the kiln air stream seems to reduce, or prevent, the formation of undesirable nitrosamines in malt. High applications of bromates at the start of malting, or spraying or steeping green malt with dilute nitric acid before kilning are other ways of reducing the formation of nitrosamines.

Respiration and growth (and modification under some conditions) can be checked by restricting oxygen supplies to a mass of grain, while allowing carbon dioxide to accumulate. In the old-fashioned Kropff malting system the germinating grain, initially steeped to an unusually low moisture content of about 41%, is placed in a deep bed in a closed well-insulated container (Fig. 5.5) [12, 13, 46, 47]. Oxygen is consumed, carbon dioxide accumulates and respiration is halted, metabolism is reduced, and the heat output of the grain is halted after a temperature rise of 3·5–4·5°C (6–8°F). As a result of the anaerobic fermentative processes that occur, ethanol and possibly acids accumulate and damage the embryo, so checking subsequent root growth and reducing malting losses by 2–3% (Table 5.1). The extract is elevated by 0·5–0·8% dry matter, but the enzyme content of the malt is reduced. Provision is made in the construction of the equipment for the periodic ventilation of the grain, to maintain its viability. The results of Kropff malting were apparently variable, possibly owing to lack of exact control and uniformity of technique. Kropff malts yield much soluble nitrogen and have a high cold-water extract, and take much colour on the kiln. This process was useful when it was originated, when malting losses of about 11–13% were common and produced dark malts suitable for the traditional dark lager beers of Munich. However, in recent small-scale trials carried out against efficiently malted controls, using gibberellic acid and with 6–8% malting losses, carbon dioxide treatments or periods of anoxia merely reduced malt quality [41]. Nevertheless schedules for giving green malts 'rests' in moist atmospheres enriched with carbon dioxide have been described [18, 48].

The idea of using chemical growth regulators in malting is widely disliked, and efforts have been made to find 'physical' treatments that may be used instead. Experimentally, embryos may be selectively damaged by freezing wetted grain [49] or by battering chitted grain with mechanical agitation during the germination period, having induced modification with added GA_3 [50]. Malt made with agitation and applications of gibberellic acid was

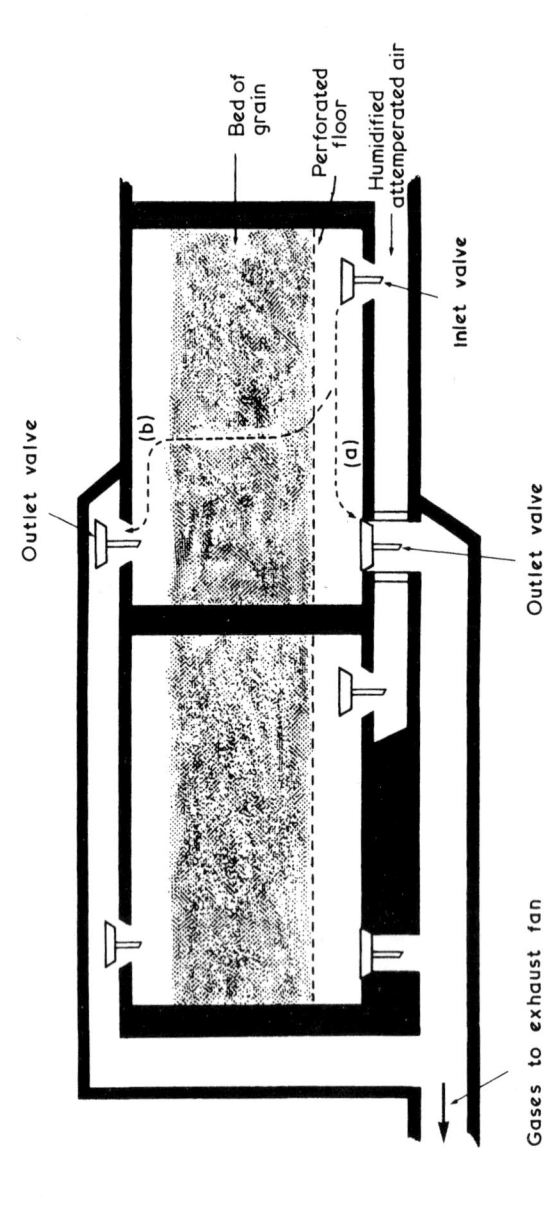

Fig. 5.5 Diagram of two Kropff boxes. The structures are thermally insulated and gas-tight when the valves are closed. Route (a), course of air during limited ventilation. Route (b), air stream during more vigorous ventilation and attemperation.

produced with only a 3·5% malting loss, and had an extract exceeding that of the controls. This type of process might be applicable to production malting if turners were designed to deliberately bruise the grain, the converse of what they are normally intended to do. The use of hot steeps for comparatively short periods rapidly hydrates grain and damages the embryos. If the correct conditions are chosen, modification can be induced by subsequent applications of GA_3, and, experimentally, excellent quality malts are produced in very high yield [51, 52] (Table 5.5). This process has not been adopted commercially, apparently because of the difficulty of rapidly mixing many tons of grain and hot water to achieve 'instantaneously' the chosen temperature of 40°C. In addition, enzyme development is sometimes insufficient, and longer 'hot steeps' at lower temperatures have been proposed as an alternative method of causing embryo damage [51].

In the resteeping process, in its simplest form, barley steeped to 40% moisture was germinated for 3 days, then the roots were killed by 'resteeping', i.e. reimmersion in an anaerobic steep at 18°C (64·5°F) for 24 hr [52–54]. The steep was drained and a flush of water was applied to remove the materials leached from the dead roots. The wet grain, at 50–52% moisture, was returned to the germination compartment for a further period ('autolysis') before kilning. Root killing reduced the malting loss to about 4% and the high moisture content of the grain after the resteep resulted in rapid modification and saving of some 15% of germination time (Table 5.6). However, care is needed to ensure an adequate level of killing of roots and root initials, otherwise roots may grow again, acrospires will bolt, and the malting loss will actually be increased. Furthermore microbial growth on the dead roots and materials leached from them may result in the production of discoloured, smelly and offensively flavoured malts. It is the need to minimize this problem that makes it essential to flush the grain at the end of the resteep. Wet malt is expensive to kiln, and the extra effluent produced by the resteep has a high BOD. Furthermore resteeping can only be carried out in special equipment (Chapter 6). Nevertheless the process has been used on a production scale, and is particularly attractive in countries like Germany where the use of chemical additives is forbidden (Tables 5.2 and 5.6) [17]. Resteeping schedules have been elaborated to take advantage of air rests, additions of GA_3 and hot-water resteeps, which have a more certain killing action in a shorter time. Using some of these schedules it is possible to start kilning the malt 3·5–4 days from first wetting the grain [56]. The inclusion of dilute, formaldehyde (0·1%) in the resteep liquor prevents microbial growth, results in more certain root killing, and leads to a clean malt which releases few anthocyanogens and less soluble nitrogen into the wort during mashing, and so gives rise to beer with a low haze-forming potential [22, 23].

TABLE 5.5

Analyses of malts prepared by steeping barley for 8 hr at 40°C (104°F), then treating with different quantities of gibberellic acid and germinated at 15·5°C {Data of POOL (1964) [52]}

Gibberellic acid (mg/kg of barley)		0·125			0·25			0·5		
Germination time (days)	4	5	6	4	5	6	4	5	6
Hot-water extract (litre°/kg at 20°C)	279·1	294·7	297·4	279·7	299·6	307·7	290·4	302·6	311·1
(lb/Qr.)	94·0	99·5	100·4	94·2	101·3	104·1	98·0	102·3	105·3
Cold-water extract (%)	9·9	13·6	15·2	11·4	14·3	17·9	11·8	15·1	19·1
Total soluble nitrogen (%)	0·39	0·50	0·54	0·39	0·51	0·65	0·40	0·52	0·63
Malting loss (% dry matter)	2·3	2·9	3·1	2·3	2·9	3·6	2·4	2·7	3·5
HWE (lb/Qr.) × malt yield (%)	92·3	96·9	98·0	92·4	98·6	100·7	96·0	99·1	102·0

TABLE 5.6
*Analyses of commercial malts prepared by conventional and resteeping processes
Malting loss reduced by 4%. Malting time reduced by 20%*
{RIVIERE (1961) [55]}

	Hot-water extract		Diastatic power	Cold-water
	(litre°/kg at 20°C)	(lb/Qr.)	(°Lintner)	extract (%)
Conventional malt	299·8	101·4	34·5	17·0
Resteep malt	300·5	101·6	38·0	20·0

Another proposed technique for reducing malting losses is 'double malting' [57]. In this process partly germinated malt is kiln-dried, derooted, rewet and put to 'autolyse' until the final kilning. Extract is said to be elevated and malting loss reduced by 5–7%, but in view of the extra drying and handling costs this process will never be economic.

5.9 *Kilning* [3, 12, 14, 18, 47]
Chemical changes that occur during kilning are discussed in Chapter 4, and the physical principles involved are outlined in Chapter 6. Kilning produces a dry product which is stable during storage, and from which the brittle rootlets can readily be separated. It also adds 'character' to the malt, altering its colour and flavour, and reducing its enzyme potential. Particular malts are made by bringing green malt of a controlled degree of modification to the kiln and processing it in a chosen way (next section). Enzyme survival is greatest at low kiln temperatures, and is greater at elevated temperatures when the malt is less moist. Since malt colour is developed least at low temperatures, in dry malts having low cold-water extracts, and maximal colour formation occurs at high temperature in wet malts that are rich in cold-water extract and preformed soluble nitrogenous substances, these conditions dictate the types of recipes needed for making pale highly enzymic malts and dark aromatic poorly enzymic malts respectively. With increasing final curing temperatures, generally 75–105°C (167–221°F), there is little change in fine-grind extract, malt hardness, wort viscosity, or malting losses, but the coarse-grind extract, wort pH and diastatic power fall, while fine-coarse extract difference and wort colour increase. In the interest of fuel economy, attempts are being made to use malts prepared using lower curing temperatures (Chapter 6). In addition, the fuel used in direct kilning influences malt character. Thus sulphur dioxide in the drying gases acidifies malt, minimises the formation of nitrosamines, raises its soluble nitrogen level, and partly bleaches it, reducing its colour. Smoke, e.g. from hardwood or peat in distilling malts adds particular aromas to 'special' malts.

5.10 Different malt types [3, 12, 18, 47, 58–61]

The brewer sets specifications for the types of malt he needs, and the maltster buys barley and processes it to malt hoping to meet these specifications. Clearly the tolerances must not be drawn so unnecessarily tight that they are impossible to meet regularly. Further, the specified combinations of analytical values must exist in one malt. Maltsters usually blend malts of a similar type to meet a specification as closely as possible. Brewers, in turn, usually blend malts before mashing with them, to keep the grist as uniform as possible. Specifications normally indicate a minimum hot-water extract, a colour range, a maximum moisture content, a maximum or particular total nitrogen content (TN), a level of soluble nitrogen (TSN or PSN), and a diastatic power (DP). Sometimes α-amylase contents, β-glucanase contents and other analyses are mentioned, such as an upper limit on the number of malt grains less than 2·2 mm in width (often 2–3% depending on the season's crop). Also upper limits are set on the quantity of residual arsenic from the kiln and the levels of nitrosodi-methylamine, NDMA. Typical analyses of malts are given in Tables 5.7, 5.8 and 5.9.

TABLE 5.7
Representative analyses of high-quality floor malts (I.o.B. analyses)
{NORTHAM (1965) [2]}

Type of malt	Pale ale	Mild ale	Standard	Lager
Hot-water extract (litre°/kg at 20°C)	307·0	296·0–304·9	301·0–306·2	307·0
Hot-water extract (lb/Qr)	103·7	100·0–103·2	101·8–103·6	103·9
Moisture content (%)	1·7	1·8–2·0	1·6–2·0	2·4
Colour (EBC units)	5·0	6·0–7·0	5·5–6·5	2·5
Cold-water extract (%)	18·0	17·0–19·0	18·0–19·7	18·6
Diastatic power (°Lintner)	35	33–48	39–47	63
Total nitrogen content (%)	1·35	1·44–1·56	1·48–1·51	1·65
Total soluble nitrogen (%)	0·49	0·50–0·58	0·55–0·60	0·66

Alterations in the type of malt produced may be achieved in various ways. To a limited extent raw materials may be varied. Thus malted rye may be used or oats can be malted for use in the manufacture of stout. Malted wheat is produced, particularly in Germany where it consititutes 75% of the grist for Weissbier. Wheat malt gives beer with outstandingly good head retention. A high extract of about 318·8 litre°/kg (20°C) (108 lb/Qr.) is obtained from wheat malt because the absence of husk (which does not contain potential extract) means that the malt contains about 8% more starch than an equal weight of barley malt. This lack of husk makes it difficult to malt the grain without damaging the acrospire, since this is unprotected.

Generally, as already discussed, the variety of barley, its germinative capacity, and its nitrogen content decide what grade of malt will be made from it. By adjusting the steeping and germination schedules, the chosen

TABLE 5.8

Representative analyses (ASBC methods) of some North American malts

Barley Crop	USA				Canada‡	
	Klages* (2-rowed)	Piroline* (2-rowed)	Karl† (6-rowed)	Midwestern† Larker (6-rowed)	(6-rowed)	(2-rowed)
Moisture (%)	3·9±0·2	3·9±0·2	4·0	4·1	3·8±0·2	3·8±0·2
Extract (%, dry weight, fine grind)	80–81	78·5–79·5	81·7	77·0	78–79	79·5–80·5
Fine-coarse extract difference (%)	1·5–2·1	1·5–2·1	1·7	1·7	1·3–2·2	1·8–2·5
α-Amylase (Dextrinizing units)	35–40	33–38	33	40	35–45	30–40
Diastatic power (degrees)	110–120	100–110	102	156	120–145	90–120
Total Protein (%)	11·5–12·5	11·3–12·3	10·4	13·3	11·5–12·5	11·0–12·0
Soluble N/total N (ratio)	39–43	38–42	45	39	38–42	38–42
Colour (°Lovibond)	1·45–1·75	1·55–1·85	1·68	1·74	1·40–1·60	1·20–1·40

* Courtesy of VOGEL, R., Great Western Malting Co., Vancouver, USA.
† Courtesy of ZYTNIAK, R., Great Western Malting Co., Vancouver, USA.
‡ Courtesy of SISLER, W. W., Brewing and Malting Barley Research Institute, Winnipeg, Canada.

TABLE 5.9

Some average analyses of conventional German malts (Analytica EBC) {OLOFF and PIENDL (1978) [61]}

Malt type	Pale lager	Pilsen lager	Vienna-type lager	Dark lager	Diastatic malt	Wheat malt
Extract (coarse grind, g/100 g, dry wt.)	79·1	78·9	79·3	77·5	77·3	82·2
Fine–coarse extract difference (%)	1·6	1·8	1·6	2·0	1·5	1·5
Moisture (%)	4·4	4·6	4·5	3·8	7·6	5·7
Conversion time (min)	10	10	10	20	5	15
Colour (EBC units)	3·4	3·0	7·1	17	2·6	4·1
Total protein (g/100 g dry wt.)	11·0	11·4	11·0	11·5	12·1	13·3
Soluble protein (mg/100 g dry wt.)	734	716	783	714	811	808
Protein modification ratio (%)	41·9	39·4	44·4	39·0	42·1	38·1
pH (EBC)	5·76	5·74	5·63	5·54	5·82	5·94
Viscosity (cP)	1·57	1·60	1·57	1·61	1·54	1·80
Hardness (Brabender)	347	358	371	365	510	694
Diastatic power* (°WK/100 g dry wt.)	289	307	215	145	433	317
α-Amylase (ASBC, D.U./100 g dry wt.)	44·0	46·0	40·0	30·5	64·0	47·0

* °WK – Windisch-Kolbach units.

level of modification is achieved, and by kilning to adjust the final diastatic power and colour, the correct malt is produced.

At present, malts of a particular grade are made in the UK using a wide variety of schedules. The traditional method of making a pale ale malt took 17–18 days from wetting to taking from the kiln. By 1968 the average figure was 10–12 days, and now periods of 5·5 days (including a 24 hr kilning cycle) are not uncommon. Such quick malting may involve the use of multiple steeping schedules, air rests, additives, high processing temperatures, and so on. However, considerable quantities of 'traditional' malt are still made, particularly for use in the older small breweries. In addition, there has been a rise in the average nitrogen contents of barleys available to maltsters and malts having higher moisture contents are now being accepted by some breweries. Many ale malts might have a total nitrogen content of 1·6–1·65% dry matter, a hot-water extract exceeding 299 litre°/kg (20°C; dry; 100·8 lb/Qr.), a moisture content of 1·5–2·5%; a colour of 4–6·5 EBC units, a total soluble nitrogen value of 0·5–0·7%, and a soluble nitrogen ratio 'index of modification' (TSN/TN) of 31–41%. British lager-type malts are now often made in a rather similar fashion, except that the use of bromates is avoided, so that the levels of dimethyl sulphide precursor are not depressed. But whereas the ale malts are likely to be kilned from 60 to 95°C (140–203°F), the lager malts are likely to be kilned at temperatures varying from 55 to 70°C (131–158°F). Such lager malts may have extracts in excess of 301 litre°/kg; (20°C; dry; 101·9 lb/Qr.), moisture contents of 4–6%, low colours not exceeding 2·5 EBC units, nitrogen contents of 1·65–1·75%, total soluble nitrogen values of 0·5–0·7% and soluble nitrogen ratios of 31–41%.

In North America the production of malts from six-rowed Californian barleys has practically ceased [3, 58], having been largely displaced by the use of Western two-rowed barleys. Nevertheless a major proportion of N. American malts, perhaps 85%, are still made from six-rowed barley varieties. In the USA the ranges of some analyses encountered in Mid Western malt types, prepared from six-rowed barleys, are: moisture (%), 3·7–5·3; thousand grain weight (g dry wt.), 27–32; growth (% acrospires 3/4–1 corn length ratio), 70–90; mealy (%), 87–98; extract (%; fine-grind, dry basis), 74·5–80·0; fine–coarse extract difference (%), 0·9–3·5; colour laboratory wort (°Lovibond), 1·2–2·9; diastatic power (degrees Lintner), 100–190; total protein (%, dry basis), 11·7–14·3; soluble/total nitrogen (ratio), 34–45. It seems that many commercial batches of North American malts are blends of malts prepared from two- and six-rowed barleys.

Coloured and 'special' malts constitute about 1% of the malt made in Britain. The great majority are manufactured by using extreme variations of the basic procedures. Thus the green malt may be drastically modified, stewed on the kiln, kilned at very high temperatures, or finally roasted. Small percentages of special malts are used in grists for some beers, to which they impart colour, special flavours and aromas, high anti-oxidant power, and 'palate fullness'. To preserve their special characters they should be used while fresh. Graded series of malts of different colours and other properties are usually available, ranging from lager and pale ale malts to chocolate and black malts. The darkest malts have little or no enzyme potential. To determine the extract of dark malts it is necessary therefore to mash them with enzymes from another source, as occurs in brewery mashes. For analytical purposes it is usual to mash coloured malts with a cold-water extract of a pale ale malt which contains the necessary enzymes, or mixed with a small proportion of a highly enzymic malt (compare with analyses of adjuncts; Chapter 8).

Chocolate or black malts are prepared from plump uniform barleys having a moderate nitrogen content (1·5–1·6%), germinated for 4–6 days, kiln-dried to about 5% moisture at which stage they may be stored. Subsequently they are roasted at a high temperature [221–233°C (420–450°F)] for 2–2·5 hr in a slowly rotating metal cylinder heated by burning coke or gas. As heating is continued, acid fumes are produced, and there is a loss of about 15% of the dry weight. The fumes must not be allowed to escape. Care is needed, since the grain is converted to charcoal at temperatures over 250°C (480°F), and may actually catch fire. Frequent colour checks indicate when roasting should be stopped by the addition of water. The water instantly turns to steam, cooling the grain and thereby checking the roasting process and causing the grain to swell. The product should be uniform in appearance without charred grains, should have a hot-water extract 254–268 litre°/kg (20°C; dry; 85–90 lb/Qr.) with a colour of 900–1100 EBC units (chocolate malt) or 1200–1500 EBC units (black malt), and should contain 1·5–3·0% moisture. The product has a sweet or acrid flavour and is used to give character to stouts.

Roasted barley is prepared in a similar way to chocolate and black malts except that raw unsteeped grain is roasted in the drum. The product has a hot-water extract of about 269 litre°/kg (20°C; dry; 90 lb/Qr.) and a colour of 1000–1500 EBC units. It has a characteristic flavour, more dry and sharp than dark malts, and is used in making some dark stouts.

European 'brumalt' is made from highly steeped 8-day malts which are placed for 24 hr in a bed about 20 cm deep under a cover so that the temperature rises sharply to a maximum of about 50°C (122°F) and then falls as oxygen in the air under the cover is consumed and carbon dioxide accumulates rather as in a Kropff box. The product, after kilning at less than 100°C (212°F), is very dark and yields a wort rich in simple sugars, consequently having a high reducing power.

Crystal or caramel malts are widely used and have delicious flavours. A fully modified malt, 1·4–1·5% N, rich in sugars, is wetted or rewetted after light kilning, and the wet material (up to 50% moisture) is loaded onto a kiln where it is warmed at 65·5–76·7°C (150–170°F) for 1·5–2 hr with minimal ventilation, so that evaporation is restricted, and 'stewing' occurs. The interior of the grain is effectively 'mashed', the starch is hydrolysed, and the endosperm liquefies to a syrup. Amylolysis rather than proteolysis is favoured. Enzyme action is limited by a lack of water, and the optimum temperature for the production of reducing sugars is 65–70°C (149–158°F), compared with 62·8–64·0°C (145–147°F) in the mash tun, owing to the increased stability of the enzymes in the drier system. The kiln temperature is then increased, with frequent colour checks being made on the malt, to about 121°C (250°F). When the malt cools, the liquid interior sets to a pale-brown hard caramel-like mass. The product, which is produced at the expense of a high malting loss, has a colour of 80–220 (generally near 140) EBC units, an extract of 254–268 litre°/kg (20°C; dry; 85–90 lb/Qr.) and contains about 3% moisture. A pale crystal malt called 'cara-pils' is prepared that yields an extract having a low fermentability.

Amber malts are prepared by kiln-drying well-modified malt to 3–4% moisture and then by 'ambering' in the kiln or a roasting drum by heating rapidly to about 93°C (200°F), in 15–20 min and then gradually to 138–149°C (280–300°F). The higher temperature is maintained until the correct colour, 35–100 EBC units, is attained. The product has a 'warm biscuit flavour', an extract of 262–282 litre°/kg (20°C; dry; 88–95 lb/Qr.) and contains little diastase. In contrast, diamber malts are/were prepared so as to retain the extract and diastatic power, although they have a deep colour and fine aroma. Brown malts are normal malts dried in a fierce heat from a fire of hardwood faggots made from oak, hornbeam, ash or beech. A final kiln temperature of about 177°C (350°F) for 2·5 hr gives a product with a colour of 100–200 EBC units, 1–2% moisture, and an extract of 262–267 litre°/kg (20°C; dry; 88–90 lb/Qr.) and a characteristic flavour derived from the wood smoke. This material is now used rarely, if at all.

In Britain, transactions involving coloured malts have in the past been complicated by the range of units of weight that were used. Thus coloured malts and roasted barleys were sold by the malt quarter of 336 lb, or 280 lb or 252 lb, or quarters of 8 or 6 bushels. Metric units of weight are now being employed.

Acid malts, such as Dixon's patent malt, are prepared in various ways and are used to reduce the pH in the mash tun. The growing malt is variously sprayed and steeped with solutions of biologically prepared lactic acid – for example, it may be sprinkled twice with 10% lactic acid and steeped in a solution of lactic acid before kilning. In addition to the effect on mash pH, the product has an elevated cold-water extract and gives a wort with a high level of permanently soluble nitrogen and formol nitrogen. Thus it provides a good source of amino acids for supporting yeast growth.

Chit malt, produced on the mainland of Europe, is normally steeped to a low moisture content (about 40%), then grown for a short period of about 3 days at less than 18°C (approx. 64°F), and then lightly kilned in a very undermodified state. Various other slightly more modified 'short-grown' malts are also prepared. These products retain much of the character of raw grain and are used at less than 10% of the grist. Beer from such material is said to have a marked hop flavour and has improved head retention, as do beers prepared from grists containing a proportion of raw wheat or raw barley flour. Chit malt was originally prepared to circumvent laws forbidding the use of raw grain in brewing. Similar laws (repealed about 1880) combined with edicts restricting the efficiency of mills explain the desire of traditional English brewers for completely mellow easily crushed malts.

Malts for whisky production may be made from barley having a more variable nitrogen content. Malt whisky is made from malt having a high extract – and hence is well modified and has a moderate or low nitrogen content. Peat is burnt on the kiln so that peat smoke condensate (reek) settles on the malt, increasing its phenol content and giving it a characteristic flavour. The phenols present include phenol and various isomeric cresols, xylenols and guaiacol. For reasons of tradition, gibberellic acid is not used in the preparation of this malt. Malt used for the production of grain whisky may have a high nitrogen content of 1·8% or more and may be made from six-rowed barley. This is essential to provide adequate levels of enzymes in the mash tun where the bulk of the extract will be provided by unmalted adjunct (maize) which supplies no enzymes for conversion, and practically no soluble nitrogenous substances to support yeast growth. For distilling purposes as much carbohydrate as possible is fermented to alcohol. To achieve this, maximum enzyme development and survival is required during the preparation of the malt. So the grain may be steeped and sprayed to a moisture content of 48–50%, and is traditionally given a long period of low-temperature germination to yield the maximum quantity of enzymes. These are preserved by drying the malt at a low temperature [53–54·5°C (128–130°F)]. For whisky manufacture the mash is made at a low temperature. The 'wort' is not boiled, so that the breakdown of dextrins to fermentable sugars is continued by the surviving malt enzymes acting in the presence of the yeast during fermentation.

Brewers of malt vinegar may also use high-nitrogen highly diastatic malts. Their mashing and fermentation procedures resemble those of the whisky distiller but their fermented 'beer' is subsequently acetified with highly aerobic bacteria.

Malts with high enzyme contents are of interest to the brewer for the conversion of adjuncts in the mash. This has always been the case in North America, where highly diastatic malts (DP approx. 200°Lintner) made from six-rowed barleys having nitrogen contents of up to 2·2% are used to convert mashes made with as much as 60% unmalted adjuncts.

Considerable savings in fuel and processing time may be made by kilning malts at a low temperature to the hand-dry stage (5–8% moisture) and then omitting the curing stage. By British ale-brewing standards such malt is 'slack', that is, it has a high moisture content. However, it can be stored, milled in a conventional way, and contains a large proportion of its initial content of enzymes. Moisture contents of 4–6% are common in European lager malts (Table 5·9) and even higher values are sometimes accepted in North American malts. The removal of rootlets may be difficult if the moisture content exceeds 8%. Hand-dry malt and even green malt have been used in experimental mashes in the UK. Green malts have exceptionally high enzyme contents but may give flavour problems in the final beers, and cannot be stored. Hand-dry malt is a practical alternative. British brewers are now accepting slightly higher malt moisture contents. In Belgium small quantities of 'wind-malt' are prepared by thinly spreading green malt from six-rowed barley on screens, in special lofts, and leaving it to dry. This material is used to prepare 'old Louvain' beers, and presumably has a high enzyme content.

Conventional malt extracts are prepared by mashing normal malt as a brewer does, and concentrating the sweet wort so obtained under reduced pressure (Chapter 8). Such syrupy extracts are used in the manufacture of some sweets and foodstuffs and in home brewing. Very occasionally they are used by brewers as a 'stand-by' to provide extract when there is a shortage of mash-tun capacity, and as the source of nearly all the extract in some very small remote breweries including temporary breweries. An alternative procedure is for maltsters to mash green malt directly, with or without unmalted adjuncts, concentrate the wort by flash-evaporation and sell the resulting syrup as a sweet-wort replacement or an adjunct. Because the cost of kilning is replaced by the cost of concentrating the syrup, the process is more economic than the preparation of conventional malt extracts. The products are sometimes quaintly known as 'liquid malt', and the wort-concentration step as 'liquid kilning' (Chapter 11).

5.11 *Rootlets* [3]

Dried rootlets (coombes; culms; cummins; malt sprouts) are separated from the malt after kilning. They are extremely hygroscopic, and must be stored out of contact with the air. They have a cash value that is less, on a weight basis, than the malt itself, and are used for feeding cattle. Some 3–5% of the original barley is recovered as rootlets in traditional malting,

TABLE 5.10
Kilned malt rootlets; approximate ranges of composition
{BRIGGS (1978) [3]}

Fraction	Proportion (%)
Moisture	Less than 7.0
Crude protein (N × 6.25)	25–34
Fat (oil)	1·6–2·2
Minerals	6–7
Nitrogen-free extractives	35–44 (or even 50)
Pentosan	15·6–18·9
Cellulose	6–10

but the yield and quality vary widely (Table 5.10). Roots are rich in B vitamins (Chapter 4), especially pantothenic acid (85) and vitamin E (83), and in peptides, amino acids and 'protein'. About three quarters of the 'crude' protein is 'true' protein, and it contains an adequate amount of lysine from a nutritional point of view. Many basic substances are present in rootlet extracts, including choline, (64), betaine (66), tyramine (59), hordenine (61), candicine (62) purine bases and nucleotides, and allantoin (58) (Chapter 4). Not surprisingly, during kilning they are apt to accumulate high levels of undesirable nitosamines owing to interactions between the nitrogenous components and oxides of nitrogen from the furnace. Root extracts provide a rich source of growth factors for micro-organisms, such as yeasts or the mould *Eremothecium ashbyii*, which is used to produce riboflavin.

REFERENCES

[1] NORTHAM, P. C. (1962). *Brewer's Guild J.*, **48**, 259, 296.
[2] NORTHAM, P. C. (1965). *Brewer's Guard.*, **94**, 97.
[3] BRIGGS, D. E. (1978). *Barley*, Chapman & Hall, London, 612 pp.
[4] BISHOP, L. R. and DAY, F. E. (1933). *J. Inst. Brewing*, **39**, 545.
[5] BISHOP, L. R. (1948). *J. Inst. Brewing*, **54**, 330.
[6] BISHOP, L. R. and MARX, D. (1934). *J. Inst. Brewing*, **40**, 62.
[7] HUDSON, L. R. (1960). *Development of Brewing Analysis*, The Institute of Brewing, London, p. 15.
[8] HOGGAN, J., COMPSON, D. G., SPENCER, E. and JENNINGS, P. (1962). *J. Inst. Brewing*, **68**, 39.
[9] HOGGAN, J., SPENCER, E. and JENNINGS, P. (1963). *J. Inst. Brewing*, **69**, 28.
[10] MEREDITH, W. O. S., ANDERSON, J. A. and HUDSON, L. E. (1962). *Barley and Malt* (ed. COOK, A. H.), Academic Press, London, p. 207.
[11] PIENDL, A. and OLOFF, K. (1980), *Brauwissenschaft*, **33** (4), 85.
[12] DE CLERCK, J. (1957). *A Textbook of Brewing*, Vol. 1, Chapman & Hall, London.
[13] LÜERS, H. (1922). *J. Inst. Brewing*, **28**, 192.

[14] POLLOCK, J. R. A. (1962). *Barley and Malt* (ed. COOK, A. H.), Academic Press, London, p. 303.
[15] SHANDS, H. L., DICKSON, A. D., DICKSON, J. G. and BURKHARD, B. A. (1941). *Cereal Chem.*, **18**, 370.
[16] SHANDS, H. L., DICKSON, A. D. and DICKSON, J. G. (1942). *Cereal Chem.* **19**, 471.
[17] NARZISS, L. (1975). *E.B.C. Monograph II. EBC Barley and Malting Symposium Zeist.* p. 62.
[18] NARZISS, L. (1976). *Die Bierbrauerei* (6th edn.). *I. Die Technologie der Malzbereitung*, Ferdinand Enke, Stuttgart, 382 pp.
[19] PIENDL, A. (1976). *Master Brewers Am. Assoc. Tech. Q.*, **13**, 131.
[20] POLLOCK, J. R. A. and POOL, A. A. (1979). *J. Am. Soc. Brew. Chem.*, **37**(1), 38.
[21] PIRATZKY, W. (1958). *Die Nahrung*, **2**, 239.
[22] WITHEY, J. S. and BRIGGS, D. E. (1966). *J. Inst. Brewing*, **72**, 474.
[23] WHATLING, A. J., PASFIELD, J. and BRIGGS, D. E. (1968). *J. Inst. Brewing*, **74**, 525.
[24] MACEY, A. (1977-Sept.). *Brewer's Guard.*, **106** (9), 81.
[25] BRIGGS, D. E. (1963). *J. Inst. Brewing*, **69**, 244.
[26] AULT, R. G. (1961). *J. Inst. Brewing*, **67**, 405.
[27] SMITH, M. T. and BRIGGS, D. E. (1979). *J. Inst. Brewing*, **85**, 160.
[28] SPARROW, D. H. B. (1965). *J. Inst. Brewing*, **71**, 523.
[29] BLOCH, F. and MORGAN, A. I. (1967). *Cereal Chem.*, **44**, 61.
[30] MACEY, A., SOLE, S. M. and STOWELL, K. D. (1969). *Proc. Eur. Brewery Conv. Congr. 12th. Interlaken*, p. 121.
[31] PALMER, G. H. (1969). *J. Inst. Brewing*, **75**, 536.
[32] PALMER, G. H. (1974-Feb.). *Brewer's Digest*, p. 40.
[33] URION, E. and BARTHEL, C. (1961). *J. Inst. Brewing*, **67**, 453.
[34] MACEY, A. and STOWELL, K. C. (1961). *Proc. Eur. Brewery Conv. Congr., Vienna*, p. 85.
[35] BROOKES, P. A. and MARTIN, P. A. (1974). *J. Inst. Brewing*, **80**, 294.
[36] DUDLEY, M. J. and BRIGGS, D. E. (1977). *J. Inst. Brewing*, **83**, 305.
[37] JONES, M. and PIERCE, J. S. (1963). *Proc. Eur. Brewery Conv. Congr. Brussels*, p. 101.
[38] JONES, M. (1969). *Brewer's Digest*, **44** (3), 60.
[39] ANDERSON, R. G., BROOKES, P. A. and MARTIN, P. A. (1979). *Proc. Eur. Brew. Conv. 17th Congr. Berlin (West)*, p. 689.
[40] SPILLANE, M. H. and BRIGGS, D. E. (1966). *J. Inst. Brewing*, **72**, 398.
[41] PONTON, I. D. and BRIGGS, D. E. (1969). *J. Inst. Brewing*, **75**, 383.
[42] LAMBERT, J. G. (1953). *J. Inst. Brewing*, **59**, 324.
[43] PAULA, G. (1932). *Woch. Brau.*, **49**, 329, 339.
[44] POOL, A. A. and O'CONNOR, T. N. (1963). *J. Inst. Brewing*, **69**, 382.
[45] COLE, W. J. and HEMMENS, W. F. (1977). *J. Inst. Brewing*, **83**, 93.
[46] SCHÖNFELD, F. (1913), *Woch. Brau.*, **30**, 354, 374.
[47] VERMEYLEN, J. (1962). *Traité de la fabrication du malt et de la bière*, Vol. 1. Assoc. Roy. des Anciens Élèves de l'Institut Sup. des Fermentations, Gand, Belgium.
[48] SOMMER, G. and ANTELMANN, H. (1967). *Brewer's Guild J.*, **53**, 7.
[49] SMITH, L. F., LINKO, M., ENARI, T.-M. and DICKSON, A. D. (1964). *Proc. Annu. Meet. Am. Soc. Brew. Chem.*, p. 86.
[50] SIPI, M. I. and BRIGGS, D. E. (1968). *J. Inst. Brewing*, **74**, 444.

[51] HOME, S. and LINKO, M. (1977). *Proc. Eur. Brewery Conv. Congr. 16th, Amsterdam*, 91.
[52] POOL, A. A. (1964). *J. Inst. Brewing*, **70**, 221.
[53] POLLOCK, J. R. A. (1960). *J. Inst. Brewing*, **66**, 22.
[54] KIRSOP, B. H. and POLLOCK, J. R. A. (1961). *J. Inst. Brewing*, **67**, 43.
[55] RIVIERE, M. V. B. (1961). *J. Inst. Brewing*, **67**, 55, 387.
[56] LUBERT, D. J. and POOL, A. A. (1964). *J. Inst. Brewing*, **70**, 145.
[57] LINKO, M. (1961). *Proc. Eur. Brewery Conv. Congr., Vienna*, p. 99.
[58] HIND, H. L. (1938). *Brewing Science and Practice*, Vol. 1, Chapman & Hall, London, p. 274.
[59] VALENTINE, G. (1920). *J. Inst. Brewing*, **26**, 573.
[60] REED, R. M. (1965). *Brewer's Guard.*, **94**, 55.
[61] OLOFF, K. and PIENDL, A. (1978-Mar.), *Brewer's Digest*, p. 39.

Chapter 6

THE TECHNOLOGY OF MALTING AND KILNING
(including the physical principles of barley drying)

6.1 *Floor Malting* [1–5]

Floor malting, the oldest method for making malt, is still carried out to a limited extent in the UK. The traditional method is described first, then ways in which the process has been modified are indicated. Originally steeping was carried out in rectangular cisterns or tanks. At intervals the water covering the grain was drained away and was replaced with fresh well water. After the last drain the grain was moved manually on to the floor. 'Floors' are of smooth water-proof concrete or similar material, usually arranged above each other in special buildings to form long rooms having low ceilings, and whitewashed walls with small shuttered windows along the sides. Typically the structure is strengthened by rows of cast-iron pillars. Normally grain from the steeps is delivered to one end of the floor, and the chutes or other devices which deliver green malt to the kilns are at the other end. Initially the grain was heaped into a 'ruck' or 'couch', that might be restrained by a wooden frame. When the temperature had risen sufficiently the grain was spread on the floor and turned at intervals by men using wooden shovels, or was raked or 'ploughed' by drawing pronged wooden devices through the piece. Temperature was controlled by opening or closing the windows at the sides of the floor, and by thickening or thinning the piece. As germination progressed, successive 'pieces' were moved, in sequence, along the floors. At the end of germination the grain was gathered up and moved in barrows to a simple kiln, where the process was completed. Such floor maltings had a low productive capacity, were costly in terms of manpower used, and only operated in the winter-time, when ambient temperatures were low. Existing floor maltings have all been mechanized to greater or lesser extents. The old cisterns are frequently replaced by self-emptying steeps and fans are often installed to ventilate the grain between immersions (steeps; waters; tides). Elevators and helical-screw-conveyors are used to move grain and malt, in place of the old baskets, hoists and barrows and the malt is turned by mechanical devices, usually either hand-guided or 'ride-on' devices (rather like grass mowers), but in one or two cases the grain on each floor section, or bay, is worked by machines which are supported on rails running each side of the bay. Such machines may be set to turn, thicken or thin a grain piece, or to move it to a conveyor leading to the kiln.

The use of air conditioning, the forced circulation of air of a known relative humidity and temperature, combined with some insulation of the buildings, now allows floor malting to take place throughout the whole year, regardless of the weather. In addition, old-type kilns have been modernized in various ways (see below).

6.2 *Trends in malting equipment* [5-10]

Many believe that the most uniform, attractive and well-modified malts are made in floor maltings. If true, this may be due to the details of the conditions under which the grain germinates. Floor maltings survive because of the demand for their products, because mechanization has cut down manpower requirements, and because there is virtually no depreciation and little cost in maintaining an old floor malting that is structurally sound. However, floors cannot produce large quantities of malt, nor would it now be economical to build a floor maltings. Newer equipment is built to reduce manpower requirements, be more flexible in its operation, be suitable for automation and capable of handling large batches of grain, and allowing it to malt rapidly. While recently the processing of 150 tonne batches in 7 days from first wetting to kilning was considered a good arrangement, plant being erected (1979) is expected to handle 500 tonne batches in 4·5 days. Larger plant is cheaper to build and operate, in terms of cost per unit weight of malt made, than smaller plant. To achieve these results requires careful planning. Thus in plant in which steeping, germination and kilning are carried out in separate vessels, the design of each part can be optimized, but all the sections must work according to a closely integrated schedule which takes account of manpower requirements, the time taken to transfer large bulks of grain and allows for unexpected difficulties in the production process. On the other hand, in 'combined' plant in which germination and kilning or steeping and germination, or indeed all three stages, are performed in one vessel, the problems of transferring large quantities of grain or of having inflexible production schedules are reduced as are the problems of automation, although at the expense of design compromises, which may lead to higher water-usage or a reduced kiln efficiency for example. The maltster is able to choose from a wide range of equipment. Many types of maltings are now in use, and the descriptions given below are only of a representative selection.

6.2.1 ABRADERS

Where barley is to be abraded, the treatment is usually carried out on the way to the garner in which the grain is held immediately before steeping. At least three types of abraders have been used. In the entoleter-type equipment grain is thrown outwards from a fast-moving rotor against an abrasive surface (Fig. 6.1) [11]. The impact causes abrasion, and the dust generated is removed by aspiration. In another type of machine the grain is abraded when stirred by an impeller in a bowl coated internally with an abrasive, while in a third type the grain stream passes horizontally through a chamber with walls lined with an abrasive, while it is stirred by rotating paddles. The degree of abrasion must be moderated so that grains remain intact and large quantities of husk do not become detached from the grain.

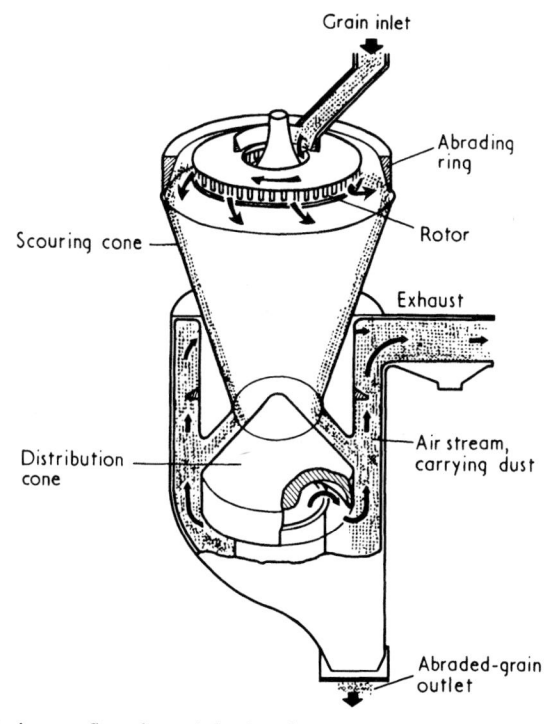

Fig. 6.1 A single-impact flow-through barley abrader {Northam and Button (1973) [11]}

6.2.2 STEEPS [5, 6, 12, 13]

From time to time cleaning devices are described in which grain is mechanically agitated immersed in water, to wash the grain, so removing dust and soluble materials, and allowing stones to settle and separate. If such washers or destoners are used they contribute to the steeping process. Grain is usually steeped by immersion. Many steeps, made of steel, are cylindrical or rectangular in cross-section, with conical bottoms which allow the grain to run

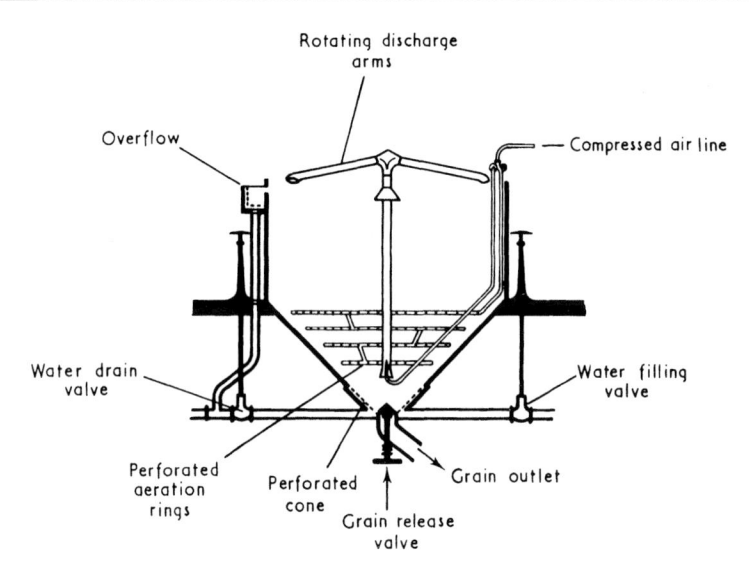

Fig. 6.2 A self-emptying conical steep tank fitted with a mixing 'geyser' with rotating hollow arms as well as with aeration coils.

out under gravity (Fig. 6.2). Barley is usually loaded into the steep by running it in from above, into a steep tank partly filled with water. To minimize the dust hazard, the spout delivering the grain from the garner or overhead conveyor may be surrounded by water sprays which dampen the falling grain. These 'conical-bottomed' steeps are convenient, but are not ideal for grain aeration or the even initiation of germination. Steeps are usually housed within buildings, but only recently have many been either insulated, or provided with means for warming the steep water to a fixed temperature. In most cases temperature 'control' is the thermal inertia of a large bulk of grain mixed with well water at its natural temperature. Generally steeps are mounted high up in the maltings' building so that the grain may be discharged by gravity to lower steeps or germination vessels. Overflow ducts are provided, and 'floaters' (light grains) and other light particles are washed out of the grain, overflow with the wash water, and are collected by sieving. They are then dried and sold for use as cattle food. Conical-bottomed steeps may be aerated by compressed air entering through internal aeration rings (Fig. 6.2) or, better, with aeration nozzles which fit into the sides of the base and leave the internal face of the cone smooth. In addition, or alternatively, the grain may be mixed and aerated by the use of a rouser ('fountain' or 'geyser') in which compressed air lifts the grain–water mixture from near the base of the steep, and discharges it above the water level, sometimes through rotating arms (Figs. 6.2 and 6.3). Grain may also be aerated by 'pumping over' a

THE TECHNOLOGY OF MALTING AND KILNING 149

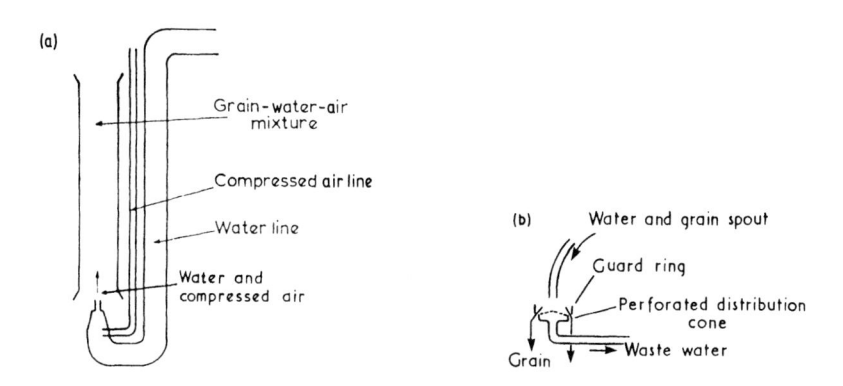

Fig. 6.3 (a) A portable steep-rouser. (b) A water separator and grain distributor for spreading a grain–water stream over the surface of a receiving steep tank while removing some of the water.

grain–water mixture either from the base of a steep back to the top or alternatively into another vessel. Part of the water may be removed from the mixture using a 'water separator' (Fig. 6.3b) or alternatively the mixture may be spread by discharging it on to the apex of a metal cone which lacks perforations. Aeration is also achieved by 'air resting' the grains between immersions (Chapter 5). It is desirable to draw air down through the bed of grain during air rests ('carbon dioxide extraction'), ensuring that fresh air reaches all the grains, and the film of surface moisture on the grains is disturbed and is encouraged to drain downwards. Blowing fresh air into the base of steeps is disliked because air compressors tend to heat air, which causes local drying in the grain, and may contaminate the grain with oil fumes. Further, this is a less efficient way of displacing the carbon dioxide, which is a heavy gas, from the mass of grain. In order to increase the uniformity of aeration, some special steeps have an inner perforated cone which allows air to be withdrawn from the grain mass at the sides as well as at the base of the cone. Normally the air in steep rooms is neither attemperated nor humidified, and so special steeps that use carbon dioxide extraction may be fitted with spray bars which allow the surface layers of grain to be wetted if they become too dry. In some spray-steeping vessels air is drawn downwards continuously, and attemperated steep liquor is sprayed on to the surface of the grain bed when temperature differentials between chosen points exceed a certain value. The spray-steep liquor is recirculated, but is wholly replaced with fresh water two or three times during the steeping period. Enclosed 'secondary' flat-bed steeping vessels, in which the incoming air is 'conditioned' (attemperated and humidified) have been described [14]. The initial stages of germination may conveniently be carried out in such vessels.

Conical steeps may be discharged wet, that is the grain–water mixture is moved away by gravity or using pumps. Alternatively dry-casting is used, that is the water is drained away, and the moist grain is conveyed away separately. The dry-discharge arrangement avoids the problem of grain arriving too wet in the germination place, and so favours the rapid onset of germination. An old patent describes the removal of the unwanted film of surface water by blowing warm air over the drained grain as it is conveyed away from the steep. In addition, dry-casting simplifies the addition of additives. Wet-casting is inconvenient with chitted grain because of the rapid and almost uncontrollable uptake of water that can occur. Furthermore wet grain may easily be damaged by the pumps used to move grain–water mixtures.

Large simple conical steeps [which may be 6 m (approx. 20 ft) deep] are disadvantageous because of the pressure occurring on the grain in the lower layers, and the difficulty in obtaining even treatment for all the grains in the bulk. To reduce the problems caused by excessive pressure some large steeps have been built with 'multiple' hopper bottoms which retain the advantage of being 'self-emptying'. Such steeps are shallower, and so the grain is subjected to less extreme pressures. Increasingly large flat-bed steeps are being built which are unloaded mechanically (Fig. 6.4). Grain depth in flat-bed steeps may be around 2 m (about 6 ft 6 in), and need be no deeper, however great the vessel's capacity. Steeps of 158 tonnes capacity are in use.

Spray-steeping works well in continuous 'Domalt' plant [15] (see below), but it does not work well in deep conical steeps, even when combined with downward aeration, and has not been widely adopted. It seems to be highly advantageous to wash grain by at least one immersion steep, preferably with agitation, before 'topping up' the grain moisture content by subsequent spray-steeping. Water can be sprayed on to grain in some steep vessels, while it is moving along in modified conveyors, and in germination vessels when the turners are equipped with 'sprinkler bars'.

The effects of water quality and additives have been discussed (Chapters 3 and 5). While alkalis or hydrogen peroxide may be used dissolved in steep liquor, it is more usual and economic to apply gibberellic acid and potassium bromate at 'steep-out'. This is usually achieved by spraying at a controlled rate and thoroughly wetting the damp grain as it is conveyed away from a dry-cast steep. The spray liquid is normally collected in a sump and recirculated.

Waste steep water is expensive to dispose of, because of its high BOD and high content of suspended solids and the large volumes involved. In addition, payment must often be made for the water taken in at a maltings. There is much interest in reducing the volumes of effluent, by using minimal volumes in steeping and, where possible, reusing steep liquors (Chapter 7).

THE TECHNOLOGY OF MALTING AND KILNING   151

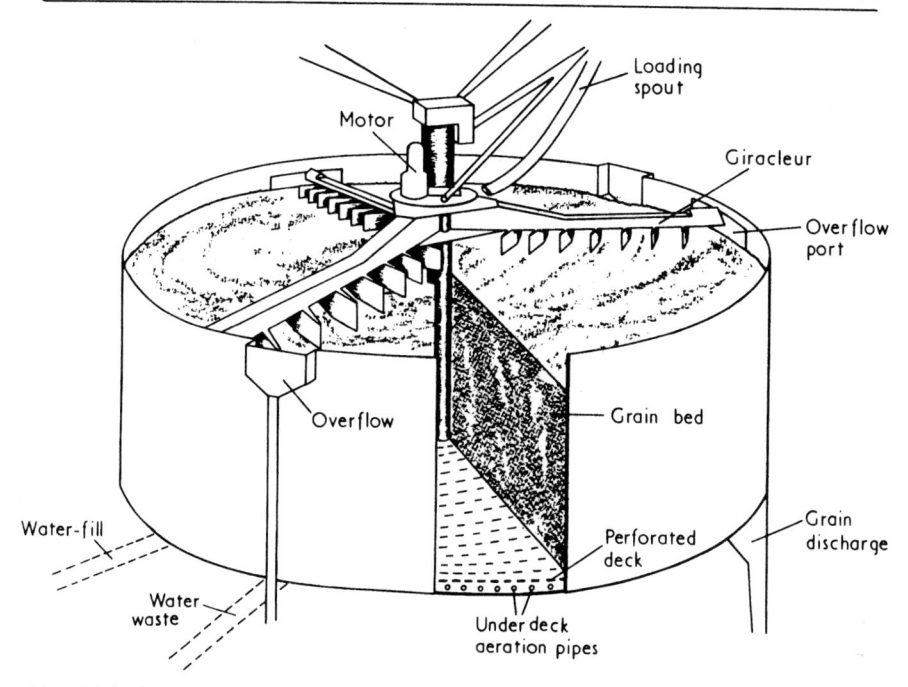

Fig. 6.4 A diagram of a 'Nordon' type of flat-bed steeping tank. Provision is made both for aeration from below the perforated deck, and for carbon dioxide extraction from beneath the grain bed during air rests. Such steeps may be large, e.g. have capacities for 200 tonnes of barley.

6.2.3 PNEUMATIC MALTING SYSTEMS [5, 6, 12, 13]

Increasingly malt is germinated in deep beds that are turned mechanically, and are attemperated by the forced passage of a strong flow of humidified and attemperated air. Such a 'pneumatic system' allows the production of large batches of malt in extensively automated plant, controlled by relatively few men. Large-scale operation is needed to make such maltings economic, e.g. 22 000 tonnes of malt p.a. Many types of plant are in use, but certain general principles of operation apply. In them high-quality malt may be made according to regular processing schedules, throughout the year, and with pleasant working conditions. Such malt factories are expensive to build and maintain, and they use large quantities of power (electricity) and fuel, especially on kilns. Consequently increasingly ways are being sought of reducing power consumption as well as water consumption and effluent production.

In early pneumatic germination equipment, air, humidified and cooled (conditioned) by drawing it through a bed of coke wet by sprays of well water, was passed downwards through the grain bed, then to waste. To cool grain faster the rate of air flow was increased. This system was inflexible and inefficient in that (*i*) large quantities of water were wasted, (*ii*) there was no control over the temperature of the well water, and (*iii*) the air was often incompletely humidified, so that the grain dried, growth ceased and the malt was incompletely modified. At present water-saturated and attemperated air is usually blown up through a bed of grain, about 1 m (39 in) deep, at a rate of say 3·7 m (about 12 ft)/min, giving about three air changes a minute. Rarely the air flow may be downwards or may be reversible. The air is conditioned by passage through many fine sprays of water that has been heated or cooled as required. The spray water drains to a sump and is recirculated through cooling equipment. The air from the grain bed may be recirculated or may be wholly or partly replaced with fresh air. The average temperature differential across the piece is kept to about 1°C (about 2°F), but differences of 1·5–2·5°C (3–4°F) may develop. The piece is cooled by reducing the 'air-on' temperature, but maintaining a fixed air flow rate. Germination temperatures are usually in the range 16–21°C (61–70°F), whereas they used to be 13–16°C (55–60°F). Temperatures may be altered during the germination period (Chapter 5).

The germinating grain is periodically turned to 'lighten the piece', and so (*i*) allow an easier passage of air, (*ii*) stop roots matting and binding the grain together, and (*iii*) break up local 'hot spots', and so (*iv*) by mixing the grain increase the eventual uniformity of the product. Depending on the stage of germination and the vigour of growth the grain may be turned up to three times in 24 hr. The capacity of maltings is often limited by the feasible depth of the grain bed. In the older types of maltings, beds deeper than 1·1 m (about 3·5 ft) were diffcult to turn, but beds of 2·45 m (about 8 ft) can be turned in some newer plants.

As a precaution provision is usually made for spraying water on to grain at a controlled rate, while it is being turned. Additives may be applied in the spray water. Inevitably the germinating grain loses water in a forced air flow. Even when the 'air on' entering the grain bed is saturated with moisture it emerges still saturated, but because it is warmer, it has removed moisture from the grain (p. 120). In any case the 'air on' is frequently not completely saturated, although this is difficult to check, because high levels of relative humidity are difficult to measure with precision. Figures quoted for a pneumatic Saladin box, with 'air on' at a relative humidity of 100% show that the moisture loss from the grain was about 0·5% of the fresh weight daily. When the relative humidity of the 'air on' fell to 95% the moisture loss rose to 2·5% daily, sufficient to stop modification in 2 days. Thus the attemperating air should be as nearly as possible saturated with moisture. Applications of moisture to germinating grain (sprinkling) are disliked, and attempts to avoid them are made by using initially high grain moisture contents (compared with those in floor-malting practice), and by germinating as rapidly as possible, so that modification is complete before the grain becomes too dry. The evaporative losses occurring in pneumatic malting account, at least in part, for the differences between pneumatic and floor malts, and for the differences in the 'best' conditions found to apply in each type of plant.

Since the temperature of the grain may be altered during germination, it is best to ensure that each germination unit has its own air flow and conditioning system so that each 'piece' is treated individually, and recirculated air has to be cooled at most about 2°C (3–4°F) per cycle. The air space over the grain should be minimized to keep the volume of circulating air needing to be conditioned as small as possible. The high humidity in germination compartments can cause rapid deterioration of paintwork and plant, which must be made of suitably resistant materials. Algae and some moulds grow on ceilings and walls, particularly near lights. This difficulty is reduced by (*i*) using paints containing toxic substances, (*ii*) by irradiating air ducts with ultraviolet light (which can constitute a risk to eyesight), and (*iii*) by metering small quantities of chlorine gas into air streams. This last procedure should be used with care, since chlorine is poisonous, and catalyses the breakdown of gibberellic acid. Using built-in electrical heaters to keep ceilings warm and free from condensate is uneconomic and has been discontinued.

6.2.4 COMPARTMENT MALTINGS [5, 6, 12–28]

Compartment or box maltings, often called Saladin boxes, are used very widely, and can vary greatly in detailed structure. They consist of rectangular open-topped containers, with a base of removable slotted metal plates on which the grain bed rests; the grain bed is usually 0·9–1·0 m (about 3 ft–3 ft 3 in) deep (Fig. 6.5), but much deeper in some newer plants (see below). Usually each box contains about 40 tonnes (200 Qr.) of steeped barley, but boxes of much larger capacity exist. Conditioned air is forced through the piece from below. The vertical long sidewalls of the box are exactly parallel, and the tops carry rails and a rack. These give support and traction to various pieces of equipment, including turners (usually with vertically mounted helical screws and a spray bar) or a plough or bulldozer, which may be used either to level the steeped grain during loading or to push the green malt to one end of the box and discharge it through a port during stripping. Turners and ploughs may serve several compartments, being moved from one to another on trolleys supported by rails running across the ends of the boxes. Boxes with end walls are likely to be equipped with contra-rotating helical-screw turners activated from above. Alternatively boxes with no end walls may use turners which in passing pick up grain and drop it behind, so

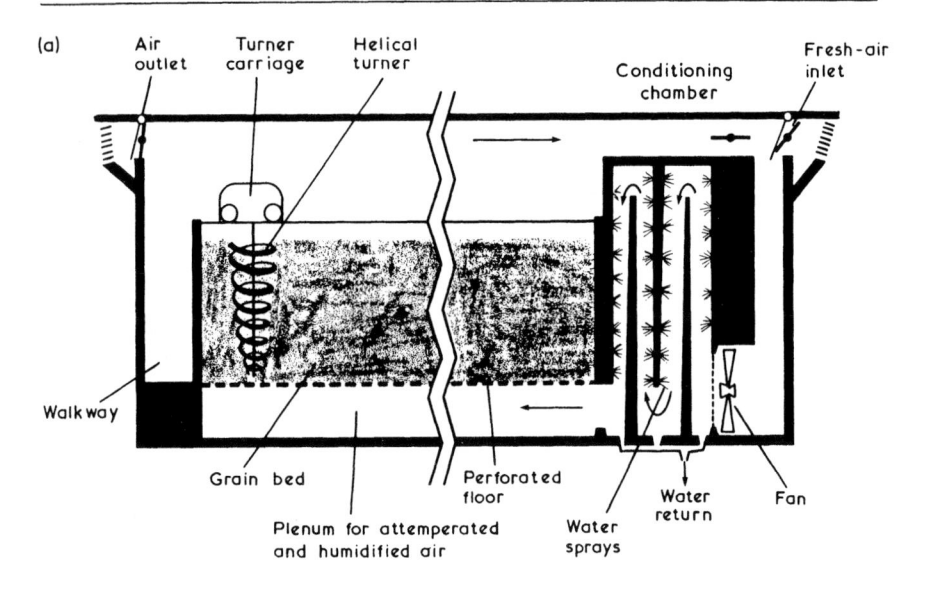

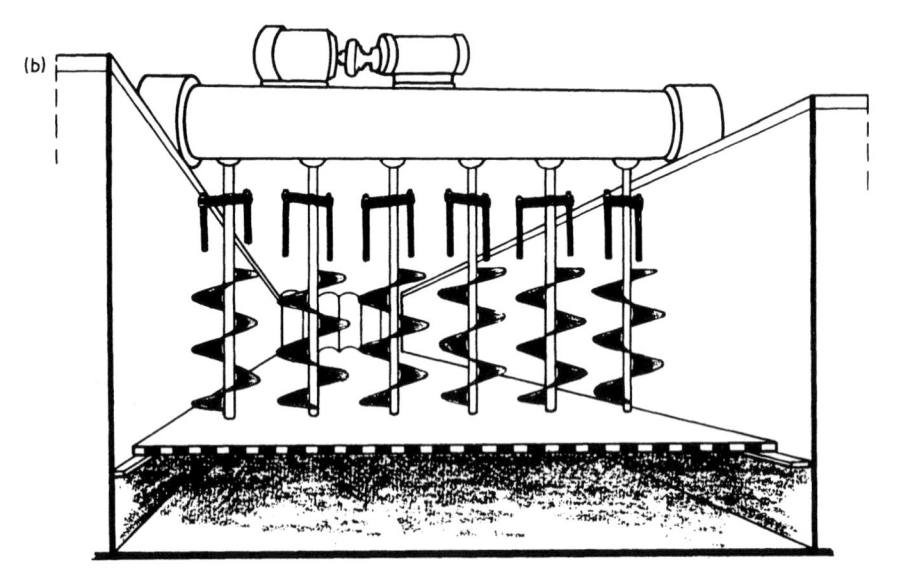

Fig. 6.5 (a) A Saladin box or compartment germination vessel. (b) A helical-screw turner of a type commonly used in germination boxes.

gradually moving the grain mass along a little (see Fig. 6.11). With this equipment, successive turns must be from opposite ends of the box, so that the piece is moved to and fro, and a small area of slotted floor, at one or other end of the box, is not covered by grain. To conserve attemperated air, valves are used to cut off the flow to these areas when they are not covered with grain. Turners must 'lighten the piece', separate matted rootlets and allow the air to pass freely. Good turners can increase the volume of the piece by one-quarter to one-third. Turners should not damage the rootlets or crush grains, otherwise secondary rootlets develop and may loosen the husk, spoiling the appearance of the malt. Also, moulds may grow on the dead rootlets and crushed or broken grains and may spread to infect the entire batch of grain.

Boxes may be loaded 'wet' or 'dry' from overhead spouts delivering steeped grain from conveyors or grain and water from the steeps. Several ways of unloading malting boxes are in use. Usually grain is 'bulldozed' to the end of the box and into a conveyor. However, discharge may be a tipping floor into a conveyor in the underfloor air duct, or the turners may be set to pick up the green malt and deliver it over the side of the box into a conveyor, or it may be delivered into a flexible pipe leading into a pneumatic conveyor system.

In some compartment maltings the grain is kilned in the box, by replacing the flow of attemperating air with a flow of hot dry air from a furnace. In specially constructed boxes, with shallow underfloor spaces and a built-in weir, fan and furnace, it is possible to dry barley and carry out steeping, germination, resteeping and kilning operations (Fig. 6.6; [18]). This plant is comparatively cheap and simple to build. Originally it was designed to hold a bed of grain about 1·5 m (about 5 ft) deep, and by using multiple-steeping and resteeping procedures at controlled temperatures root growth was to be restricted, and so turners were not needed. However, in newer versions of the plant, grain beds are deeper and turners are installed to allow other modes of operation. This was the first type of plant in commercial use that could be utilized in these various ways. By having all the production stages in one vessel, process programming is simplified, and the need to convey grain between the steeping and germination stages and the germination and kilning stages is avoided. However, engineering compromises are involved, so that in many respects design details are less than ideal. Using a germination box for kilning dries out the structure and should reduce corrosion and infection problems. On the other hand, the structures must withstand considerable changes in temperature. When large batches of malt are to be made, conveying begins to take up a disproportionate amount of the processing time and, where possible, should be avoided. Because of the undesirability of either increasing the depth of the grain bed, or extending the box width or length, some newer malting plants utilize very large multipurpose vessels of circular cross-section. For example, in a 'Clova' malting plant (Fig. 6.7), four vessels are grouped to allow the use of common services and ducting, exchange air during kilning, and so on. Each vessel, which may be used for barley drying

(600 tonnes/12 hr), steeping, germination, resteeping and kilning, is equipped with a rotating annular wedge-wire floor which carried the grain bed past loaders, strippers, turners and spraying equipment. An ingenious arrangement of ducts and louvres allows furnace air to be directed through the grain bed to waste, or to recirculate, or to pass to another vessel. Fan speeds, and hence air flows, are variable.

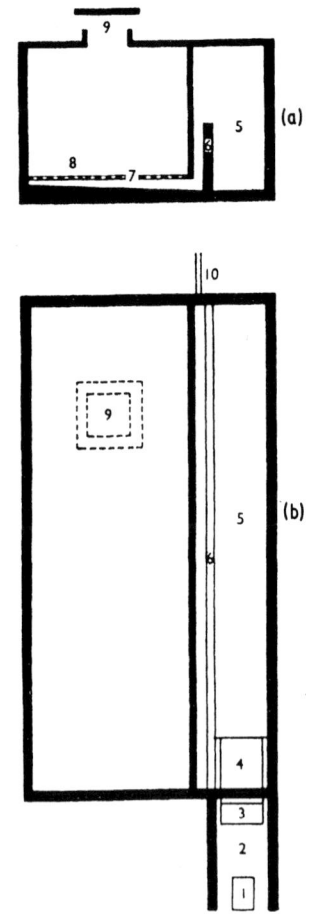

Fig. 6.6 A multipurpose box which may be used for steeping, germination, resteeping and kilning during malting, or may be used for barley drying. {Griffin and Pinner (1965) [18]}. (a) Vertical section. (b) Plan view. 1. Oil burner. 2. Furnace. 3. Silencer. 4. Fan. 5. Air duct. 6. Wall to retain water. 7. Perforated floor of wedge wire. 8. Position of bed of grain. 9. Air vent. 10. Drain cock.

Following small-scale studies, a 40 tonne pilot plant was built that allowed malt to be made with no turning, by taking advantage of the ability of resteeping to suppress rootlet growth [19]. The plant consisted of one or more

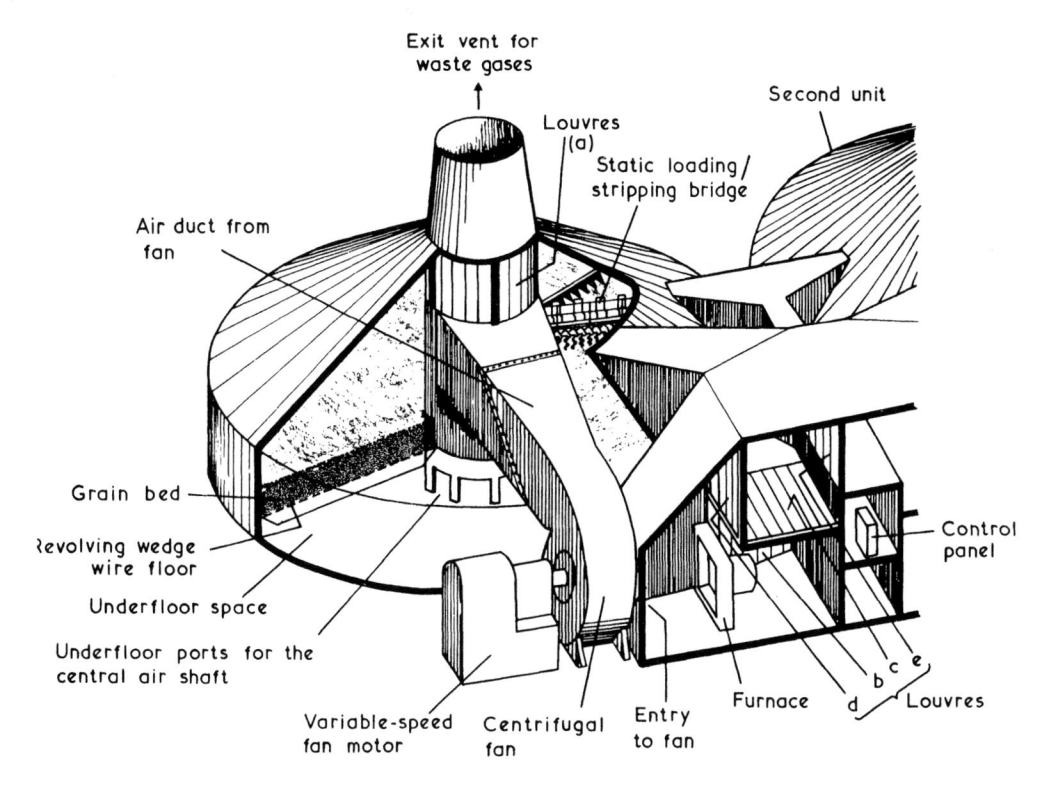

Fig. 6.7 A 'Clova' malting unit, one of a group of four, in which barley drying, and the steeping, resteeping, germination and kilning operations of malting can be carried out. The perforated floor is rotated during loading, stripping, turning. The bridges carrying the turners and the loading/stripping machinery are stationary. By adjusting the settings of the louvres, (a–d) it is possible to arrange air flows to dry barley, to germinate, or kiln with fresh air, to recirculate air during kilning, or to take hot 'curing' air from one unit and direct it to reinforce the fresh air stream in an adjacent unit that is in the drying stage of kilning. Water usage and the volume of steep effluent are high. [Courtesy of Griffin (1979) Moray Firth Maltings, Ltd.]

hopper-bottomed metal tanks, with a perforated conical false base that could be loaded with grain from above, ventilated with attemperated air, flooded with water at a chosen temperature, and from which water and grain could be separately withdrawn. To overcome the failure of the damp grain to flow, as a result of 'arching', a metal bar bearing short rods was mounted on a universal joint situated in the base of the cone. Such 'arch breakers' work by moving around the conical base of the hopper, dislodging the grain which falls away under gravity. In this equipment grain could be turned by emptying it from a vessel and conveying it either back to the top, or into a second vessel. A similar type of conical-bottomed steep, with arch breaker, is now used in conjunction with another type of free-standing flat-bed germination

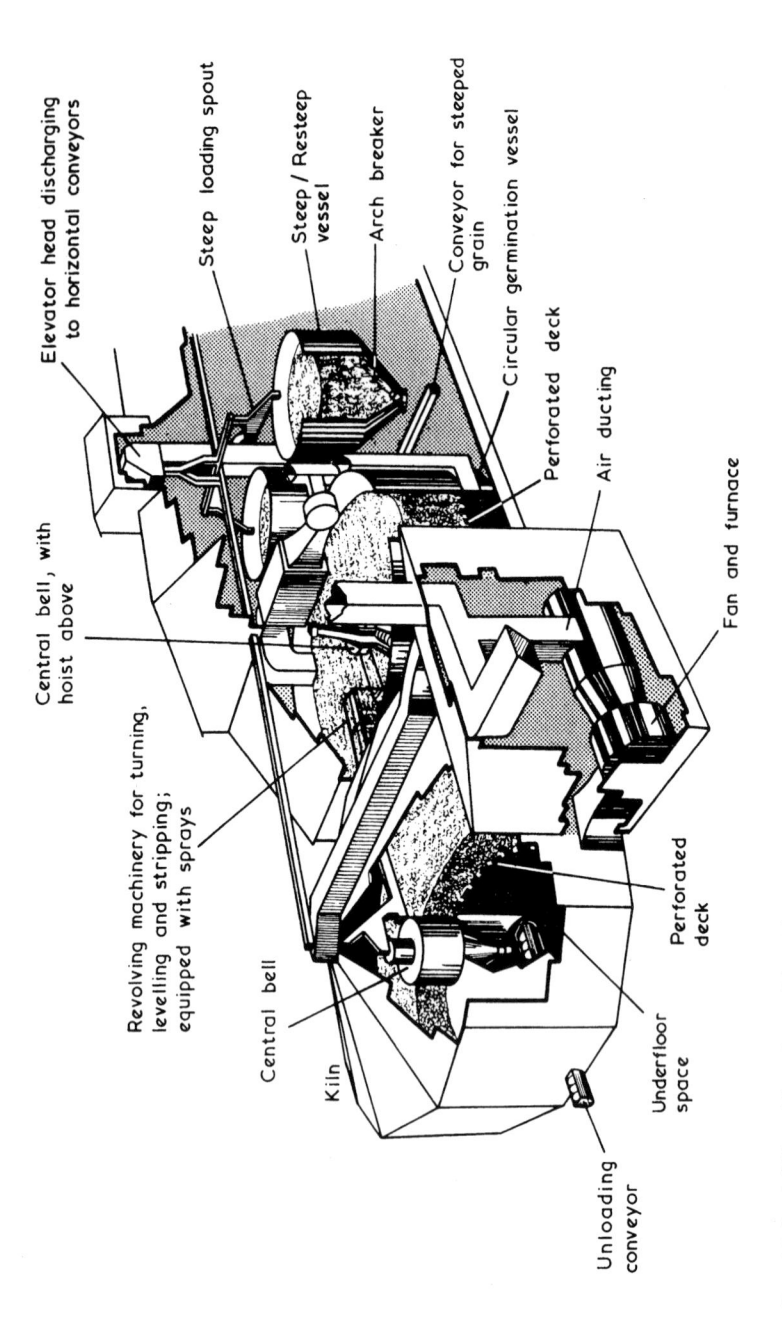

Fig. 6.8 An MTI malting plant with conical-bottomed steeps fitted with arch breakers (for steeping or resteeping), a circular flat-bed germination unit with rotating machinery for loading, levelling, turning or stripping and a flat-bed kiln. In a commercial plant several units would be used. [Courtesy of Rockley (1969) Vickers Ltd., Malting Division.]

THE TECHNOLOGY OF MALTING AND KILNING

vessel and kiln (Fig. 6.8). The vessels, which were originally designed as multifunctional germination and resteeping units [5], have stationary floors, but the handling and turning equipment rotates horizontally about a central point. To strip the floor, a central 'bell' is lifted, which allows the grain to fall into an underfloor conveyor.

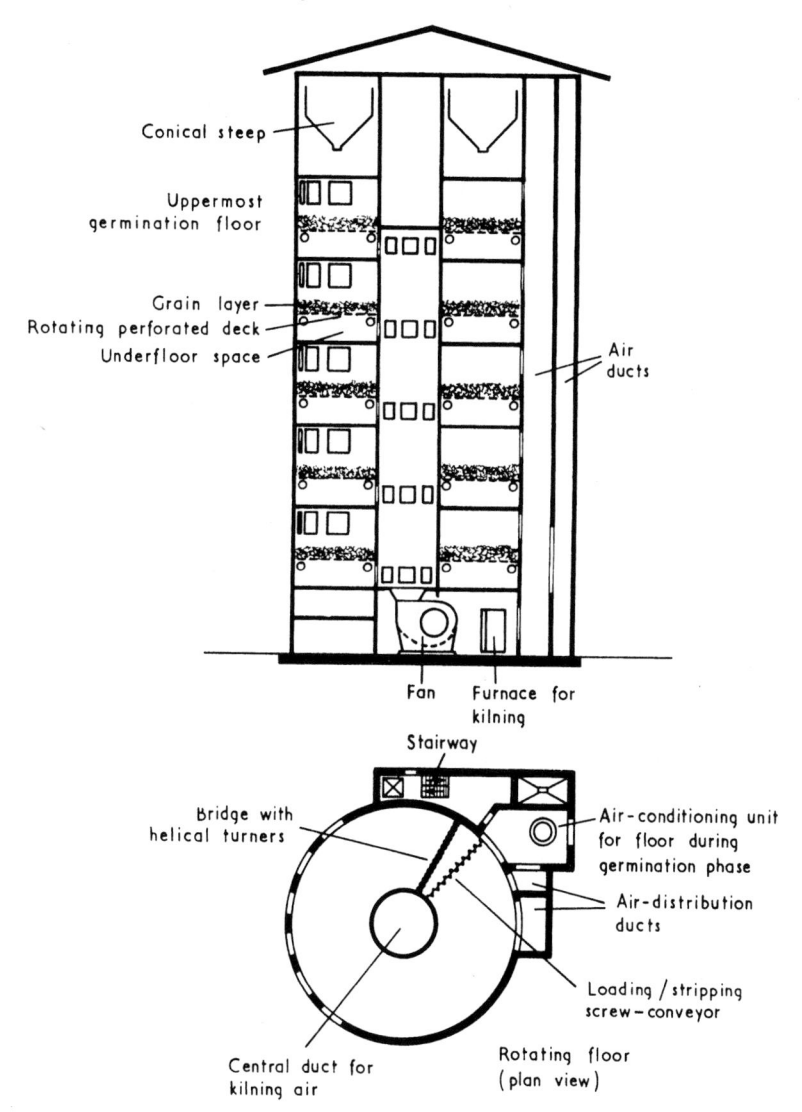

Fig. 6.9 Diagram of a Bühler-Miag tower malting, with the steeps mounted at the top and the kiln in the base. The annular floors can rotate, taking the grain from the loader, or past the turners, or to the stripping machinery.

Numerous other variations on the compartment-type of malting have been described. Various arrangements in which boxes are built-in columns above each other, in towers, are finding favour. In some such towers a series of rectangular boxes are mounted above each other, while in others annular germination chambers, either with moving turners or moving floors are employed (Fig. 6.9) [6]. An arrangement in a tower has the advantages of saving ground space, and allowing very efficient distribution of services and comparatively easy heat recovery. Such towers may have steeps at the top, or may be accompanied by a separate steep house. Each floor may be operated as a separate compartment, or some proposed plants could be run 'semicontinuously', the germinating grain being conveyed or discharged through tipping floors to a lower level at successive stages of germination [6]. In some designs the kilns are housed in the base of the tower, in others they are in a separate building.

The Popp malting plant is, in principle, a deep-loaded compartment with a novel method of turning the grain [13, 22–25]. Steeped barley is loaded into a well-lagged cylindrical steel vessel, connected to a supply of attemperated air, to the unusually large depth of 2·1–2·4 m (about 7–8 ft; Fig. 6.10). A temperature differential of about 2°C (3–4°F) occurs across the grain bed. The grain rests on a perforated metal floor, which may be tipped in two sections to discharge the green malt into the hopper. The grain is lifted and turned by suddenly releasing compressed air beneath the piece. The air blast is provided by a compressor, which charges a pressure vessel, and a special pressure-discharge valve. If the grain is 'blown' too infrequently, the rootlets bind together and mat, whereas if it is 'blown' too frequently the blast passes through the uncompacted grain with too little resistance, causing channelling. In either case inadequate turning occurs.

6.2.5 SEMICONTINUOUS AND CONTINUOUS MALTING PLANTS

Various systems have been proposed in which grain moves discontinuously down towers [6], entering successive sections in which the conditions are adjusted to suit the stage of germination that has been reached.

The Ostertag 'Wanderhaufen', or 'moving piece' system is, in principle, an extension of the open-ended malting compartment [13, 22, 27]. Various forms of this plant are used, but in each the steeped grain is discharged on to one end of a 'street', a long box with a slotted metal base on which the grain rests (Fig. 6.11). The passage of a special turner moves the grain one place down the street. At intervals new loads are levelled at the starting end, a series of pieces move down the street in sequence, and finished pieces are off-loaded into a kiln. The kiln may be separate and of a conventional type, or may be in two sections (the drying and the curing sections) level with the street, but separated from it and from each other by sliding heat-proof

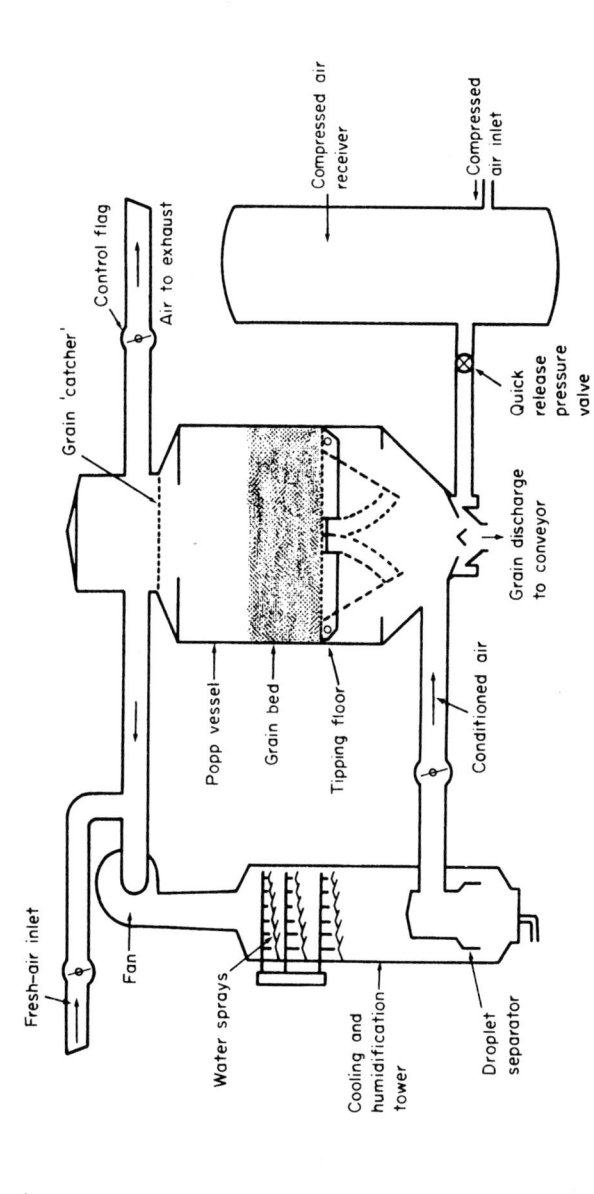

Fig. 6.10 A Popp germination vessel, in which turning is achieved by the sudden release of compressed air beneath the grain bed.

(a)

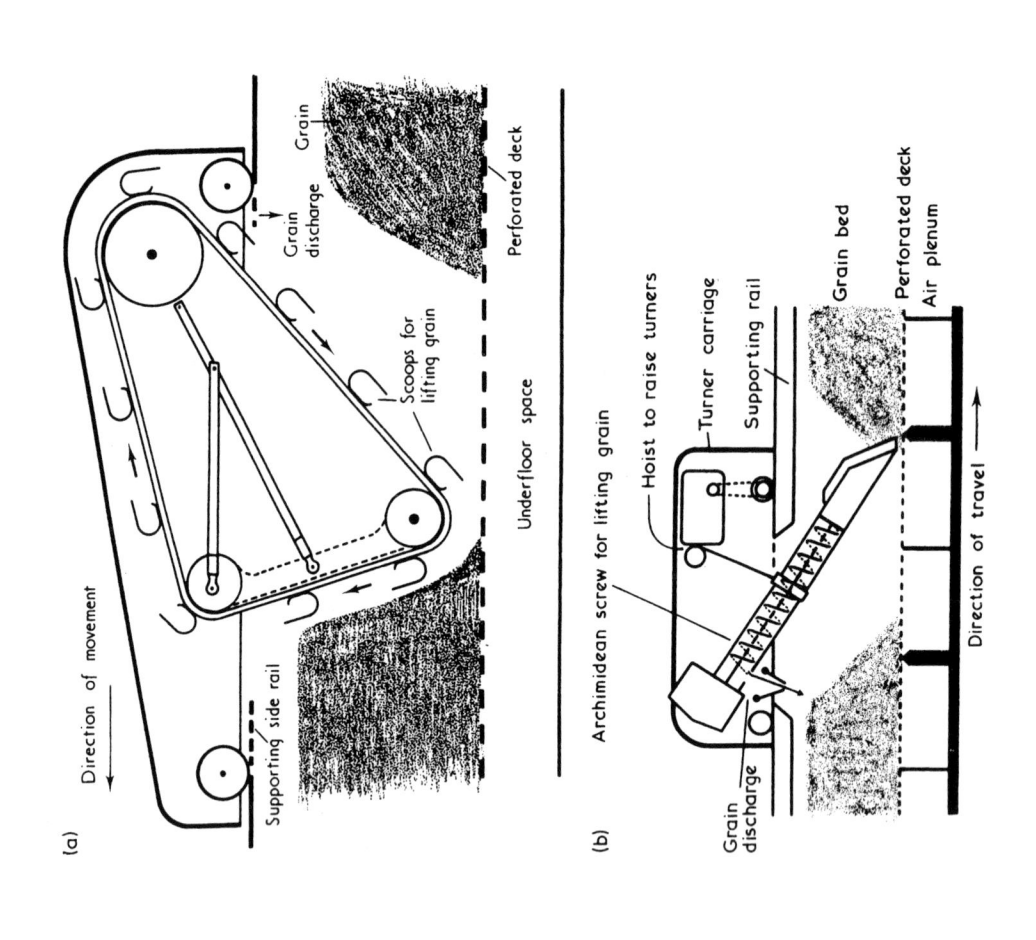

Direction of movement

Grain discharge
Grain
Scoops for lifting grain
Perforated deck
Underfloor space
Supporting side rail

(b)

Archimidean screw for lifting grain
Hoist to raise turners
Turner carriage
Supporting rail
Grain bed
Perforated deck
Air plenum
Grain discharge
Direction of travel

(c)

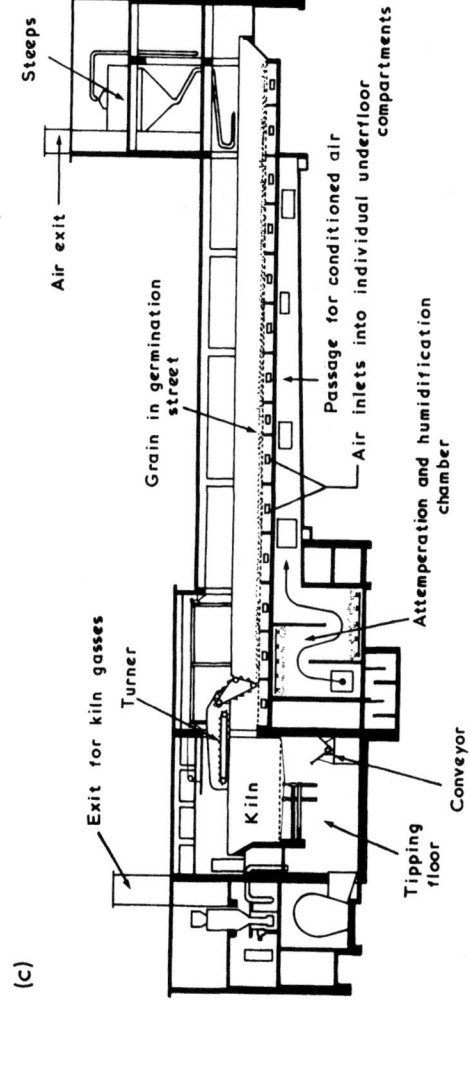

Fig. 6.11 Wanderhaufen malting equipment. (a) An Ostertag turner in its working conformation. The turning mechanism can be retracted into its carriage so it may be moved above the bed of grain. (b) A Seeger turner in its working conformation. (c) A section of a complete Seeger Wanderhaufen malting plant, with a kiln (Ostertag patent).

doors. The pieces may be turned and advanced two or three times each day. Different sections of the street may be conditioned to different temperatures, and air flows and some sections may be used for spray-steeping. The key to this system is the use of one of several patterns of special turners which pick up grain and deposit it in a more advanced position, with a minimum of mixing between different lots of grain – as little as 5% mixing has been claimed (Fig. 6.11). Each turner normally serves several streets. Since it must always move from one end of the street, provision is made to retract the working parts so that it may be returned above the grain to the starting end. The floor is cleaned by brushes on the turner. Also a 'space', created by not adding a load of grain at an appropriate time, may be allowed to move up the street to allow inspection, cleaning and repair operations. The Lausmann transposal system is similar in concept, except that grain is moved forward along a street divided into physically separated compartments (Fig. 6.12) [28]. The floor of each compartment may be raised or lowered. To move grain forward the floor of a full compartment is slowly raised, while the adjacent floor of an empty compartment, which begins in the fully raised position, is slowly lowered. At the same time the 'turner', a comparatively light structure which spans two compartments, gently skims the green malt from the top of the emptying box and desposits it into the box being filled (Fig. 6.12).

Some advantages of continuous or semicontinuous systems are (*i*) they are easily automated, (*ii*) they can be adjusted to provide 'optimum' malting conditions for a given grain type by simply maintaining steady-state conditions in the different sections, (*iii*) their manpower requirement, relative to their capacity, is low by traditional standards (*iv*) they have reduced variability in demands for water, power and other services, (*v*) saving of heat consequent upon avoiding cooling and reheating kilns. From time to time numerous 'near-continuous' malting plants have been described. Only a few will be considered here.

In the modern Czeck 'Solek tunnel' malting, grain is passed through an agitated immersion wash, then is subsequently borne along on endless belts, being spray-steeped, germinated in conditioned air and kilned in turn. The grain is mixed as it tumbles from one belt to the next (Fig. 6.13). The Domalt plant [13, 15, 22–24, 26, 29] provides a fully continuous and automated system, which is in use in several parts of the world (Fig. 6.14). Screened dry barley is metered from a hopper into a flow of water, using a vibrating feed. The grain–water mixture is pumped to an inclined spiral washer, consisting of a large rotating Archimidean screw of perforated metal plates in a tubular housing which carries the grain upward, at an angle, against a flow of attemperated water, sprayed in at the upper end. The return water drains from slots in the lower end of the housing, and is returned into the system.

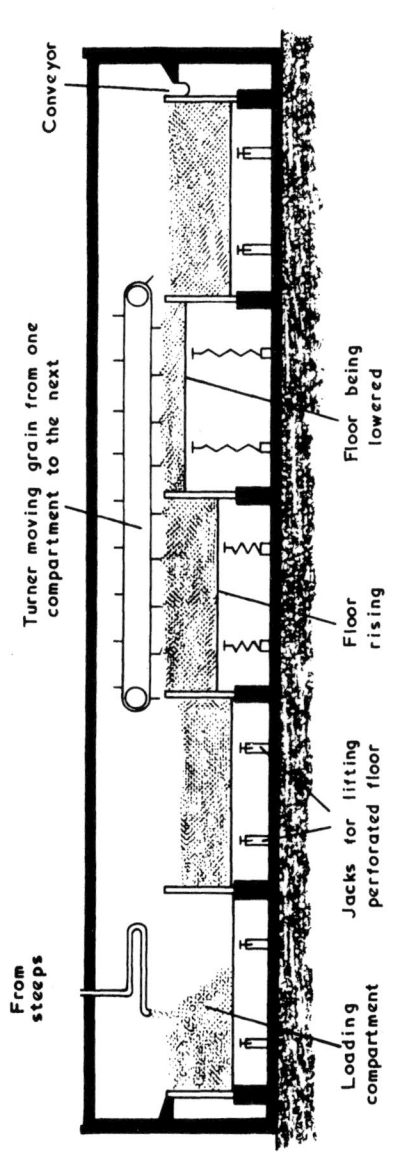

Fig. 6.12 Lausmann transposal germination street, showing germinating grain being moved.

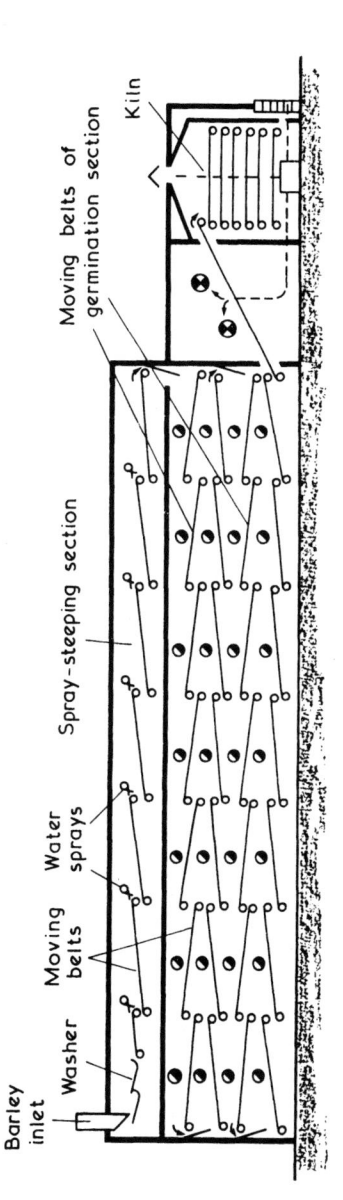

Fig. 6.13 Solek 'tunnel' continuous malting system.

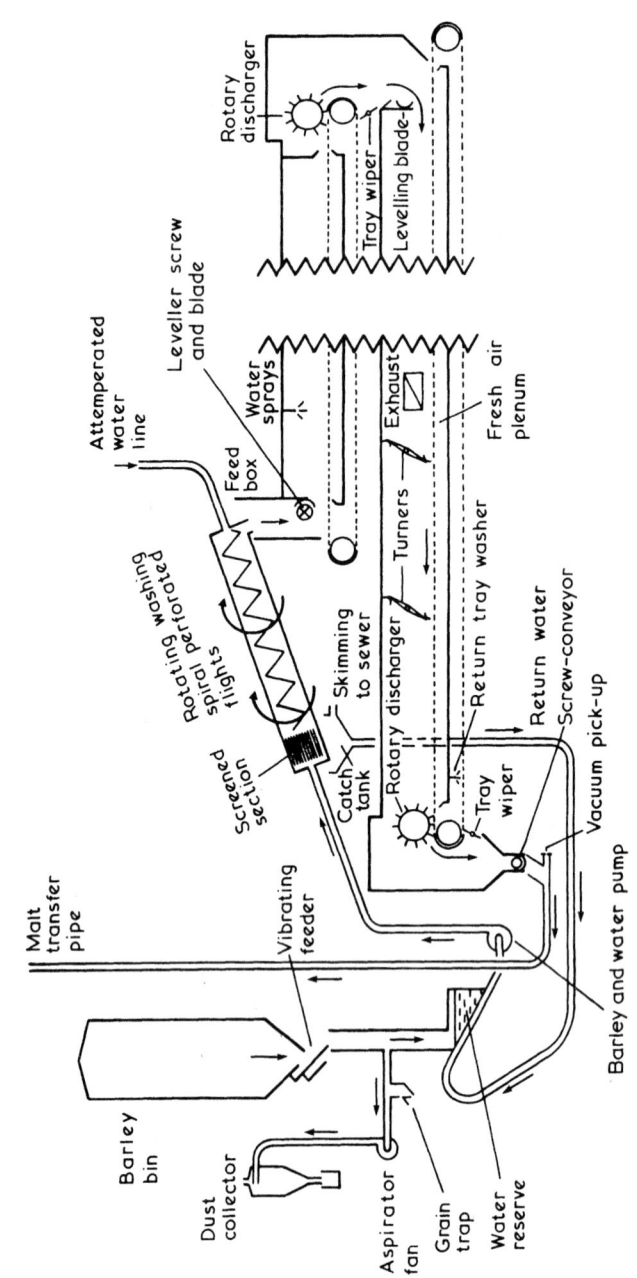

Fig. 6.14 Diagram of some parts of a 'Domalt' continuous malting plant. After Stoddart, et al. (1961) [15].

The grain emerges at the top about 100 min later washed and wetted under good conditions of aeration. The discharged grain is spread and levelled into a bed about 0·9 m (36 in) deep, on an apron or floor of perforated metal sheets slowly moving forward in an enclosed tunnel at about 0·7 m (27 in)/hr. Attemperated air is passed up through the piece from below. As the grain is moved forward, it is spray-steeped in the initial stages, and at intervals it meets simple rotating turners which gently lift and turn it. Grain may be sprayed at intervals as necessary, and different sections may be adjusted to different temperatures. Eventually the malt is carried through a continuous kiln, to a derooting plant. The equipment is self-cleaning, and has built-in safety devices which sound alarms and stop the plant if any faults develop. The cost of installation is higher than that of a Saladin box plant of similar capacity. A Scottish Domalt plant has produced malt from Scottish two-rowed barleys as well as fast-growing Canadian six-rowed barleys. The excellent aeration during the first wetting and washing stage and the spray-steeping results in a rapid start to germination. In Canada processing times were reported as about 80 hr. compared with the 240 hr. needed to produce similar malt by a batch process. The Domalt and Solek malting systems both use an intial immersion wash and steep, followed by spray-steeping. In view of the water uptake/time curves for steeping grain the initial wetting should provide a high percentage of the final malt moisture content (Chapter 5). In the proposed Geys system the grain is spray-steeped with ozonized water [26].

At least two malting systems have been proposed in which grain was to be held in trucks with slotted bases, which were moved to various 'stations' designed to provide attemperating or kilning air. In practice the sliding seals gave difficulties, and in the Saturne system as now used the grain is moved on a rotating annular floor of large diameter, progressing around the outside of the building which houses the servicing plant and grain stores. Between the germination 'ring' and the control buildings is an annular kiln, in which the grain is continuously kilned [30].

6.2.6 DRUM MALTING PLANT [5, 6, 12, 13]

At present, most drums are used to hold germinating grain, but about seventy years ago a drum was designed in which steeping and germination could be carried out, and another recent proposal was for a drum which could be used for all of the stages of the malting process, and would allow resteeping. A limited number of drums are used for barley drying or for kilning green malt. All drums are long horizontal metal cylinders, supported on rollers, in which the grain may be mixed, levelled and turned by rotation about the long axis. In general, germination drums are not insulated. The grain may be automatically sprayed while being turned. A great many types of drum have been tested, and several sorts are in operation. The grain is conditioned 'pneumatically' by the forced passage of humidified and attemperated air. In the perforated and Galland types of drum (Fig. 6.15), the conditioned air is directed to pass through the grain from a central duct to the periphery of the drum or vice versa. In these installations the loaded drums are about three-quarters full with green malt. In decked or box-type drums the grain is aerated only when the drum is stationary in a pre-set position when the perforated deck is horizontal. Air is forced up through the deck and out of the top of the grain bed, and is recirculated through the water sprays of the air-conditioning unit.

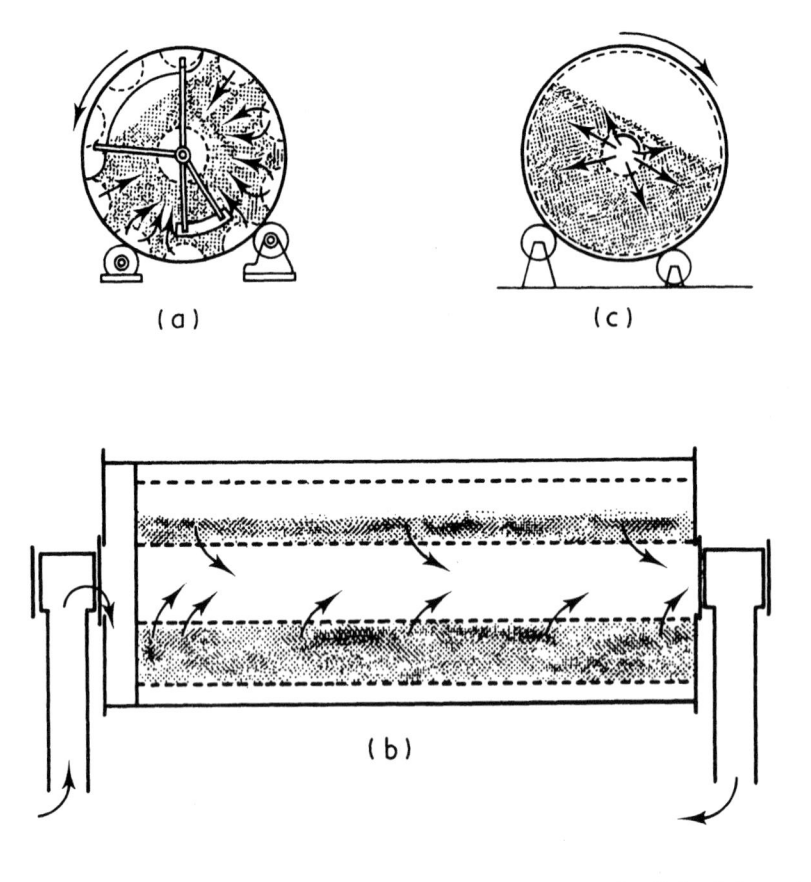

Fig. 6.15 (a) and (b) Transverse and longitudinal vertical sections of a Galland drum, in which air from perforated ducts situated around the periphery is exhausted through a central duct. The position of the sliding valve which cuts off the air supply to the ducts above the grain surface is indicated. (c) Transverse section through a 'fully perforated' drum in which the air moves from the central duct.

Modern decked drums are self-filling, levelling and self-emptying and have provision for under-deck in-place cleaning [17, 31, 32] (Fig. 6.16). Typically, each can hold about 30 tonnes (150 Qr.) of steeped barley. In some patterns continuous helical blades are fitted, extending from the ends of the drum to the middle. During filling, steeped grain is fed in at one end, while the drum is turning and the grain levels itself and forms a bed about 1·4 m (4·5 ft) deep. During turning the grain bed forms at an angle to the horizontal. Consequently, to ensure that the perforated deck is level with the grain surface, and to encourage an even air flow, the drum is turned until the deck is beneath the grain, and parallel to its surface. It is then reversed until both are horizontal, and parallel with the ground. When the drum is to be emptied a ring of doorways, 'ports' or 'hatches' is opened. These are arranged around the centre of the curved sides of the drum. The drum is turned and, as it rotates, grain falls out of the hatches into a collecting hopper leading into a conveyor. The spiral blades continually guide the green malt from the ends of the drum to the middle and so out of the hatches until the drum is empty. In other types of drum the ports are arranged in a line along the length of the drum. During discharge the green malt falls into a hopper which extends for the full length beneath the drum. Filling, turning, emptying and cleaning are fully mechanized. The temperature of the air may be varied automatically during the germination sequence. Thus the quantity of grain that one man can malt with this type of equipment is governed almost entirely by the size of the plant.

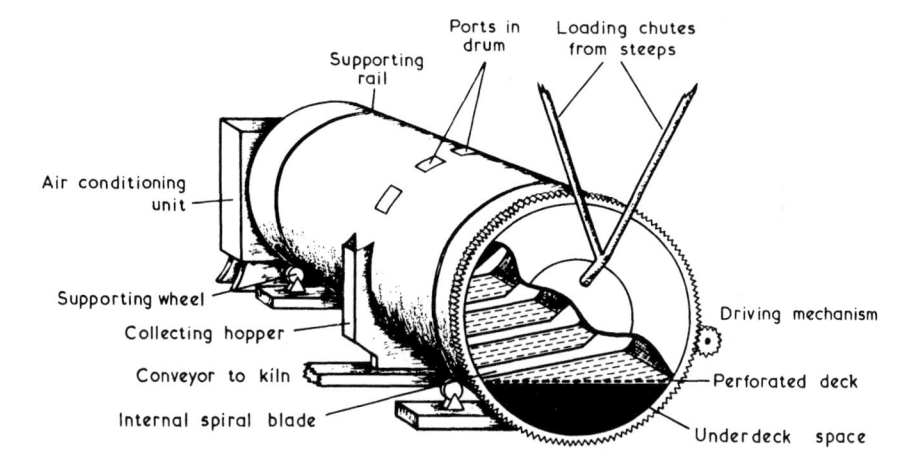

Fig. 6.16 A Boby type of decked drum. The air-conditioning unit, which is situated at the end away from the loading spout, humidifies and attemperates the air stream, which subsequently moves up through the grain bed from the underdeck space.

Other drum-malting systems have often been suggested. For example, to overcome the drying caused by attemperating the grain with a forced air flow and to produce conditions more similar to those found on a malting floor, a drum was designed in which much of the heat generated by the growing grain was conducted away through the walls [33, 34]. To achieve this, the area of the internal surface of the drum, and hence the area of contact with the grain, was increased with numerous metal protuberances. The drum, which was evidently mechanically complicated, apparently worked well.

Although new malting plants do not always introduce new principles into malting, they (*i*) try to fully utilize well-known principles to reduce labour requirements to a minimum, (*ii*) are as fully automated as possible, (*iii*) are designed to conserve energy, (*iv*) are designed to minimize water usage and effluent production, and (*v*) to be sufficiently flexible to allow new processes to be employed without the need for major structural alterations. (*vi*) In addition, conveying is often reduced to a minimum by using multipurpose vessels. Many of these objectives led to conflicting design requirements, and the compromises reached in plant design depend largely on the current economic situation. Success of a plant also depends on installation, maintenance, running and replacement costs relative to the gain in value of converting barley into malt, as well as its reliability. Virtually no data on the comparative economies of different plants are available.

6.2.7 GRAIN CONVEYORS [6, 12, 13, 35]

Some types of conveyors have already been mentioned briefly (Chapter 2). Dry grain and malt must be moved with as little abrasion, breakage and dust production as possible, and green malt should be moved gently to minimize skinning, that is the removal of the husks, crushing and the separation of rootlets. Skinned grains darken on the kiln and separated rootlets become mouldy. In order to minimize damage, green malt is sometimes moved by jog- or endless belt-conveyors, but screw-conveyors, chain- and link-conveyors and elevators, bucket elevators and pneumatic systems are also used. Jog-conveyors cause the grain to advance in a series of small hops. Conveyors should remain as clean as possible to minimize infestation with insect pests. Elevating grain to a high level and processing it subsequently at progressively lower levels has the advantage that powered conveyors are needed only for a fraction of the time in which the grain is moving. The power consumption for the volume of grain moved varies greatly between different conveyor systems. Thus pneumatic systems, in which grain is 'fluidized' and swept along in a rapid current of air, are relatively costly to install and operate. They are, however, practically dust-free and are flexible in operation. Grain dust may be a health hazard (e.g. farmers' lung), and is certainly a danger in that mixtures with air are explosive. It is important therefore that elevators and other grain-handling equipment should not generate sparks.

6.3 Kilning malt and the physics of barley drying [5, 6, 12, 13, 36–41]
Malt is kilned to produce a friable readily milled stable product that may be stored for long periods, and from which roots may easily be removed. In the transition from green to finished malt, 'green-grain' flavour is removed and characteristic odour, flavour and colour are developed (Chapters 4 and 5). The physical principles involved in drying barley and the early stages of kilning malt are the same, but the viability of the barley must be maintained so that lower processing temperatures are used, although warming grain helps to ensure that any dormancy is minimized (Chapter 2).

At any temperature, a given sample of barley or malt has a characteristic water-vapour pressure. This vapour pressure increases sharply with increasing temperature. Relative humidity (RH) is the quantity of moisture present in the air expressed as a percentage of the quantity of moisture needed to saturate the air at the same temperature. Thus, with an RH of 0% the air is perfectly dry, while with an RH of 75% it is three-quarters saturated. However, at higher temperatures air with a given RH will contain much more water vapour than air of the same RH at a lower temperature. It is usual to express the water-vapour pressure of the grain as the RH of the air in equilibrium with the moisture in the grain at the prevailing temperature. This is helpful because, as the temperature increases, the water-holding capacity of the air increases as well as the water-vapour pressure of the grain, and so the equilibrium RH of the system alters much less than the equilibrium water-vapour pressure. Measurements of the vapour pressure of grain have been made under static conditions and at fixed temperatures where the moisture in the grain is approximately in equilibrium with the moisture in the surrounding air. Many apparent equilibria are in fact only approximate because of the excessive periods of time that may be needed for true equilibria to be attained. If non-equilibrium or 'dynamic' conditions are considered, as with a stream of air with an RH below the equilibrium RH of the grain through which it is passing, then the grain will dry until its equilibrium RH at its reduced moisture content equals the RH of the air stream. Conversely, grain will pick up moisture from air having an RH above the equilibrium RH of the sample. Equilibria are approached faster at higher temperatures.

The evaporation of water from the grain cools it by an amount equal to the latent heat involved in the evaporation (539 cal/g; 970 B.Th.U./lb, at 100°C; 212°F). Incidentally the latent heat of steam rises with decreasing temperature. This cooling is important in maintaining the viability of wet barley during drying because to be safe the calculated temperature of individual grains should probably never exceed about 38°C (approx. 100°F).

Higher air temperatures are in use in driers, and if evaporation is restricted, for example by an insufficient air flow, the grain temperature will rise to a dangerous extent, and viability may be reduced. In kilning green malt, the removal of moisture at low temperatures allows the maximal survival of enzymes and the least development of aroma and colour. In experimental malts, the full complement of diastatic enzymes is believed to survive if the green malt is dried in a rapid air flow at 40°C (104°F) to not less than 6% moisture [42].

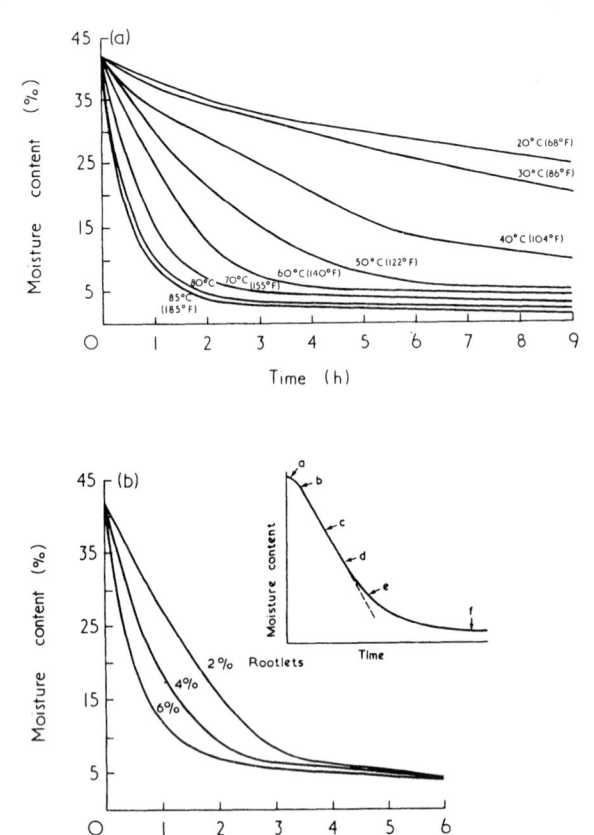

Fig. 6.17 (a) Experimental drying curves for barley in a fixed air flow, at the different temperatures indicated, (b) Drying malt grown to different extents in a fixed air flow at a fixed temperature, 70°C (158°F). Inset: the components of an 'idealized' isothermal drying curve. a, The initial phase in which the grain bed is warming and coming into equilibrium with the air stream. b, Onset of the period of steady-state drying. c, The approximately linear steady-state stage in which 'free water' is removed. d, Transitional stage, leading to e, the intermediate stage of drying and f, the slow drying period in which bound water is being removed. {Schuster and Grünewald (1957) [43]}.

The higher the temperature of each grain, the faster the moisture will diffuse from the interior to the surface. From the surface the water evaporates rapidly into the warm air stream, so that faster drying occurs at elevated temperatures (Fig. 6.17). As the grain dries, it shrinks and reduces the distance the residual moisture must travel from the interior to the surface, so tending to hasten drying. On the other hand, the associated reduction in surface area slows the drying rate (Table 6.1). The interesting suggestion has been made that 'root-free' malts (made using resteeping for instance) should be lightly crushed before kilning, to facilitate the drying process. The presence of roots on malt appreciably accelerates the drying process, presumably because of the increased surface from which evaporation can occur,

TABLE 6.1

Estimates of the effect of moisture content on the surface area of malt
{Data of SCHLENK cited by DE CLERCK (1957) [12]}

Moisture in malt (%)	Surface area of dry malt (m^2/100 kg)
45·0	360
40·0	330
30·0	250
20·0	200
10·0	178

(Fig. 6.17). It is possible, but not quite proven, that another reason why malt is more readily dried than barley is because the water it contains is less firmly bound to the grain components [38]. When the vapour pressures of barley at ambient temperature (and malt at kilning temperatures), expressed as percentages of the vapour pressure of water, are plotted against moisture contents, curves of the form shown in Fig. 6.18 are obtained. Unfortunately, as indicated in Fig. 6.18, there are discrepancies in some of the available data (including those for malt). At high moisture contents, the equilibrium vapour pressure of samples is the same as that of free water. But below characteristic moisture contents (about 30% for barley and possibly a lower value for malt), the vapour pressure is progressively restricted, and this 'restriction' becomes serious below moisture levels of about 25% [40, 43]. This is because below these moisture contents all the water is physically or chemically bound and is associated to varying degrees with the molecular structure of the grain. It has been proposed that malts of different degrees of modification have different equilibrium RH values at lower moisture contents, and this causes them to dry at different rates on the kiln. It is also possible that the enhanced rate of drying of better modified malts is due to an increased ease of movement of the moisture to the grain's surface, including

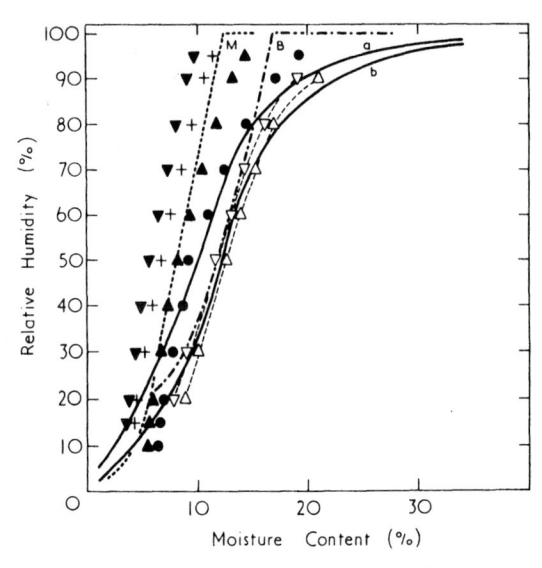

Fig. 6.18 Relationships between equilibrium relative humidities (RH) and the moisture contents (fresh wt. bases) of barley and malt. The heavy curves are based on values of barley, at normal temperatures, starting at 4% moisture, a, and 5% moisture, b. These curves limit numerous sets of experimental results. The lack of coincidence of these curves is caused by the failure to attain true equilibrium between the grain moisture and the water vapour within the experimental period. Measurements were made with barley at 30°C (86°F) (▽) and at 10°C (50°F) (△) respectively. Curves B (–.–.) and M (– – – –) for barley and malt (at 82°C; 179·6°F) probably represent cases in which measurements were made on rewetted samples in which some of the added moisture had not equilibrated with the grain's interiors. Other curves for malt, at 20°C (68°F), more or less coincide with the barley values enclosed by lines a and b. Other data are given for green malt held at 30°C (86°F) (●), 40°C (104°F) (▲), 60°C (140°F) (+) and 80°C (176°F) (▼). After Briggs (1978) [5]; Pensel (1967) [4]; St. Johntson (1954) [38]; Abrahamson (1978) [40]; Tuerlinckx and Goedseels (1979) [41].

the rootlets, and evaporation from it. The factors affecting the rate of drying on a kiln include (*i*) the volume of air passing through the grain bed, (*ii*) the depth of the grain bed, (*iii*) the quantity of moisture to be removed, (*iv*) the temperature and moisture content of the air entering and leaving the grain, and (*v*) the hygroscopic state of the barley or malt. Other related factors that are of importance are (*vi*) the temperature within the grain, and possibly (*vii*) the degree of shrinkage and (*viii*) the quantities of roots that are present. In practical kilning the rate of evaporation of moisture from malt begins to fall appreciably when the moisture content is about 30%. This is more striking at 23% and the decline in the evaporation rate becomes serious at moistures of 20% or less [40, 43].

Usually the temperatures of the air immediately below and above the bed of grain are measured with remote indicating electrical thermometers, and frequently the data are continuously recorded on a chart. The temperature of the outside air is also important. The temperatures indicated by thermometers in the bed of grain, while of interest in showing whether or not the load is being heated uniformly, are of no value in drying calculations. In critical studies, the rate of air flow through the grain bed also needs to be measured, using a sensitive anemometer. The relative humidities (RH) of the outside air and the air leaving the malt should also be measured, although, at high curing temperatures in particular, measuring the RH of the air leaving the grain bed to a sufficient accuracy is difficult. It is usual to make approximate determinations of the moisture content and colour of malt during kilning. Because the answer must be obtained quickly to be of use in controlling the process, various rapid but slightly inexact techniques are employed.

Invariably malt (or barley) is dried by passing heated air through the grain bed. In older shallow-loaded kilns the bed is turned at intervals to ensure uniform treatment. Heating air is costly in terms of fuel. Unless the heated air removes as much moisture as possible from the grain, or in other words unless the air emerges from the grain bed saturated with moisture, heat is being wasted. In practice, during the initial stages the air must not reach saturation within the bed of grain, otherwise the top layers will become wetter. No cooling by evaporation will occur, and the grain will overheat and 'stew'. Later, when the moisture content is reduced, the RH of the air passing through the grain bed must be below the equilibrium RH of the grain. Further, the air above the grain must not become saturated or else condensation occurs, the kiln becomes foggy and 'reeks', and 'drip-back' on to the grain occurs from condensate separating on the kiln structure. To overcome this, the air flow is adjusted so that the air emerging from the bed of grain has an RH of 90–95% for as long as possible. Inevitably there is an unavoidable waste of heat when air leaves the bed of grain at less than 100% RH. Other causes of waste are (*i*) imperfections in the combustion of fuel, (*ii*) heat used in heating the kiln structure, (*iii*) heat lost from the kiln and (*iv*) heat used in warming the malt dry matter together with its associated moisture, and not evaporating moisture. Attempts are now being made to reduce these losses, and to recover some of the heat from the wet air leaving the kiln (see below).

The process of moisture removal from malt during kilning may be divided, for the reasons already mentioned, into four merging phases (Fig. 6.17). At the beginning there is an initial phase in which the grain and kiln are warmed, and the air flow through the grain bed is established and the rate of water removal builds up. In the free-drying stage the air flow, with an 'air-on' temperature of 50–60°C (say 112–140°F) removes water from the malt without restriction. Evaporation occurs at a steady rate so that the air flow can be adjusted such that the 'air off' the grain has a consistent RH of 90–95%. When the moisture content of the malt has fallen from its original level of

43–48% to about 25%, that is when about 60% of the total water initially present has been removed, the rate at which moisture reaches the grain surface and the bound nature of the residual moisture begin to seriously restrict evaporation, and the intermediate stage of drying begins. As a result, the rate of drying begins to decline. The rate of air flow is now restricted, because moisture is now leaving the grain more slowly and it takes longer to establish equilibrium between the grain and the air. To accelerate water removal the temperature of the ongoing air may be increased. The intermediate stage lasts until the moisture content of the green malt has fallen to about 12%, when all the water in the corns is firmly 'bound'. Thus to increase the water-vapour pressure of the grain and to accelerate drying, the temperature of the 'air on' is increased to about 65–75°C (say 149–167°F), and the air flow may be reduced still further. If the kiln is equipped to be able to recirculate some of the air, then recirculation may begin shortly after the beginning of the bound-water phase, as the RH of the 'air off' is beginning to decline still further in spite of the higher temperature and the reduced rate of air flow being used. The air flow must not be reduced to too low a value, as draughts of less than about 1·2 m (4 ft)/min probably will not allow even drying of the malt. The RH of the 'air off' should be slightly less than the equilibrium RH of the malt at all stages of kilning. When the malt is 'hand dry', that is when the moisture content has been reduced to 5–8%, the sample is ready to be cured, so the temperature of the 'air on' is increased still further, to 'cook' the malt, add to its character and reduce its moisture still more (Chapters 4 and 5). In 'traditional' malting the temperatures attained in curing will be about 80–105°C (176–221°F) for pale malts or 95–105°C (203–221°F) for dark malts. In some deep-loading kilns, 75% of the air may be recirculated at this stage, as it is not doing much drying. Consequently a major saving in heating fuel is achieved. These kilning stages are recognized by convention, but are not distinct, and merge into each other.

The type of malt required determines (*i*) the duration of curing, (*ii*) the intensity of curing, (*iii*) the degree of modification when the green malt is loaded on the kiln, (*iv*) the moisture content of the malt at which curing is begun and (*v*) the temperature programme which precedes the curing stage. In general the more modified the green malt, the moister it is at the onset of curing and the higher the temperature at which curing begins, the greater will be the final colour, flavour and aroma, but the enzyme complement will be less. Final moisture contents of conventional British brewing malts are now generally in the range 1–4%, although sometimes higher moisture contents may be accepted. Malts which contain more than the specified amount of moisture are termed 'slack'. North American malts may contain 4% or more moisture, and some European brewers will use malts with 5–6% moisture (Chapter 5). Interestingly, in continental three-floor kilns the

THE TECHNOLOGY OF MALTING AND KILNING 177

alterations in the moisture contents of green malt on the top, middle and bottom floors may be 42–20%, 20–10% and 10–5%, roughly corresponding to the conventional free water, intermediate and bound-water drying stages respectively. Examples of the temperature, and other, changes occurring in continental malts in a two-floor shallow-loading kiln, and the temperature changes in a deep-loaded Winkler kiln are shown in Figs. 6.19 and 6.20. The details of kilning schedules in use vary widely [6].

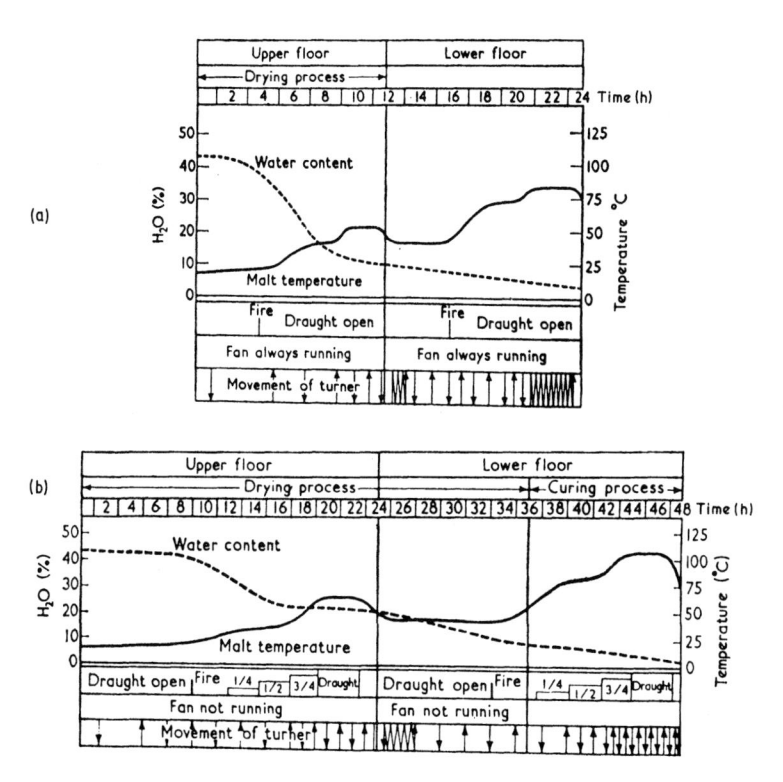

Fig. 6.19 Kilning conditions on two-floor kilns (a) for a pale malt being kilned on a 24 hr cycle, and (b) for a dark malt on a 48 hr kilning cycle. After Schuster (1962) [45].

In the past, mathematical treatments of the dynamics of the removal of water (12, 38, 41] have not been particularly helpful at the practical level, but with increasing sophistication of kiln construction, instrumentation and operation this may be changing. Nevertheless estimates of the efficiency of kiln drying can be made comparatively easily, on the basis of the relationship between the heat produced by burning a given amount of fuel, the maximum quantity of moisture that can theoretically be evaporated by this heat and a knowledge of the water actually removed. During the free drying stage about

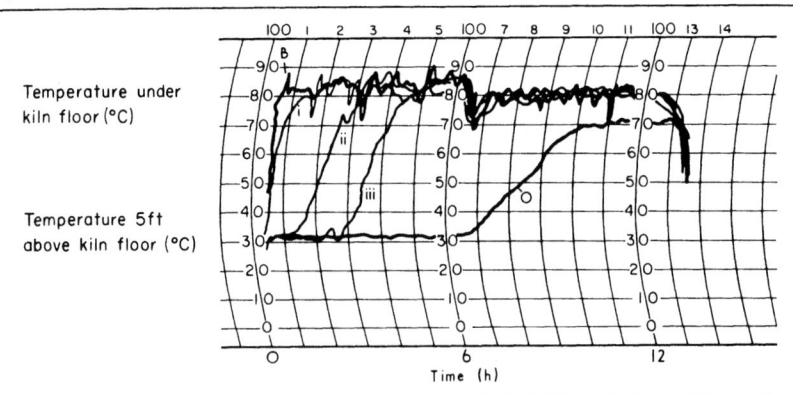

Fig. 6.20 Temperatures at different depths in the couch (i, ii, iii) and above (O) and below (B) the grain bed during drying on a Winkler kiln, using a 14 hr kilning cycle. After Fernbach (1928) [46].

two-thirds of the total moisture initially present when the kiln is loaded is removed. Within this period it is possible to use a knowledge of the temperature and RH of the air entering and leaving the bed of malt to calculate the 'in-grain' temperature. Similar calculations are also feasible if allowance is made for the progressive restriction of the water-vapour pressure as malt dries. Rates of drying may also be calculated if the volume of air passing in a given time, the draught, is known. Calculations may suggest economies that can be achieved by alterations in kilning practice. They also allow a more reasoned appreciation of a kilning cycle from the readings of the recording thermometers and hygrometers used to monitor the process. Fig. 6.21 shows a modified 'Mollier diagram' in which the relationships between air temperature, moisture content, RH and heat content (enthalpy) are summarized. The use of this chart is best understood by reference to examples.

EXAMPLE 1, KILNING MALT
Air enters a kiln at 11°C (52°F) saturated with moisture (RH 100%). From the chart its moisture content is 0·0080 kg of water/kg of dry air (point A, Fig. 6.21). It is heated to 60°C (140°F) by indirect heating, and its moisture content remains the same, since there is no addition of moisture from the combustion of the fuel. Its enthalpy (heat content) is now 81.4 kJ/kg (35 B.Th.U./lb of dry air) and its RH is 6% (point B, Fig. 6.21). The air passes through the bed of green malt and in so doing it gains moisture and is cooled by the latent heat of evaporation of the water. But its enthalpy remains the same, so that as the RH of the air rises, the state of the air is described by points along the enthalpy line. This continues until it becomes equilibrated with the malt moisture, that is until its RH equals the equilibrium RH of the malt. If the malt moisture-vapour pressure is not restricted (RH 100%),

THE TECHNOLOGY OF MALTING AND KILNING 179

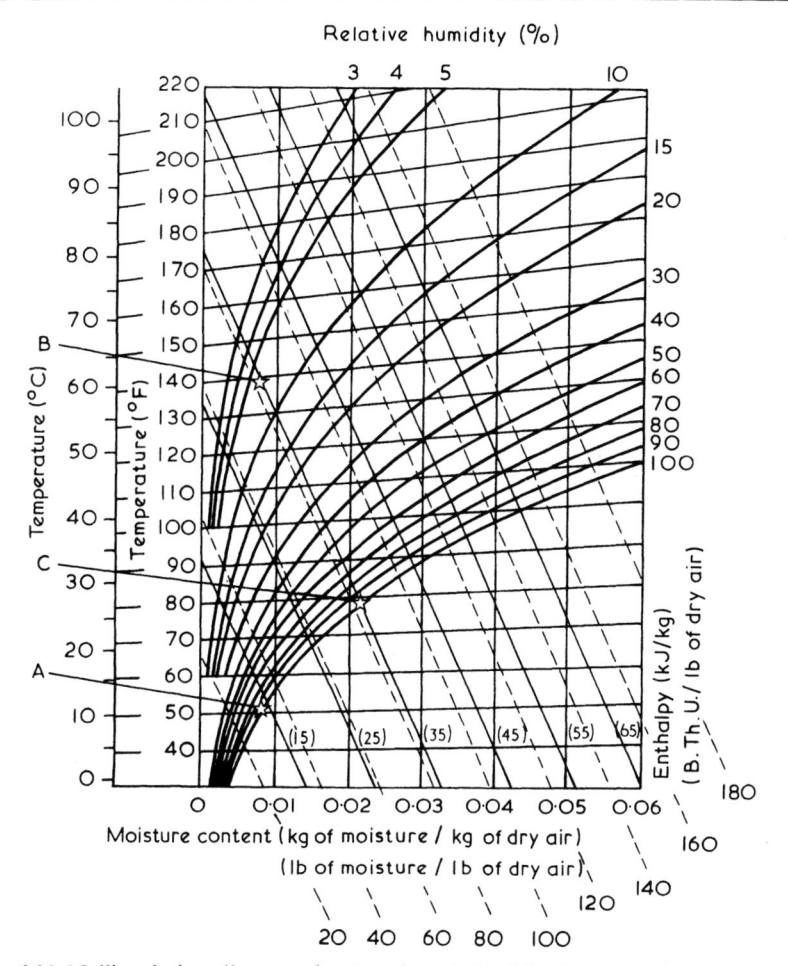

Fig. 6.21 Mollier drying diagram showing the relationships between air temperature, moisture content, relative humidity and heat content (enthalpy; 1 B.Th.U./lb of dry air equals 2·326 kJ/kg of dry air). The temperatures shown are 'dry bulb' values. To find the 'wet bulb' temperature for any set of conditions find the intersection of the equivalent enthalpy line with the '100% relative humidity' curve, and read off the equivalent dry bulb temperature. The points A, B and C refer to example 1 in the text. Modified from St. Johnston (1954) [38].

then the state of the air that leaves the bed of grain in a saturated condition is defined by the intersection of the enthalpy and 100% RH lines (point C, Fig. 6.21). Thus at this stage the air has a temperature of 25·5°C (78°F), and a moisture content of 0·0215 kg of water/kg of dry air. In practice, the air flow is adjusted so that the air leaves the grain 90–95% saturated in the free-water stage, and so the temperature of the 'air off' will be

26–27°C (79–81°F). Thus at an 'air off' temperature of 25·5°C (78°F), the air flow removes 0·0215–0.0080 = 0·0135 kg of moisture/kg of dry air (or lb/lb of dry air – numerically these units are interchangeable).

EXAMPLE 2, DRYING BARLEY

The barley has a moisture content of 17%, and therefore its moisture-vapour pressure is partially restricted and its equilibrium RH is about 85% (Fig. 6.18). Air enters the kiln at 0°C (32°F) saturated with moisture (RH 100% and has a moisture content of 0.0040 kg/kg of dry air. It is heated to 49°C (120°F), when it has a relative humidity of 8% and an enthalpy of about 63 kJ/kg (27 B.Th.U./lb of dry air). The air passes through the bed of barley and is in equilibrium at the surface of the grain. Thus the temperature in the grain is defined by the intersection of the enthalpy, and 85% RH lines at 23°C (approx. 73°F). The moisture content of the 'air off', at 85% RH, is 0·0140 kg of water/kg of dry air, therefore 0·014–0·004 = 0·010 kg of water is removed per kg of dry air passing through the kiln. At an in-grain temperature of 23°C (70°F) barley viability is entirely preserved, and hence these conditions are suitable for barley drying. As the grain dries and its water-vapour pressure falls, the conditions will move along the appropriate line of equal enthalpy. The temperature at each point is defined by the intersection with the line of equilibrium RH appropriate for barley at its particular moisture content. Thus if the grain is particularly well-dried, to 10% moisture, its equilibrium RH will be about 30% (Fig. 6.18) and the in-grain temperature will be 34°C (approximately 94°F). In more precise calculations allowance is made for the moisture added to the air from the combustion of the fuel used in the kiln. It is worth noting that different fuels burn to produce different quantities of moisture, and this must be allowed for in the calculations.

As malt (or barley) dries, so in time the water-vapour pressure becomes restricted and the relative humidity of the 'air off' falls. 'Air on' temperature is increased and the air flow is reduced until the RH of the 'air off' is a little below that of the equilibrium RH of the sample. Also, as the rate of evaporation decreases, the air passing through the malt is cooled less, so the 'air off' temperature, i.e. that of the air leaving the grain, rises towards the 'air on' temperature, the underfloor temperature on a kiln, thus indicating the changing state of the sample. For example, a malt with a moisture content of 4% will have an equilibrium RH of about 10% and so the drying air must have an RH of less than this. As the grain dries and the relative humidity of the 'air off' falls, it takes longer for the air passing through the grain bed to come into equilibrium with it. It is not possible to reduce the air flow through the bed of malt indefinitely because of the characteristics of the fans used to drive the air, the need to maintain combustion in the furnaces, and because the safe practical lower limit to air flow through a bed of grain to maintain uniform treatment is about 0·9–1·2 m (3–4 ft)/min. Thus, when the RH of the 'air off' has fallen to 25%, fuel may be saved by recirculating a proportion of air through the grain. Such air requires little reheating and so fuel is saved. Up to 75% of the air may be recirculated at the end of curing. Consideration of the example shown in Table 6.2 indicates how great the saving in fuel may be.

TABLE 6.2

Effect of draught control and air recirculation on the fuel consumption of a kiln. The results are adjusted to a constant ambient air temperature of 7°C (approx. 45°F)
{ST. JOHNSTON (1954) [38]}

Kiln operating conditions	Fuel used, as equivalent coal consumption + 5 lb for loss in ash (lb of fuel/Qr. of malt kilned)
Kiln used in a traditional manner	125·8
Saving when draught regulated, monitored by anemometer	21·1
Saving due to air recirculation during curing	25·1
Average fuel consumption when kiln is used with draught regulation and recirculation	75·0

The indirect heating of air by combustion gases is relatively inefficient. However, the combustion products of the fuel do not come in contact with the malt, and so impurities, such as arsenic and sulphur or any oxides of nitrogen formed during the combustion process, are of little consequence other than in terms of corrosion of the heat exchangers. Hence cheap low-grade fuels may be used in kilning when indirect heating is used. Continental three-floor kilns require less fuel/unit weight of malt produced than two-floor kilns, but they are complicated to manage [47].

The steady forced-draught through deep 0·9–1·5 m (3–5 ft) beds of grain, as used without turning in Winkler or Müger kilns, results in short 18–24 hr kilning cycles as compared with 3–4 day cycles in the old natural-draught shallow-loading kilns. Thus in Winkler kilns heat can only be lost from the structure for a fraction of the time available in old kilns, and so this loss is reduced. It is economically important to kiln for as short a time as possible to get maximum usage from the kiln.

To varying extents the drying performance of kilns, especially old natural-draught kilns, is dependent on the weather [35–39, 42, 48]. The higher the moisture content of the incoming air the less is the kiln's drying power. Also the higher the temperature of the outside air the higher its moisture-carrying capacity the wetter it may be and therefore the lower its drying capacity after heating (Mollier diagram, Fig. 6.21). In warm weather less draught is generated in the kilns, since this is related to the temperature difference between the air inside and outside. If the draught in a kiln at 57·2°C (135°F) is taken at 100 units when the outside air temperature is 0°C (32°F), the draught will be 89 and 72 units when the outside air temperatures are 10°C (50°F) and 21°C (70°F) respectively [38]. Air saturated with water at 0°C (32°F) contains only a fraction of the moisture it holds when saturated at 21°C (70°F). Therefore the lower the external temperature, the higher the drying power of the kiln will be.

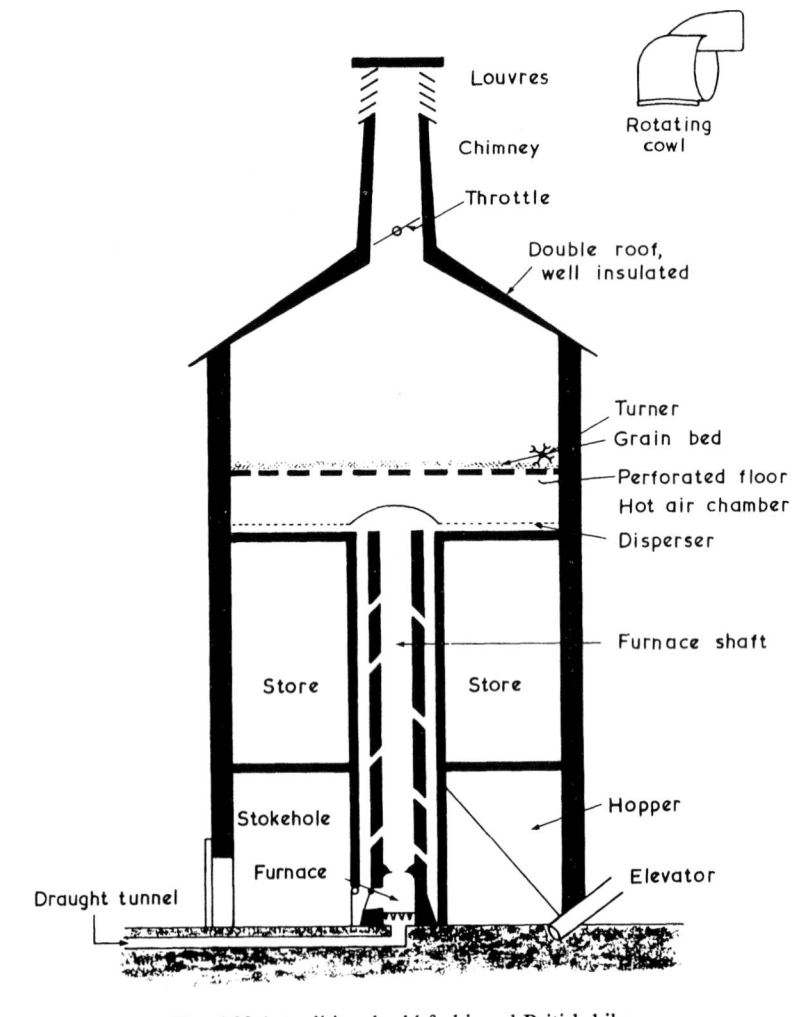

Fig. 6.22 A traditional, old-fashioned British kiln.

In old natural-draught kilns the regulation of air flow was inadequate (Fig. 6.22). In warm moist weather the draught was poor and the air sometimes emerged saturated from the grain. The air cooled in the space over the grain, condensation occurred, and sometimes so much moisture settled on the malt that the top layer began to sprout. Conversely in cold frosty weather the draught was often excessive, air emerged comparatively dry from the grain bed and fuel was wasted. The use of (*i*) fans to obtain at least a minimum draught, (*ii*) dampers to throttle the air flow, (*iii*) cowls to check down draughts caused by wind and (*iv*) thermostatically controlled furnaces has largely overcome these problems. However, the range of malts that are produced is so wide and the performance of kilns varies so much that theoretical knowledge is only used to supplement and check practical experience.

6.4 *Types of malt kiln*

The detailed designs of kilns vary greatly, and some types have already been mentioned and illustrated (Figs. 6.6–6.9, 6.11, 6.13, 6.14). However, it is convenient to consider kilns in a sequence from shallow loading to deep loading. This approximates to the order in which they have been used in the UK. The simplest kiln of which there is a clear description consisted of a fire-basket situated in a lower room, the dunge or dungeon [2]. The ceiling of the dungeon (the floor of the room above) was of perforated tiles or other material, even inflammable wooden slats covered with a cloth of woven horsehair. The green malt was barrowed on to the floor. During kilning the malt was turned with forks, shovels or wooden 'ploughs' and was levelled with rakes. The kiln floor was roofed over and the aperture in the roof, through which the wet air escaped, was covered to prevent the entry of rain. To reduce the risk of roots falling through the floor and on to the fire, and so causing the whole structure to catch alight, a metal sheet or disperser was suspended over the fire-basket.

With the desire for generally better control of kilning more sophisticated kilns were evolved. Exact instructions were evolved for the proportions to be used in their construction [2, 48] and for positioning them on rising ground away from trees or other obstructions that might cause down-draughts, and with the lower door 'facing the prevailing wind', to pressurize the furnace (Fig. 6.22). The fire-basket was replaced by a more easily regulated furnace. Sometimes air was led to the furnace through an underground duct. From the furnace the hot air passed up a double shaft, possibly being mixed with a controlled proportion of cold air at the same time. The hot air would be distributed throughout the hot air chamber beneath the perforated kiln floor, by an elaborate hollow brick or metal disperser, sometimes consisting of a series of radiating tubes. The air would escape from the disperser, then rise through the floor of perforated fire-clay kiln tiles, perforated metal sheet or woven wire, through the grain bed, into the body of the kiln and up a chimney to waste. In the best kilns the roof was of double thickness, being well-insulated and draught-free. The chimney might be finished with a waterproof cap having louvres at the sides or, better, with a rotating metal cowl which moved with the wind and prevented down-draughts and minimized the effects of changing wind direction. A damper was often situated in the throat of the chimney to regulate the draught (Fig. 6.22). The space within the kiln structure around the outer hot-air shaft was often used to store both malt and fuel for the furnace. Following an arsenic poisoning scare at the turn of the century, analysis for arsenic in fuel became, and still is, a routine measure. To trap fumes which contained arsenic it was suggested that the hot furnace gases be passed through a maze of channels in a stack of chalk blocks [49]. The heat converted the chalk to lime, which retained 90% of the arsenic and much of the gaseous sulphur oxides. This device was claimed to work well, but it is not now used.

In unfavourable weather kilning might take 5 days to complete. To ensure a more reliable draught most kilns were fitted with power-driven fans. Sometimes a fan was placed in the throat of the chimney to aspirate air through the kiln, but often it was installed below the floor to pressurize the hot-air chamber [50, 51]. As a result, steady draughts are maintained and kilning can be completed in a more regular manner. The fan also overcomes the resistance of the grain bed to air flow, so the depth of the grain on the floor may be increased from 10–40 cm (4–16 in) to about 60 cm (24 in). Further, by installing external or internal well-insulated ducts from above the kiln floor via the fan to the hot-air chamber, it is possible to recirculate air in these old kilns. Another modernization, usually carried out long before the installation of fans, was the fitting of paddle or screw turners to replace manual turning of the grain. Loading may be by gravity-fed chutes or by steerable devices equipped with rapidly moving endless belts which throw the green malt out in a fan-shaped pattern and so spread it evenly on the kiln. Discharge is now mechanically aided with power-assisted shovels, or the turners may be set to strip the kiln. Kiln tiles have usually been replaced with wedge-wire floors having a large free area, and so offering minimum resistance to the flow of gases. Such metal floors may be made to tip either in one piece or in sections, allowing the malt to fall into a chute. Such 'self-stripping' kilns are convenient, but are more expensive to build.

On the mainland of Europe kilns of a similar type but of different appearance were developed [5, 6, 12, 13, 45, 46]. Traditional continental kilns have two, or more rarely, three floors situated one above the other, but one-floor kilns are also used. Provision is made to supply fresh or hot air below each floor, as desired where, in the cases of the upper floors, it mixes with the air from the lower grain beds. Green malt is loaded on to the top floor and drop-

TABLE 6.3
Fuel consumptions of several kilns making pale malts
{FERNBACH (1928) [46]}

Kiln type and operation	Fuel used (coal, lb/Qr. of malt kilned)
English kiln, operated with minimal draught, no curing	42
Continental 2- and 3-floor directly heated kilns*	39–57
Continental 2- and 3-floor indirectly heated kilns*	43–89
Winkler kiln, 24-hr cycle	23 (coke)

* Lower values – three-floor kilns.

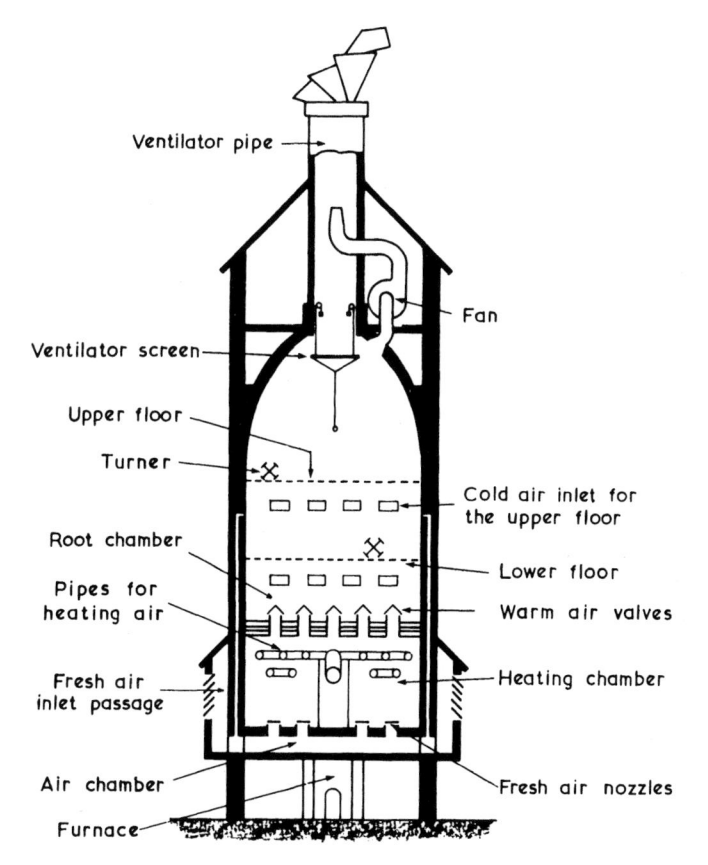

Fig. 6.23 A traditional two-floor indirectly heated Continental (European) kiln.

ped on to the floor beneath for the next and higher temperature stage in the kilning process. Fuel consumption, per unit of malt dried, is less in three-floor than two-floor kilns which, in turn, are more efficient than simple single-floor kilns (Table 6.5). The relatively low volume of hot dry curing air rises from the lower floor and is supplemented with other cool air before rising through the grain on the upper drying floor. In many continental kilns the air is heated by contact with a system of pipes of pear-shaped cross-section that are warmed by either the furnace gases or a stream of hot water. Such indirect heating has the advantage that cheap fuels may be used, because the furnace gases do not come into contact with the malt. On the other hand the fuel is used less efficiently because of the losses involved in the transfer of the heat to the drying air stream. A two-floor indirectly heated kiln is shown in Fig. 6.23. However, such systems are relatively costly and complex to build and have high maintenance costs. Directly heated one-, two- and three-floor kilns are also used. These types of kiln are loaded to a small depth and are equipped with mechanical turners.

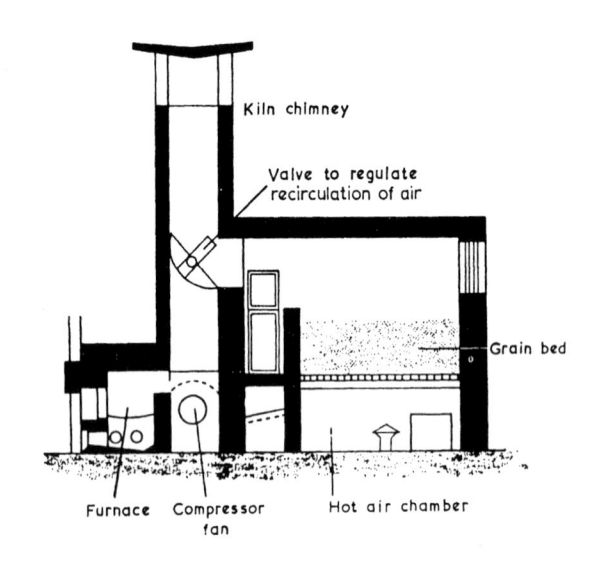

Fig. 6.24 A deep loading box kiln of the Winkler type.

A different type of single-floor deep-loading kiln of high capacity [5, 6, 12, 13, 45, 46], in which the deep bed of grain is not turned at all during kilning, is associated with the names Winkler and Müger (Fig. 6.24). This type has gained in popularity because of (*i*) its simplicity, (*ii*) its relative cheapness to build, (*iii*) its small size and (*iv*) its high efficiency of fuel utilization. Such kilns are used in this country for the production of the paler types of malt. On traditional kilns with a shallow grain bed and regular turning, the malt may be thought of as kilning evenly. In an unturned deeply loaded kiln, where the grain bed depth is 0·9–1·5 m (3–5 ft) and is exactly level, the hot air forced from below moves the drying zone progressively from the bottom of the grain bed to the top (Fig. 6.20). Draught and temperature are regulated so that the air emerges 90–95% saturated during the free-water evaporation phase; drying therefore occurs throughout the bed of grain, but clearly the product is slightly uneven. The forced draught of 12.2–15.3 m (40–50 ft)/min cools the wet malt by sustained evaporation and hence the 'air on' temperature may be relatively high, without damaging the product. As the RH of the air emerging from the grain bed falls below 90%, so the drying efficiency of the kiln falls. The temperature of the 'air on' is increased, the air flow may be reduced and later recirculation is started. Subsequently, the temperature is increased still more in order to cure the product. In the final stages, the air may be 75% recirculated. The low head-space over the malt adds to this efficiency by reducing the volume of air to be circulated. Such a kiln is practically independent of the weather. However, it is dependent on the continuous running of a fan, so there is a steady consumption of power. Such kilns are often equipped with tipping floors for stripping, but other stripping devices may be used. They appear better suited for kilning lager and pale ale malts rather than some of the darker highly cured malts.

Some other kiln configurations have been suggested, and some that are used in various other malting plants have been noted already. The vertical type of kiln, which may be modified to operate as a continuous kiln, is used to a small extent on the Continent of Europe, but not in Britain. In this type of structure, the green malt is held between perforated walls, and hot gases are forced through the grain bed from side to side [6, 13]. Frequently the upper, middle and lower sections are separated by doors, but when these are used the shrinkage of the malt may leave a gap at the top of each column, and this allows gases to pass freely, without touching the grain. Alternatively these doors may be kept open, green malt is added from above, and the kilned product is withdrawn from the bottom either in batches or continuously.

To a limited extent malt is kilned in insulated drums equipped with either ducting or decks as in germination drums, and which may be rotated to turn and mix the contents. Many 'special malts' (Chapter 5) are roasted in small metal drums. These used to be fired with coals, turned by hand and quenched by throwing water into the drum [52–54]. No attempt was made to conduct the extremely acrid smoke and steam out of the building. Now these drums are equipped with flues to carry away the combustion gases which may pass through a 'scrubber' to trap the fumes or an 'afterburner' to consume the organic residues that are present. They are mechanically turned, and they have sprinkler bars inside for quenching. For convenience they are now usually heated with gas burners in the UK, giving more complete temperature control.

6.5 Kiln furnaces

Kilns were originally fuelled with hardwood logs or faggots and small quantities of a few special malts are still made using such fuels, as the smoke from burning hardwood gives the products particular flavours. Similarly Scotch whisky malt is flavoured with peat smoke, i.e. it is 'peated'. In Britain the most important fuels used in kilns are low-arsenic hard coals, low-sulphur oils and natural gas. Experimentally, barley dust is being used to fuel barley driers with a considerable reduction in running costs [55].

Solid fuel furnaces used to be wasteful in that up to 10% of unburned coal might be lost with the ash. Stoking was a dirty process requiring much labour and skill. Simple coal furnaces are unsuitable for exact automatic thermostatic control. Newer furnaces may be automatically stoked with small low-arsenic anthracite which is burnt in a fan-driven flow of air. The fuel may be metered into the furnace chamber from above or may be delivered from below, i.e. 'underfed', by screw or ram systems [3, 4]. Such furnaces are easy to run semi-automatically, to control and clinker. Water circulating in a cooling jacket round the combustion cylinder and passages within the hollow furnace door is used to preheat incoming air by warming a suitably positioned radiator or heat exchanger. Compared with oil, coal is low in sulphur and so the flue gases and dust are less acid. Also, less water is produced by the combustion of coal and therefore less moisture is added to the heated gases. On the other hand, natural gas is 'pure', lacking sulphur and arsenic, but it burns to produce a comparatively large quantity of water vapour, more than is produced by burning oil. It seems possible that malt 'picks up' significant quantities of oxides of nitrogen (NOX) during kilning operations that use direct heating with natural gas. NOX, reacting with basic nitrogenous substances in the malt, may give rise to carcinogenic nitrosamines (Chapter 5). Specially designed 'low-NOX' fuel burners are being designed to minimize the production of oxides of nitrogen. Sulphur dioxide, present in the combustion gases of many fuels, may overcome the problem of nitrosamine formation. Sulphur burned at the air inlet serves as a source of sulphur dioxide, but may cause some increase in air pollution. Both oil- and gas-fired furnaces are easily thermostatted.

Oil-fired furnaces work by blowing a spray of oil into a burner together with a fan-driven air stream. Oil fuel is convenient, since it is clean to handle and is readily stored, controlled and measured. Such furnaces need to be carefully adjusted to prevent the formation of smoking flames, especially when the furnace is running at a reduced heat output during periods of air recirculation. The oil used is usually a light fraction having a low sulphur content, otherwise the sulphur trioxide produced during combustion may produce speckled unsightly 'magpie' malt [56, 57]. The brewing quality of magpie malt seems, however, to be unimpaired. The acid nature of the combustion products contributes to the deterioration of kilns. New furnaces with double toroidal burners will burn heavy fuel oils completely and produce satisfactory malts. In view of the policy of avoiding high-sulphur fuels, it is surprising that in the past in Britain (and still in the USA) malt was 'sulphured', that is sulphur was burned and the fumes were deliberately led through the malt, so bleaching or 'brightening' the product [58]. The malt treated in this way produces a wort having an elevated level of soluble nitrogen and a reduced pH. In view of the risks of nitrosamine formation, 'sulphuring' is being re-introduced in the UK, at least experimentally.

A modern automatically controlled kiln must be able to hold a number of air flows and temperatures and to achieve these states with the greatest economy of fuel. It may be more efficient to have several furnaces feeding one kiln, so that some furnaces can be completely shut down during recirculation or other periods of reduced heat output. This is often preferable to partially closing one large furnace. Some successful small-scale tests have also been made on drying malt with infra-red irradiation, but other tests have proved to be less satisfactory [59].

Many efforts are being made to improve the efficiency of kiln operations and reduce fuel consumption [6, 9, 10, 60, 61]. Experimentally solar panels have been used to preheat incoming kiln air. Kiln insulation has been improved, kiln and furnace operations are often checked, and malt with unusually high moisture contents (e.g. 3–5% against 1·5–2·5%) have become acceptable in the UK. This trend is likely to continue, since rootlets are easily separated from malt having less than 8% moisture, and malt with 10% moisture can be stored satisfactorily in temperate climates. To save fuel final curing temperatures are also being reduced, and so in the UK some pale Pilsener-type malts are being finished at 60°C (140°F), lager-malts at 75°C (167°F) and ale malts at 90–100°C (194–212°F). Recirculation is now routinely used on modern kilns, and to gain the advantages of the 'two-floor' configuration kilns may be linked, to allow 'curing' air from one to pass into a larger cooler volume of drying air used in the other, as in 'double-kiln' arrangements (Fig. 6.25). 'Continuous' and two-stage semicontinuous kilns also prove to be thermally efficient, partly because cooling and reheating the kiln structure is avoided. Heat generated by diesel or gas engines, used to drive the fans, has been used to preheat incoming air. More efficient variable-pitch fans are being used, and microprocessors are being used to regulate all the kilning variables in the most efficient way. Also economic

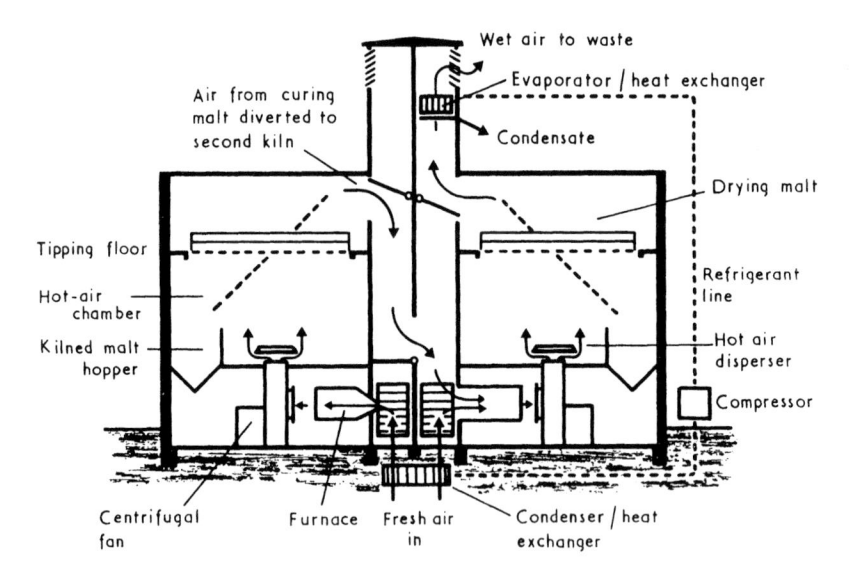

Fig. 6.25 A 'double' box kiln with tipping floors arranged to use dry 'curing' air from one compartment to help dry green malt in a second compartment. A possible heat-pump installation is also indicated.

ways are being sought to use the 'low-grade' heat of the gases from the kiln outlet to preheat incoming air. Simple heat exchangers made of large areas of plate-glass sheets or glass tubes work well, but are bulky and difficult to install in old kilns. Dust and volatile materials in the outgoing air dirty the exchanging surfaces, which are wetted with condensed moisture, and these must be cleaned. The increased resistance to gas flow, caused by the heat exchanger, increases the power needed to drive the fans. More efficient metal heat exchangers situated in the kiln outlet may be connected to radiators at the air intake by a recirculating liquid, e.g. a water–glycol mixture. The most expensive and complex type of installation currently under evaluation is the heat pump. This is installed as indicated in Fig. 6.25. The 'heat sink' withdraws heat from the kiln effluent gases by using it to evaporate a refrigerant such as Freon. The heat is released into the incoming air stream by circulating the refrigerant and condensing it under pressure in a radiator. A possibility that has apparently not yet been exploited is to use 'heat pipes' to carry heat from the kiln outlet to the incoming air [60]. Low-grade 'waste' heat may be used to heat green-houses or in other enterprises not otherwise linked to malting or brewing.

6.6 Separation of culms

After kilning the furnaces are shut off and usually the fans are run to reduce the temperature to 65°C (150°F) for example. Less usually the malt is 'rucked-up' (piled into a heap) to keep it warm and allow further development of colour and aroma without the expense of using more fuel. When the malt is cool enough the kiln is stripped. It is more efficient to strip the kiln quickly, and to reload it before it has had a chance to cool. The malt may be cooled separately. In principle, the heat might be recovered and used to preheat incoming kiln air. To break and remove rootlets the cooled malt is agitated with beaters and is screened [6, 12, 13]. The malt may be passed through a revolving perforated drum that retains the corns and allows the roots to pass out through the perforations. At the same time, the malt is aspirated with dry air to cool it and to remove dust. The air must be dry because moist 'slack' malt is not saleable. Also, damp roots become pliable, are not readily broken, and so are hard to separate from the malt. Alternatively the malt is derooted by battering in a centrifugal device in a strong air flow. The malt falls downwards to be collected, but the rootlets and dust are swept away in the air flow and are collected separately in cyclones. The air is filtered and recirculated. The rootlets and dust are very hygroscopic and must be stored dry, before being sold for compounding in animal feed. They are also bulky, and it may be worth while for maltsters to 'compress' and 'pelletize' rootlets before selling them. The malt needs to be cooled before storage (*i*) to reduce the chance of infestation with insects, (*ii*) to check the development of colour and flavour and (*iii*) to arrest the decline in enzyme levels. There is also a belief that malt should be stored for a minimum period of 4–6 weeks before being mashed. There is some evidence that the quality of wort obtainable from malt improves during this period of storage.

In view of the high costs of removing the last 2–5% of bound water from malt during kilning, perhaps in the future there will be a market for pale malts that are only 'hand dry', i.e. contain perhaps 5–8% moisture. Such material would be easy to use, unlike green malt, could be prepared to retain a large complement of enzymes, and would have lost its 'green' flavour.

6.7 Dispatch and transport of malt

Before leaving the maltings malt is usually screened, aspirated to remove dust, and passed over a magnetic separator. In the UK it is sold by weight. Malt used to be dispatched in sacks, and small quantities may still be so distributed, rapid delivery being relied on to prevent it picking up too much moisture, and becoming slack. The sacks now used are made relatively waterproof with a lining of a plastic film. However, with (*i*) increasing handling and labour costs, (*ii*) the risk of insect infestation, and (*iii*) the chance of grain becoming slack when sacks are used, malt is usually handled in bulk. Lorries or freight-wagons carry malt in airtight containers from which it may easily run through pipes like a fluid, with no direct handling and little contact with the air. Such 'grain tankers' may release the malt through 'hopper-bottoms' working under gravity (with or without the aid of tipping), or they may be emptied by 'sucking' the grain into a pneumatic grain-handling system.

REFERENCES

[1] STOPES, H. (1885). *Malt and Malting*, F. W. Lyon, London.
[2] LANCASTER, H. (1908). *Practical Floor Malting*, The Brewing Trade Review, London.
[3] NORTHAM, P. C. (1962). *Brewer's Guild J.*, **48**, 259, 296.
[4] NORTHAM, P. C. (1965). *Brewer's Guard*, **94**, 97.
[5] BRIGGS, D. E. (1978). *Barley*, Chapman & Hall, London, p. 526.
[6] NARZISS, L. (1976). *Die Bierbrauerei. 1. Die Technologie der Malzbereitung*, Ferdinand Enke, Stuttgart.
[7] Anon, (1976-Jan.). *Brewer's Guard.*, **105**,(1), 31.
[8] HYDE, W. R. (1975-Feb.). *Brewer's Guard.*, **104**,(2), 21.
[9] MINCH, M. J. G. (1978). *MBAA Tech. Q.*, **15**(1), 53.
[10] STOWELL, K. C. (1978-May). *Brewer's Guard.*, **107**(5), 50.
[11] NORTHAM, P. C. and BUTTON, A. H. (1973). *Proc. Eur. Brewery Conv. Congr. 14th Salzburg*, p. 99.
[12] DE CLERCK, J. (1957). *A Textbook of Brewing*, Vol. 1, Chapman & Hall, London.
[13] VERMEYLEN, J. (1962). *Traité de la fabrication du malt et de la bière*, Vol. 1, Assoc. Roy. des Anciens Élèves de l'Institut Sup. des Fermentations, Gand, Belgium.
[14] WILLIAMS, L. H. O. (1976). *MBAA Tech. Q.*, **13**(4), 261.
[15] STODDART, W. E., GRAESSER, F. R. and WESSON, J. P. (1961). *Proc. Eur. Brewery Conv. Congr. Vienna*, p. 105.
[16] CHUBB, A. R. (1954). *J. Inst. Brewing*, **60**, 205.
[17] PALMER, A. H. (1965). *Brewer's Guard.*, **94**, 17.
[18] GRIFFIN, O. T. and PINNER, B. C. (1965). *J. Inst. Brewing*, **71**, 324.
[19] POOL, A. A. and POLLOCK, J. R. A. (1967). *Proc. Eur. Brewery Conv. Congr. 11th, Madrid*, p. 241.
[20] Anon. (1968), *Int. Brewer's J.*, 74.
[21] Anon. (1969). *Int. Brewer's J.*, 47.
[22] LEATON, K. W. (1963). *Brewer's Guild J.*, **49**, 237, 288.
[23] MACEY, A. (1963). *Brewer's Guard.* **92**, 41.
[24] MACWILLIAM, I. C. (1963). *Brewer's J.*, **99**, 68, 127.
[25] GREEN, A. D. (1965). *Brewer's Guild J.*, **51**, 280.
[26] MACWILLIAM, I. C. (1965). *Brewer's Guard.*, **94**, 20.
[27] PHILBRICK, H. L. (1965). *Brewer's Guard.*, **94**, 39.
[28] Anon. (1969). *Int. Brewer's J.*, 77.
[29] Anon. (1965). *Brewer's Guard.*, **94**, 63.
[30] MAUCLAIRE, D. (1977). *Proc. Eur. Brewery Conv. Congr. 16th, Amsterdam*, p. 105.
[31] Anon. (1964). *Brewer's Guard.*, **93**, 61.
[32] Anon. (1967). *Brewer's Guard.*, **96**, 19.

[33] LUNDIN, H. (1936). *J. Inst. Brewing*, **42**, 273.
[34] JONSSON, A. E. (1937). *J. Inst. Brewing*, **43**, 205.
[35] SIMMONS, N. O. (1963). *Feed Milling*, Leonard Hill, London, p. 36.
[36] Ministry of Fuel and Power (1951). *Fuel Efficiency Bulletin*, No. 52, H.M.S.O., London.
[37] BEAVEN, E. S. (1904). *J. Fed. Inst. Brewing*, **10**, 454.
[38] ST. JOHNSTON, J. H. (1954). *J. Inst. Brewing*, **60**, 318.
[39] LOGAN, R. K. (1954). *J. Inst. Brewing*, **60**, 340.
[40] ABRAHAMSON, M. (1978). *Proc. Congr. Inst. Brewing 15th (Australia and New Zealand)*, p. 247.
[41] TUERLINCKX, G. and GOEDSEELS, V. (1979). *Proc. Eur. Brew. Conv. Congr. 17th, Berlin (West)*, p. 701.
[42] DICKSON, A. D. and SHANDS, H. L. (1942). *Cereal Chem.* **19**, 411.
[43] SCHUSTER, K. and GRÜNEWALD, J. (1957). *Brauwelt*, **97**, 1446.
[44] PENSEL, S. (1967). *Brauwelt*, **107**, 562.
[45] SCHUSTER, K. (1962). *Barley and Malt* (ed. COOK, A. H.), Academic Press, London, p. 271.
[46] FERNBACH, A. (1928). *J. Inst. Brewing*, **34**, 119.
[47] LAUFER, L. (1928). *J. Inst. Brewing*, **34**, 610.
[48] CHUBB, H. M. (1911). *J. Inst. Brewing*, **17**, 452.
[49] BEAVEN, E. S. (1904). *J. Fed. Inst. Brewing*, **10**, 520.
[50] MINCH, M. J. G. (1954). *Irish Maltsters' Conference, 1952–1961, Selected Papers*, p. 210.
[51] JAQUES, G. and MUNGHAM, H. H. (1964). *Proc. Irish Maltsters' Conf.*, p. 40.
[52] VALENTINE, G. (1920). *J. Inst. Brewing*, **26**, 573.
[53] REED, R. M. (1965). *Brewer's Guard.*, **94**, 55.
[54] Anon. (1963). *Brewer's Guard.*, **92**, 21.
[55] Anon. (1980) *Brew. Distil. Int.*, **10**(4), 26.
[56] MACEY, A. (1958). *J. Inst. Brewing*, **64**, 222.
[57] MACEY, A. and STOWELL, K. C. (1957). *Proc. Eur. Brewery Conv. Congr., Copenhagen*, p. 104.
[58] WITT, P. R. and ADAMIC, E. (1957). *Proc. Annu. Meet. Am. Soc. Brew. Chem.*, p. 37.
[59] SCHUSTER, K. (1952). *J. Inst. Brewing*, **58**, 219.
[60] SCRIBAN, R. and DANCETTE, J. P. (1979). *Bios*, **10**(4), 12.
[61] HOUTART, J. and DELATTE, J.-L. (1979), *Bios*, **10**(4), 24.

Chapter 7

BREWING WATER

7.1 *Geology of water supplies* [1, 2]

Rain and melted snow normally supply the water needed by communities, rather than sea water which requires costly desalination. A variable proportion of the fresh water drains from the land by rivers without appreciable penetration of soil and rock. In many countries, however, large volumes of water do penetrate, are held temporarily or permanently underground and help to maintain a fairly constant level of water table. By creating wells, bore holes or underground galleries, water of almost uniform composition and temperature is made available from these underground sources.

Three groups of rock formations based on physical properties are recognized, namely (*i*) unconsolidated (or recent) sediments, (*ii*) consolidated sediments and finally (*iii*) igneous and metamorphic formations. Not all geological formations furnish adequate supplies of water, despite intense precipitation of rain. The best formations in this respect are referred to as aquifers. They are usually of a coarse porous texture with either abundant fissures or weak cementation.

7.1.1 UNCONSOLIDATED SEDIMENTS

Unconsolidated sediments include recently deposited sands such as glacial sands, alluvial fans of gravel caused by erosion and river terraces of sand and gravel. Glacial sands can be almost sealed by clay and may therefore become water-saturated. The difficulty in tapping this plentiful supply of water is that the sands run easily and quickly fill the bore hole. Alluvial fans are often thick layers of gravel, and in southern Europe are important sources of water. They and the river terraces vary considerably according to the texture and depth of the gravel.

7.1.2 CONSOLIDATED SEDIMENTS

Consolidated sediments include sandstones, limestones, clays and marls. In Britain, a large proportion of the water requirements are tapped from the sandstones which are of a coarse porous texture. In contrast, limestones tend to be intrinsically impervious and only yield water if they are intensively fissured, but when a bore hole strikes a suitable ramification of fissures, water supply may be good. Clays, shales and marls are generally impervious to water and do not provide reasonable supplies.

7.1.3 IGNEOUS AND METAMORPHIC ROCKS
Igneous and metamorphic rocks are usually impermeable and they provide little water unless there is a suitable system of fissures. It is clear therefore that water supply depends on porosity of the rock and the fissures system within the formation.

7.2 *Water quality*
The quality of the water which may be withdrawn depends on many factors. The porous sandstones not only effectively filter the water but in some cases may effect base exchange. In contrast, water within fissured limestone receives little filtering. The chemical composition of the rock is also important. Thus the Permo-Trias rocks were deposited during desert-like conditions and in consequence have a high salt content so that Keuper sandstone, for example, often yields a water which is highly mineralized. At Leamington Spa the water drawn from this sandstone has as much as 1·75% dissolved matter of which 75% is common salt. Only 8 miles away, Stratford-upon-Avon has water derived from the same sandstone which has only 0·03% dissolved matter and sodium chloride represents no more than 20% of the solids. Possibly there are differences in porosity of the rock and, at Stratford, the salts have become leached away [3].

In the cases of limestone, chalks and sandstones with calcareous cementation, the waters that are withdrawn usually contain carbonates and sulphates of calcium and magnesium. Temporary and permanent hardness is common, but toxic salts of lead, copper and nickel are normally absent. Fluorides are comparatively rare except in very special situations. Iron salts are much more important because of the high iron content of many sandstones; nevertheless a great deal of the iron is precipitated or complexed in the rock.

Variations in salt content between one source of water and another near by often occur where there is a movement of water along an inclined stratum of rock, which gives rise to leaching of salts from the rock. A bore hole which taps the stratum close to where it outcrops gives water with a lower salt content than one tapping the same stratum at a deeper level. In general, therefore, deep bore holes give water with a higher saline content than do shallow ones.

7.3 *The supply of water* [2, 5]
Certain maritime regions have special problems because the sea produces, in areas close to the shore, its own saline-water table. Lying on top of this saline water and presenting an abrupt change from salt to fresh, is the fresh-water table. If great care is not taken, fresh water may be withdrawn so fast

from a bore hole that the levels are disturbed and sea water is pulled up. This will put the bore hole out of commission for several months until equilibrium between the water tables is re-established. In sandy maritime areas, especially in sand-dunes, water is either collected from numerous shallow bore holes or from a ramification of slightly inclined galleries dug into the dunes.

Surface contamination from sewage, industrial effluents or sea water flooding is often reduced by exploiting the natural filtering action of porous sandstones or gravels. In addition, it may be possible to seal off those strata in a bore hole which might, by permitting entry of unfiltered or highly saline water, endanger the quality of the water.

Supplies of water from underground sources fail to meet the demand in highly industrialized areas of Britain. There has been a continuous lowering of the underground water level over the past century and in some cases only half the water is returned by rainfall. Under such circumstances, many bore holes and wells dry up and are abandoned. Not surprisingly, water authorities restrict the rate at which water is abstracted from privately owned bore holes because excessive withdrawal will imperil neighbouring bore holes by lowering the water table. To breweries, bore holes may have especial value. The composition may be particularly suitable for brewing, the water may cost about 20% that of the publicly supplied water and the content of salts may be far more uniform. In contrast, water authorities reserve the right to switch without warning from one reservoir to another whose water has substantially different composition (Table 7.1).

TABLE 7.1

Variation in Burton-on-Trent public and bore hole water (p.p.m.) {Rudin (1976) [4]}

	Public water		Well water	
	Day	Night	1975	1962
pH	7·6	7·6	7·0	7·0
NO_3^-	5	36	18	28
Cl^-	27	32	16	31
SO_4^{2-}	87	100	820	848
HCO_3^-	123	168	320	289
Ca^{2+}	58	100	352	72
Mg^{2+}	21	14	24	72
Na^+	21	37	54	26

Other but more expensive methods of obtaining water have gradually been adopted by large communities because of water shortage. Upland catchment of surface water normally involves the building of dams and the piping of the water to the towns over large distances. It is not surprising that close attention is paid to the reuse of water in urban areas.

On average in Britain, each person contributes about 2·27 hl (50 gal) of sewage effluent daily. This has to be processed at sewage works so that content of solids, organic salts residues and bacteria are reduced to acceptable levels before discharge into rivers. Conventional treatment involves filtration, sedimentation, biological oxidation and final sedimentation. This sequence of treatments yields a clear slightly coloured liquid with an earthy odour which is bacteriologically inferior to urban river water. Quite apart from the presence of bacteria, some of which are likely to be pathogenic, the water could not normally be reused directly because of synthetic detergents and other complex organic materials, nitrates and chlorides, which remain. Nevertheless, there is the possibility of reusing the treated effluent water for irrigation of crops, and many industrial applications. Special treatment of this kind of treated effluent can yield a safe drinkable water. Similarly, canal water in Belgium is used by breweries after extensive treatment to yield a water of high purity.

7.4 Desalination [6-9]

The greater part of this planet is covered by saline water, and continued attempts are made to purify such water for public water supplies. The theoretical energy requirements to remove salts from 45·5 hl (1000 gal) of sea water by various means are 3200 kWh for distillation, 340 kWh for freezing and 3 kWh by reverse osmosis.

Up to the present time, a distillation procedure called the multistage flash (MSF) is widely used [254 plants producing about 10^6 m^3 (280 $\times$ 10^6 gal of fresh water)]. Great improvements have been made by designing the evaporation cycle, so that some of the latent heat given out by steam during its condensation is recovered in a useful way, instead of being wholly dissipated in cooling water. Cost is not related to the salt concentration – in contrast with other methods – and distillation has been used in the Persian Gulf coastal areas where the salt concentration varies from 2000 to 42 000 mg/l and where fuel is relatively cheap. The States of Guernsey Water Board installed in 1960 substantial plant to augment normal resources during seasons of high demand which produces 2700 m^3 (0·5 $\times$ 10^6 gal) per day.

Intensive research and development in various parts of the world have indicated that reverse osmosis has great promise in the desalination of water. The method is based on applying high pressures to saline water held in semi-permeable membranes made of cellulose triacetate or aromatic polyamide. The pressurizing means that prime energy is required, while, in the MSF technique, waste heat can be used. Assuming steam costs for MSF to be $25 for 10^6 B.Th.U (1·055 $\times$ 10^3 MJ), a plant producing 5 million US gal (18·9 $\times$ 10^3 m^3)/day would incur costs of about $4 per 1000 US gal. Assuming electricity to be 2·2 cents/kWh, it is suggested that reverse osmosis would cost about $2·5 per 1000 US gal [6].

7.5 Deionization [10, 11, 34]

The removal of salts from less-brackish waters is sometimes achieved by ion-exchange methods in which mineral ions are exchanged for hydrogen and/or hydroxyl ions present in special resins. These comprise charged groups on a polystyrene matrix. Anion and cation exchanges are effected in either separate resin beds or in a mixed-resin bed. An illustration of the method is given in Fig. 7.1. Where cation exchange alone is carried out, calcium ions would be replaced by sodium ions. Such waters would possibly be suitable for washing casks, bottles and so on, although unacceptable as brewing liquor. It will be appreciated that the ability of a particular resin bed to deionize will be lost after a certain weight of salts has been removed from the water. The greater the ionic concentration in the water, therefore, the faster the bed will be exhausted, assuming that complete deionization is effected. If throughput is high, the resultant water may have a significant but acceptable level of salts. In any case water treated in deionizing beds is rarely completely free from solutes. The salt content of the treated water is normally measured by electrical resistance. A good deionized water may have a resistance of over three million ohm/cm^3, but silica 'breakthrough' cannot readily be detected by this means. Silica should never exceed 125 mg/litre in boiler feed.

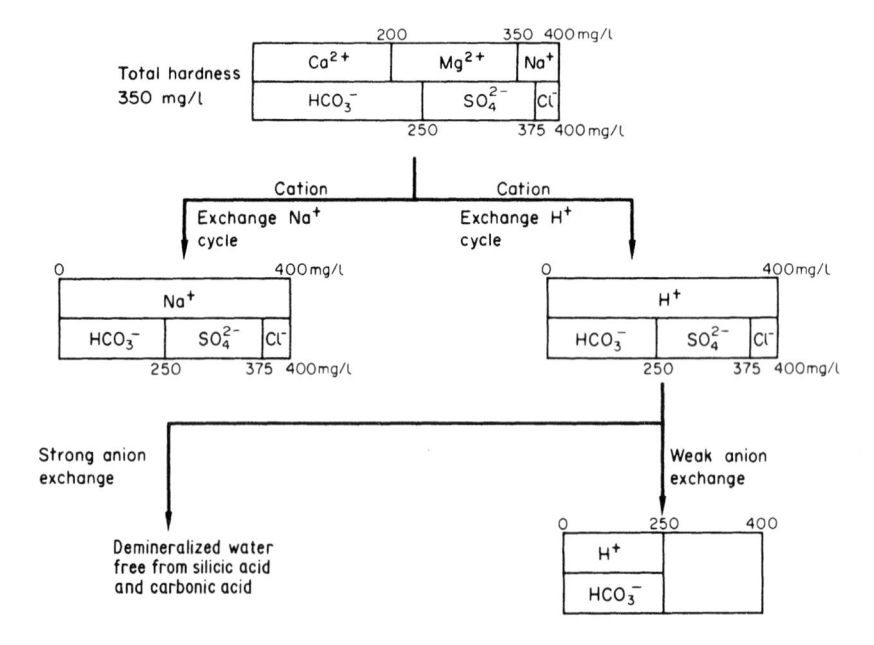

Fig. 7.1 Demineralization of a water with 200 mg of calcium ion/litre, 150 mg of magnesium ion/litre, 50 mg of sodium ion/litre, along with bicarbonate, sulphate and chloride ions.

With two-bed systems the cation bed can be regenerated with hydrochloric acid and the anion with caustic soda. The mixed-bed systems having a mixture of anion and cation resins, subject water to an almost infinite number of successive demineralizing stages. These systems cannot normally be regenerated simply like the two-bed systems and the suppliers operate an exchange service for the resin canisters. Some leakage of sodium ions may occur and lead to water of high pH. Recent developments have included the mixing of cation resin for base exchange softening with a macroreticular anion resin for scavenging organic materials. The column can be regenerated simply with sodium chloride. Continuous ion exchange has been developed to use the resin beds to the maximum. The beds are fluidized and portions of the resin withdrawn frequently, regenerated and replaced on an automatic system.

7.6 *Softening procedures* [4, 12]

Less-expensive methods than deionizing are available for the partial removal of hardness of water caused by the presence of salts of calcium and magnesium. 'Temporary' hardness is due to the presence of bicarbonates of these metals and is partly removed by boiling. Under these conditions, carbon dioxide is driven off and hydroxides are formed. In contrast 'permanent hardness' results from the presence of sulphates of calcium and magnesium, and boiling has no substantial effect on their concentrations in solution. The whole subject is bedevilled by the wide range of units to denote concentrations of salts, ions or the appropriate acids and bases (or even oxides of the elements). For example, salts such as calcium sulphate or magnesium bicarbonate may be expressed as parts per million (mg/l), as parts per 100 000, or as grains per gallon (1 grain equals 0·0648 g). A preferable method is to express concentrations in terms of ions present because ions do not arbitrarily associate as salts. While ionic concentration may be expressed as mg/l, grains/gal etc., there is much to commend expressing the results as millivals (1 millival is 1 mg equivalent/litre). Bicarbonate ions are normally estimated as carbonate so that there is correspondence with dry weight residues.

National units of water hardness give rise to some confusion. In Britain, 1 Clark degree is 1 grain of $CaCO_3$ per imperial gallon (10 lb of water at 62°F or 4·54 kg of water at 17°C). In France, a degree of hardness is 1 mg of $CaCO_3$/1000 litres, while in Germany the corresponding unit is 1 mg of CaO/1000 litres. Further complications arise from the American gallon being 3·78 kg of water at 15·5°C (8·33 lb at 60°F). One Clark degree is equivalent to 14·25 mg of calcium ion/litre, equivalent again to 0·7 millival. (Calcium has an atomic weight of 40 and being bivalent its equivalent weight is 20.) A water with less than 2·5 millival is considered soft, 2·5–5·0 as moderately soft, 5·0–7·5 slightly hard, 7·5–12·5 moderately hard and 12·5–17·5 hard; greater than 17·5 is very hard.

In softening procedures, boiling or aeration of water (or a combination of the two) is fairly expensive. By driving out carbon dioxide from solution, aeration favours an equilibrium well to the left-hand side of the equation:

$$CO_3^{2-} + CO_2 + H_2O \rightleftharpoons 2HCO_3^-$$

Aeration of boiled water, by driving off carbon dioxide, prevents the carbonate redissolving as bicarbonate. It is important to degas such water if destined for boilers because the oxygen (especially in the presence of carbon dioxide) may cause severe corrosion problems. The precipitation of calcium carbonate from the bicarbonate is very effective, but, because magnesium carbonate is considerably more soluble, the technique is far less satisfactory when magnesium ions are relatively abundant. During the boiling process, iron tends to be precipitated as ferric oxide or hydroxide.

A further method of removing temporary hardness is by the addition of carefully measured quantities of slaked lime which precipitate calcium carbonate and magnesium carbonate. The lime has also to neutralize any free acid in the water undergoing treatment, together with any carbon dioxide present, and precipitate salts or iron, silicates and also organic material. If too much lime is added, free calcium ions are again made available for hardness.

Permanently hard water may be softened by the addition of sodium carbonate (soda ash) which reacts with soluble calcium, magnesium and iron salts, but this is rare in the treatment of brewing liquor. In the older cold process involving lime and soda, it is necessary to add a precipitant such as aluminium sulphate (which hydrolyses to the hydroxide). This is mixed by paddles along with the lime and soda in a large tank with the water and, under still conditions, the flocks are allowed to settle, and the softened water is drawn off from the top of the tank. Modern lime/soda methods use warm water, rarely employ coagulants and operate on a continuous system. There is a mixing vessel, a conical sedimentation tank and a sand filter. Tenfold improvement in throughput is experienced when the water temperature is raised from 15 to 80°C (59 to 176°F), because of the reduction in viscosity of the water.

A split process is used when the magnesium content of the water is too high. An excess of lime water is added to two-thirds of the water for treatment. The calcium is precipitated as carbonate and the magnesium as hydroxide. After the precipitate has been separated, the remaining water is mixed in. Most of the calcium and some of the magnesium from this remaining volume is then precipitated by the lime water which remains in solution.

A recent survey of German breweries indicates that, for pale beers, brewing water exceeding 8 mg of CaO/m³ is softened. The 1976 costs in DM/m³ for removal of 14 mg of CaO/m³ were: lime softening 0·04, cation resin exchange (weak) 0·15, cation resin exchange (strong) 0·30, anion resin exchange 0·30, and total demineralization 0·60 [13].

7.7 Other water treatments

Acids are sometimes used to counterbalance the alkalinity of certain brewing waters, but not of course to soften them. Generally sulphuric and/or lactic acid is used but hydrochloric acid and phosphoric acid have also been employed. The question of flavour is important and lactic acid is considered to give a softer taste. Quantitatively the treatment is equivalent to titrating the bicarbonate in the water with acid, to the Methyl Orange end point at pH 4·4.

Modern boilers evaporate their charge of water in as little as 15 min, so that dissolved materials and suspended solids are rapidly concentrated [14]. Eventually the solubility products of the salts are reached and precipitation occurs, either as a loose deposit – 'sludge' – or as scale adhering to the metal surfaces. Sludges can be removed by drawing off some of the boiler water (blow back) from time to time or continuously, but heat is lost from the system. Scale formation (usually from compounds which are less soluble in hot than in cold water) is more serious because heat transfer is impaired. Overheating of the boiler tubes may occur and tubes and valves may block. Calcium sulphate is of particular importance in this respect along with calcium silicate and magnesium silicate. In low-pressure boilers, soda ash treatment suffices, but above 10 bar gauge pressure soda ash gives rise to caustic soda and carbon dioxide and thereby intense foaming and corrosion. High-pressure boilers therefore use water treated with sodium phosphates ranging from the acid NaH_2PO_4, to the progressively more alkaline Na_2HPO_4, Na_3PO_4 and $Na_2P_2O_7$. The correct treatment depends on the acidity of the water and the concentration of the scale-forming salts, but a normal dose gives 25–50 mg of phosphate/litre. Calgon and Versene treatments are also often employed. Calgon incorporates sodium hexametaphosphate and a little sodium pyrophosphate while Versene is the sodium salt of ethylenediaminetetra-acetic acid. Scale can be made softer and easier to remove by adding tannin or starch, while boiler cracking by caustic soda can be reduced by additions of lignin. Corrosion of boilers is frequently attributable to free oxygen and/or carbon dioxide and therefore boiler water is either mechanically deaerated or oxygen is taken up by additions of sodium sulphite.

7.8 Water purity [5, 15, 22]

Biochemical or biological oxygen demand (BOD) is the commonest test for indicating the degree to which water is contaminated by organic material. It measures the loss of oxygen in solution from a closed sample at 18·3°C (65°F) for 5 days. The test is often replaced or augmented by measuring the quantities of potassium permanganate or potassium dichromate reduced under standardized conditions. The values obtained are referred to as permanganate values (PV or Tidy figure) or dichromate values (DV). Various temperatures are used for the tests. The permanganate value is not totally reliable on its own as an indication of water quality, but for comparison with previous figures or with other analytical figures it serves well. In addition to organic matter, ferrous salts, nitrates and sulphides react with potassium permanganate. A further test for a given water indicates its chlorine demand and involves additions of sodium hypochlorite. This test gives a good indication of the amount of organic material present. If it exceeds the permanganate value, animal pollution is probable, whereas if it is less, vegetable pollution is likely.

Other non-biological tests which are usually carried out on a new potential source of potable water include estimation of (*i*) free and saline ammonia, (*ii*) albuminoid ammonia, (*iii*) nitrates, (*iv*) nitrites, (*v*) chlorides, (*vi*) phosphates, (*vii*) iron, (*viii*) lead and zinc salts, (*ix*) total solids or electrical conductivity, (*x*) nature and quantity of suspended matter. Physicochemical properties such as (*xi*) colour, (*xii*) turbidity, (*xiii*) pH value, (*xiv*) taste and (*xv*) odour are also examined.

Chemical analysis is important with respect to the sanitary quality of the water, especially in connection with past or remote pollution. But it does not follow that water with a satisfactory chemical characteristic is free from dangerous bacteria. Both chemical and bacterial examinations should therefore be carried out regularly. Routine bacteriological analysis is concerned with the detection of more or less harmless bacteria and in particular with those normally inhabiting the human, domestic animal and wild-animal intestines. These faecal non-pathogenic bacteria are so abundant in dejecta (10^8–10^9 cells/g) that pollution of water by extremely small traces of excrement can be demonstrated bacteriologically. The numbers and viability of pathogenic bacteria in excrement tend to be very much lower initially than those of non-pathogenic strains. The absence of the non-pathogens in a sample of water virtually precludes therefore the presence of the pathogens. If a few faecal bacteria are present in a sample of water, the sample may not necessarily be condemned, but viewed with suspicion. The human faecal bacteria include strains of *Escherichia coli*, *Streptococcus faecalis* and certain spore-forming bacilli. Pathogenic bacteria arising from human excreta include strains of *Salmonella typhi*, *Salmonella paratyphi*, *Salmonella typhimurium*, *Shigella dysenteriae* and *Vibrio cholerae* (typhoid-, paratyphoid-, food-poisoning-, dysentery- and cholera-producing organisms). But it so happens that bacteria that resemble *E. coli* in many respects live naturally in unpolluted waters, notably *Klebsiella aerogenes* (*Aerobacter aerogenes*). This means that quantitative estimations of faecal coliform bacteria have to be carried out in such a way that these non-faecal organisms are excluded. But difficulties arise from the occurrence of coliform bacteria which may occasionally be of faecal origin but are usually non-faecal. These are referred to as atypical *E. coli*.

Coliform bacteria are Gram-negative, non-sporing, sometimes motile, aerobic or facultative anaerobic organisms, 2–4μm long and about 0·5μm wide. They grow best at 37°C (98·6°F) and ferment various carbohydrates. Certain strains are commonly present in brewery wort and are referred to as

TABLE 7.2

Biochemical reactions of common bacterial contaminants of water. M.R., Methyl Red test; V.P., Voges-Proskauer test; G.C., growth on citrate test; I.P., indole-production test; E.R., Eijkman reaction test; G.S., gelatin stab test
{TAYLOR (1958) [23]}

Designation*	M.R.	V.P.	G.C.	I.P.	E.R.	G.S.	Probable Habitat
E. coli (Type I)	+	−	−	+	+	−	Human and animal intestine
E. coli (Type II)	+	−	−	−	−	−	Doubtful, probably not intestinal
E. coli (Type III)	+	−	−	+	−	−	Human and animal intestine
Citrobacter freundii (Type I)	+	−	+	−	−	−	Mainly soil
Citrobacter freundii (Type II)	+	−	+	+	−	−	Mainly soil
Klebsiella aerogenes (Type I)	−	+	+	−	−	−	Mainly vegetation
Klebsiella aerogenes (Type II)	−	+	+	+	−	−	Mainly vegetation
Klebsiella cloacae	−	+	+	−	−	+	Mainly vegetation

* The Bacteriological Examination of Water supplies (1956). Ministry of Health Report No. 71, on Public Health and Medical Subjects.

'wort bacteria' (see Chapter 12). Table 7.2 shows the range of biochemical reactions of the spectrum of common water-borne coliforms. The various tests include:

1. The Methyl Red test for acid production involves 3 days' growth at 30°C (86°F) in a liquid medium of glucose, peptone, dipotassium hydrogen phosphate and a drop of Methyl Red. A positive reaction is the production of a red colour (pH 4·5 or less), and negative a yellow colour (above pH 4·5).
2. In the modern Voges-Proskauer test for acetoin a 2-day-old glucose-broth culture is added to a 5% solution of α-naphthol in absolute alcohol plus a little caustic potash. A positive result is the development of a pink colour in 5–15 min, demonstrating the production of acetoin by the bacteria.
3. The growth of the bacteria after 3 days at 37°C (98·6°F) in a sodium citrate liquid medium is investigated.
4. For the indole-production test, a 3-day culture at 37°C (98·6°F) in peptone water is added to Ehrlich's rosindole reagent (*p*-dimethylaminobenzaldehyde in absolute alcohol and hydrochloric acid). A rose colour develops if indole is present.

5. The Eijkman reaction involves culturing the bacteria at 44°C (111·2°F) for 24 hr in a lactose medium and noting whether gas is produced.
6. The gelatin stab test is carried out with a nutrient broth–gelatin medium. A positive result is a white line of growth within the gelatin and a disc of growth on the surface. Gas spaces form in the gelatin.
7. A further test, which is widely used, concerns media containing bile salts. Thus MacConkey's medium is a peptone and lactose solid medium with sodium taurocholate and Methyl Red added. Organisms which produce acid from lactose, such as *E. coli*, form rose-pink colonies in a day at 37°C (98·6°F), while bacteria which fail to ferment lactose, e.g. *S. typhi, S. paratyphi, S. typhimurium, Shig. dysenteriae*, and *V. cholerae* produce colourless ones. Nevertheless not all the pink colonies which appear are necessarily *E. coli*.

In the majority of routine examinations the differential and confirmatory tests are reduced to: lactose fermentation at 37°C (98·6°F) (acid and gas production), MacConkey test at 44°C (111·2°F), growth on citrate, Methyl Red test and indole-production test. It will be appreciated that distinguishing and counting typical faecal coliforms is far from easy and requires considerable experience. Reference must therefore be made to modern detailed handbooks and critical descriptions of procedures [20–23]. There are instances when brewery water supplies require general bacteriological inspection regularly and the quality-control laboratory should have the facilities for this. It is extremely helpful to augment the tests described by inoculating sterile hopped wort and beer with small volumes of the water under test in order to assess the ability of water-borne bacteria to grow in the brewery. The international limits for bacterial numbers accepted for drinking water in 100 ml samples are as follows. In a given year 95% of the samples should be free of coliforms, no sample should have more than 10 coliforms, and if coliforms are present in a sample, the following sample should be free of bacteria. European and US standards differ slightly in detail. In hot climates, particular care has to be taken not only with bacterial contamination but also with other forms of biological contamination [15]. Table 7.3 gives details of internationally accepted standards on drinking water from physical, chemical and biological standpoints.

Some breweries carry out occasional checks on the presence of the beer-spoilage bacterium *Zymomonas* by inoculating water samples into glucose and yeast extract medium at pH 4. After incubation at 27°C (80·6°F), the cultures are examined for hydrogen sulphide production [4].

7.8.1 STERILIZATION OF WATER [24]
Chlorination is the most common method of sterilizing water supplies. Either chlorine gas in liquid form or sodium hypochlorite is used to get 1 mg

TABLE 7.3

International standards for drinking water (1971) plus additional limits applied to European drinking water (1970) in mg/litre.
{ANON (1971) [17, 18]}
US standards are almost identical with those listed but there are some additional limits especially in connexion with chlorinated hydrocarbon {Anon (1978) [16]}

	Permissible	Excessive
Total solids	500	1500
Total hardness (as $CaCO_3$)	100	500
Fe^{3+}	0·1	1·0
Mn^{2+}	0·05	0·5
Cu^{2+}	0·05	1·5
Zn^{2+}	5·0	15·0
Ca^{2+}	75	200
Mg^{2+}	30–150*	150
SO_4^{2-}	200	400
Cl^-	200	600
F^-	0·6–1·7†	—
NO_3^-	—	45
As	—	0·05
Cd^{2+}	—	0·01
CN^-	—	0·05
Pb^{2+}	—	0·1
Hg^{2+} (total)	—	0·001
Se	—	0·01
Anionic detergents	0·2	1·0
Mineral oil	0·01	0·3
Phenolic substances (as phenol)	0·001	0·002
Polynuclear aromatic hydrocarbons (µg/l)		0·2
Gross α emission	—	3 pCi/l
Gross β emission	—	30 pCi/l
pH	7·0–8·5	< 6·5 and > 9·2
Ba^{2+}	1	
Cr^{6+}		0·05
H_2S		0·05
NO_3^-		0·05
NH_4^+		0·05
Dissolved O_2	> 5	
Free CO_2	zero	

* Depends on the SO_4^{2-} level, the 30 being applicable to 250 mg of SO_4^{2-}/litre.
† Depends on maximum daily air temperature, highest values for 10–12°C (50–53·6°F).

of free chlorine/litre in the water. If there is less than 0·3 mg/l after 1 hr, the dose should be increased. Excess chlorine in the water can give rise to trouble in the brewery, particularly if stainless-steel or resin-bonded tanks are used. (Stainless steel corrodes in acidic chlorinated waters while certain resins are degraded to chlorinated phenols which strongly and adversely flavour beer.) Excess chlorine is therefore removed by rousing 1–2 mg of potassium metabisulphite/litre into the water; the bisulphite at this concentration acts as a harmless bacteriostat.

Other sterilizing procedures involve the production of ozone by passing dry air through an electrical field of high potential difference and then bubbling it through the water. The ozone is effective, and soon breaks down so that it does not affect the brewing process. It is, however, corrosive. An alternative is the use of a continuous ultraviolet light sterilizer emitting radiation in the range 200–315 nm to which all micro-organisms are susceptible. The cost of electricity for treating 45·5 hl (1000 gal) of water by the u.v. method is estimated at about 8 pence. Water treated in this way is used for bottle rinsing, make-up water and suspending pitching yeast. A further process employs colloidal silver (the Katadyn process) at 0·05 mg/litre, a concentration which is claimed to be toxic to most micro-organisms. The resultant water is potable.

7.9 Brewing water [4, 12–26]

During the past millenium, certain centres have become renowned for the high quality of their beers, the type of beer in each location being distinctive. Thus Burton on Trent became famous for its strong pale ale, London and Dublin for their dark ales, Munich for its dark lager beers and Pilsen for its pale lagers. These successes and their attendant distinctions can be attributed, at least in part, to the water composition in these areas. Thus Burton water is high in permanent hardness while those of Dublin, London and Munich have less calcium sulphate and more calcium carbonate. Finally Pilsen has soft water (Table 7.4). The ionic concentrations of a particular brewing

TABLE 7.4

Ionic concentrations (mg/l) of brewing water used in three brewing centres {SCHOLEFIELD (1956) [26]}

	Pale ale (Burton)	Pale lager (Pilsen)	Dark lager (Munich)
Ca^{2+}	268	7	76
Mg^{2+}	62	2	18
HCO_3^-	280	14	152
SO_4^{2-}	638	5	10
NO_3^-	31	zero	zero
Cl^-	36	5	2

water and derived wort and beer are shown in Table 7.5. Whereas concentrations of the calcium ion hardly change from water to wort to beer, there are considerable changes in the concentration of other ions. It must be emphasized, however, that these changes will vary according to the raw materials used and the brewing process employed. Ionic concentrations of various typical UK beers are shown in Table 7.6.

TABLE 7.5

Ionic concentrations (mg/l) of brewing water, wort and derived ale (12°Plato) [4]

	Ca^{2+}	Mg^{2+}	Na^+	K^+	Cl^-	SO_4^{2-}	PO_4^{3-}	HCO_3^-
Water	169	36	55	6	147	205	zero	165
Wort	165	127	101	550	450	338	846	zero
Beer	168	113	110	440	420	330	520	zero

TABLE 7.6

Ionic concentrations of various typical beers (mg/l) [Courtesy of D. Taylor]

	Ca^{2+}	Mg^{2+}	Na^+	K^+	Fe^{3+}	Cl^-	SO_4^{2-}	PO_4^{3-}
Lager (12°Plato)	80	50	70	400	0·05	200	200	400
Lager (8°Plato)	60	40	60	280	0·05	150	150	350
Ale (10°Plato)	130	60	80	350	0·1	250	400	300
Ale (8°Plato)	100	40	50	270	0·06	250	350	260

It is understandable that the brewing industry grew quickly in such centres, depending on the quality of the water and its abundance. Brewing requires a considerable volume of water and at the present time British breweries use on average between 8 and 8·5 times more water than the beer that they produce. In the past fifty years, there have been several developments which have brought about some change in the siting of breweries, including a reappraisal of water requirements. For instance, the development of softening, deionizing and purifying procedures have made suitable many sources of water previously rejected for brewing. Furthermore, appropriate addition of various mixtures of salts to water enables the brewer to match the salt composition of the favoured localities. Thus 'Burtonizing' water means the addition of calcium sulphate and possibly other salts to water so that its composition becomes similar to that of Burton well water. But only 10–20% of the water used in a brewery is actually used for wort production and the requirements for the other processes, such as washing, boiler feed water, refrigerating, and so on, appear to be for soft water. (A small degree of hardness is still desirable so that there is restricted dissolution of metals from pipes and vessels.) Thus subject to satisfactory supply of manpower and beer-distribution facilities, new breweries tend to be located where there is an abundance of fairly soft water. This means probably more treatment than used to be the case for brewing water (but very much less treatment for greater volumes of water used by the brewery for other purposes).

Other chapters describe in detail the effects of salts on mashing, wort boiling, fermentation and finally beer quality. At this stage it is sufficient to note that in mashing, enzyme activity and enzyme stability is influenced and therefore the yield of extract. At the same time, the pH level and concentration of phosphates in the mash and derived wort are strongly affected by particular salts. Extraction of hop bitter substances and precipitation of tannins and proteins are influenced by both the pH level of the wort and the concentrations of salts. Fermentation, growth and flocculation of the yeast

TABLE 7.7

Summary of the effects various ions present in water have on the brewing process

Ions	Comment
H^+ and OH^-	Always present. pH depends on their ratio.
Ca^{2+}	Important in water hardness, both permanent and temporary. In brewing, precipitates phosphate in wort to reduce pH, Ca^{2+} stabilizes α-amylase, increases total and α-amino nitrogen of worts, improves running off of worts from lauter, precipitates oxalate (haze and gushing), decreases wort colour, improves flocculation of trub and of yeast, limits extraction of certain coloured and astringent substances and also silicious materials.
Mg^{2+}	Rarely greater than 30 mg/l (Ca^{2+} may be as much as 200 mg/l). Salts more soluble than those of Ca^{2+} and therefore have less effect on wort pH and beer flavour. Important as cofactor for certain enzymes, e.g. pyruvate decarboxylase. Gives undesirable astringent bitterness to beers unless Ca^{2+} in great excess.
Na^+, K^+	Rarely present in high concentrations; Na^+ gives sour and salty taste, chloride better than sulphate. NaCl, especially at 75–150 p.p.m., gives desirable palate fullness. K^+ gives salty taste, but above 10 mg/l inhibits certain enzymes and makes beer laxative.
Fe^{2+} or Fe^{3+}	Range from traces to 30 mg/l as bicarbonate or, in peaty waters, complexed with organic material. Deposits slime in pipes and weakens yeast even at 1 mg/l. At this concentration contributes to oxidation of beer tannins and haze. Easily removed from water as ferric hydroxide by aeration followed by sand filtration.
Mn^{2+}	Traces present in malt. Important enzyme cofactor in yeast, but levels should be below 0·2 mg/l.
NH_4^+	Normally present only in polluted water.
Pb^{2+}, Sn^{2+}, Ti^{2+}	Potent haze formers, inhibitory to certain enzymes.
Cu^{2+}	Mutagenic to yeast but accumulated by cells. Copper spray for hops important source of ion. Contributes to haze formation.
Zn^{2+}	Toxic to yeast at high concentrations and also inhibitory to certain enzymes. However, at 0·1–0·2 mg/l, stimulatory to yeast and often added to wort as $ZnCl_2$.
HCO_3^-	Decomposes on heating. High concentrations give too high pH and flavour is affected. More potent in adjusting pH up than Ca^{2+} is in lowering it. Should not exceed 50 mg/l.
SO_4^{2-}	Contributes to drier more bitter flavour. Source of SO_2 and H_2S during fermentation.
Cl^-	Gives mellow palate and palate fullness. Limits yeast flocculation, improves clarification and colloidal stability.
NO_3^-	Levels above 10 mg/l may be indicative of sewage pollution. In the presence of Enterobacteriaceae, reduced to toxic NO_2^-
SiO_3^{2-}	With Ca^{2+} and Mg^{2+}, gives scales and hazes.
F^-	Up to 10 mg/l, has no adverse influence.

are similarly affected, and finally beer flavour stability, foaming and gushing are also influenced. Table 7.7. gives a summary of the effects that various ions present in the brewing water and other sources have on the brewing process; the calcium ion plays a major role at each stage.

Particularly from the standpoint of ale production, two aspects of water treatment which are of paramount importance concern the pH level of the wort. The first is the reduction in the concentration of bicarbonate and carbonate ions in the mashing liquor – and even more so in the sparge liquor. The second is to add calcium ions to these liquors so that a proportion of the phosphate present in the grist is precipitated as calcium salts (see Chapter 9).

In lager brewing, the adjusting of the wort pH is achieved largely by additions of acid to the liquor and/or by the partial removal of carbonates to give 50–100 mg of calcium carbonate/litre. (A recent specification given for a lager brewery requires brewing water with total hardness of 55 mg/litre and maximum permanent hardness of 25 mg/litre and a maximum pH of 7·5.) Bicarbonate and carbonate ions have a particular marked influence on wort pH when part of the mash is boiled during a decoction. The practice in pale ale breweries on the other hand is to pay more attention to the addition of gypsum, although removal of carbonates is often carried out. The following procedure has much to commend it. The alkalinity of the brewing water is reduced to certainly less than 50 mg/litre, and preferably less than 25 mg/litre, using lime treatment or acid treatment. Gypsum is added to the brewing water to bring the calcium ion concentration to 50 mg/litre (2·5 millival), and a further charge of gypsum is made to the copper, equivalent to 50 mg of calcium/litre because too high a level of calcium ions in the mash is disadvantageous. The final concentration of calcium in the beer will then be in the order of 60–80 mg/litre (3·0–4·0 millival). If sodium chloride additions are normally made, then these too can be made in the copper. It is impossible, however, to do more than indicate the guidelines on this topic, because of the variability of raw materials and the wide spectrum of brewing methods. Treatment is best decided on the basis of brewing trials and carefully conducted taste testing.

7.10 *Water economy in maltings and breweries* [27–30]

Quite apart from the need to protect national resources, there are three excellent reasons for conserving water in a brewery. Water, particularly from public supplies, becomes increasingly expensive. Thus in the Birmingham area over the last five years the cost has risen by almost 30% (even taking into account inflation) to £0·12 per m^3 or £0·55/1000 gal. Secondly, the greater the use of water by a brewery, the greater the production of industrial effluent which incurs charges for treatment by the local authority of some 10–20%

more than the water supplied, say £0·14 per m³. Finally, water discharged to sewers is usually warmer than ambient and so heat energy is lost. In recent years, breweries have sought methods to reduce water usage, and effluent production and improve heat conservation. The first step is to measure these parameters at each step of the process.

Water economy in maltings is centred on cleaning the barley as far as possible by dry treatments and on reducing the volume of water used in steeping, notably by spray-steeping rather than successive immersions. Cooling and humidifying water must be recirculated.

In brewing, a great deal of water is present in spent grains but it is usually too expensive to dry them. Several large breweries now resort to pressing the spent grains as dry as possible and returning the press liquor to the mashing process. Spent hops and trub, even surplus yeast and beer tank bottoms, can be pumped to the lauter tun or mash tun before the spent grains are discharged. They drain on top of the grains and an acceptable (or even enhanced) mixture results which can be sold profitably as cattlefood. Final worts from mashing can be used for mashing in the next brew.

Other economies arise from automated cleaning of vessels in which the minimum amount of water is used and the cleaning fluids are recirculated and reused. Slightly soiled water can be used as first wash for a dirty vessel, clean water only being needed for the final rinse and that can be used again as first rinse. Intermittent rinses are just as effective and more economical than continuous spraying. Where hoses are used in a brewery, they should be of rather small diameter and have a trigger valve at the end. All vessels, pipes and floors should drain readily. Big savings accrue from recovering steam condensates and from having closed circuits for water employed in cooling vessels and other equipment. This particularly applies to the cooling of wort in the heat exchanger where the hot water discharged can be used for mashing. Boiler blow-down liquid should be reduced to a minimum and water in pasteurizers should be recirculated wherever possible. Clearly there must be adequate maintenance of all equipment to reduce leaks of water.

In recent years there has been considerable interest in 'high-gravity brewing' in which wort of greater than normal specific gravity is produced, fermented and processed. At a very late stage, say before the final filtration, deoxygenated carbonated dilution water is added to the beer. This process brings about large savings in heating and cooling, with modest economies in water use. It also incurs the preparation of a special water for dilution and attention to the salt concentration of the final beer (Chapter 20).

The target must be to reduce water usage so that it represents only three to four times the amount of beer produced. Several breweries have achieved this, in one case a combined brewery and maltings. Some allowance must be made, however, for the large water demands made by washing returnable beer bottles. Many breweries, particularly those with a high proportion of their production in returnable bottles, use 8–15 times the volume of water compared with the volume of beer.

7.11 *Effluent treatment* [5, 31–33]

Reference has already been made to the need to recycle water, in other words to convert domestic sewage and industrial effluents, as far as possible into potable water. It is claimed that water in the river Rhine is used many times as it flows northwards across Western Europe. Mention has also been made of the high cost of treating effluent so that it can be discharged into waterways. The question therefore arises whether the maltsters and brewers could with advantage carry out a full or partial treatment of their industrial effluents and re-use the water or whether they are better to have the treatment carried out by the local authorities.

The impurities present in sewage usually comprise suspended solids, colloidal materials and substances in solution. Fig. 7.2 shows the techniques which are used to remove the impurities in relation to their size. Sewage screens or sieves have openings varying from 0·6 to 7·5 cm (0·24–3·0 in), and mechanical screens of the band, disc and drum types are used. In the microstrainer, which is a drum screen with finely woven stainless-steel mesh down to 20 μm, organic matter builds up a biological film which is capable of removing particles as small as 10 μm because the film reduces the holes in the mesh and also tends to adsorb or electrically attract particles.

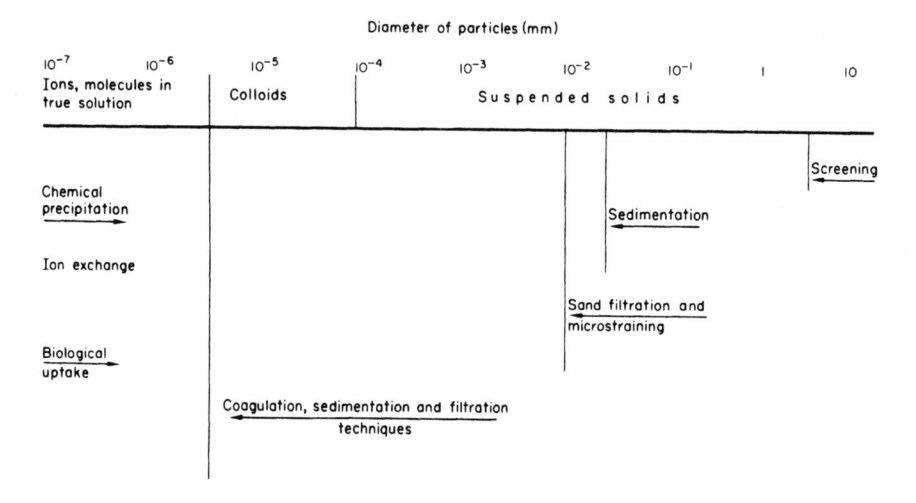

Fig. 7.2 Range of sizes of materials in dissolved, colloidal and suspended state and the methods used for removing them. {Tebbutt (1977) [5]}

Sedimentation tanks receive the strained liquid, and suspended particles settle at an increased rate until maximum velocity is reached, according to the expression:

$$(D_s - D_f)V_g = f(\mu D_f R_s d)$$
gravitational attraction   drag force

(D_s and D_f are the densities of the solid particles and the fluid, V is the volume and d the diameter of the particle, g is the gravitational acceleration, μ the absolute viscosity of the fluid, R_s the settling velocity of the particle and f is a notation for 'function of'.) Particles of sand or silt, with a specific gravity of about 2·65, will sediment at 153 mm/s (1 ft/2 s) if 1 mm diameter, 1 hr if 10 μm diameter, and 500 days if 0·1 μm diameter. The sedimentation tank should therefore be shallow, but the rate of flow through the tank should not be so rapid that particles are appreciably hindered in their sedimentation. Large storage reservoirs or lagoons with average retention times of many days give considerable improvement in water quality. Thus a typical river water held for 2 weeks in a reservoir of 2m (6·5 ft) depth may have the turbidity and bacterial population reduced by 85–95%.

Fine particles and colloids may be removed by coagulation using either inorganic materials, such as ferric sulphate and aluminium sulphate, or by living organisms. Usually the coagulating agent is negatively charged so as to attract positively charged ions, and a large flock develops with the colloid or fine particle at the centre. The larger the flock, the greater the rate of sedimentation. With aluminium sulphate treatment, the dose varies between 10 and 100 mg/litre and the optimum pH is 6·5 to 7·5. Rapid mixing for about a minute to disperse the coagulant is followed by slow stirring of the tank with large paddles.

Filtration by a trickling bed system can be regarded as a combination of straining, sedimentation and flocculation, and is normally the last process in sewage purification. The filterbed often has 0·6–1·0 m (2–3 ft) of sand topping and 0·3–0·6 m (1–2 ft) of graded gravel, below which is a drainage system to collect the filtered water. The bed provides for (*i*) screening, (*ii*) sedimentation and (*iii*) coagulation, because micro-organisms form a gelatinous coat around the sand particles to which adhere finely divided solids, as small as 10 μm diameter. Organic matter and some of the salts in solution are used as food by the micro-organisms. The filterbed becomes a complicated microcosm where protozoa, worms, larvae of midges, and moth-flies colonize the bacterial scum and use it as food. The water which filters off the bed is chlorinated to kill the organisms present and is discharged into rivers or other water ways.

For a given volume of water to be treated the trickling filter bed system uses a great deal of land and is generally expensive. 'Activated sludge' techniques are superior in these respects and therefore often replace it. Large sludge tanks or circular ditches are employed in which the sewage is aerated by diffuser plates or paddles. The sewage is oxidized and is precipitated by a mixed culture of micro-organisms, chiefly bacteria developing large quantities of mucilage. The treated effluent is discharged, but the sludge of bacteria and sewage left behind is used in part for treating a further charge of effluent. In one system, the raw effluent is screened and is then held in a holding tank before being passed on to a circular ditch. The ditch has two large aeration rotors and discharges effluent into a settling tank where the settled sludge is withdrawn. There are many methods of aeration available.

Sludge can be digested anaerobically at 30–35°C (86–95°F) in special tanks by bacterial action to produce carbon dioxide and methane. The latter can be used to heat the digestion tanks and other equipment. Sludge may be dewatered by vacuum or pressure filtration or by heating, and used as fertilizer or a fuel. The cost of dewatering is relatively high.

7.11.1 EFFLUENTS IN MALTING AND BREWING [30–40]

Many countries have recently applied stringent regulations controlling the discharge of effluents. The cost of disposal of effluents to local authority sewage works depends upon (*i*) the volume, (*ii*) the concentration of suspended and total solids, and (*iii*) the ability of the effluents to take up oxygen (the BOD test or agreed alternative). Complex formulae for calculating the costs of effluent treatment are provided by the authorities and such costs vary by an order of magnitude within Europe. In Britain, costs are based on the Mogden formula or a modified version of the formula. (The name arises from the Mogden sewage works where the formula was first introduced.)

$$\text{Cost} = A + B(ME/MS) + C(SE/SS)$$

where A is the cost of handling, pumping and overheads, B the cost of biological treatment and C the cost of sludge separation, removal and disposal. ME and MS refer respectively to the average strength of effluent discharged by the industrial plant and to the average strength of effluent at the sewage works. SE and SS correspondingly denote the concentrations of suspended solids at the two locations. Thus B is important with regard to discharge of all organic matter, particularly of high BOD, while SE concerns the discharge of all suspended solids such as spent grains, yeast, kieselguhr and label pulp. In addition to the formula, there are usually restrictions on peak BOD and suspended solids, temperature and pH. From time to time, therefore, there is a temptation to avoid peaking by diluting the effluent with clean or almost clean water. Attention to detail and long-term planning should eliminate this undesirable and expensive practice, and significantly reduce both water usage and effluent production.

TABLE 7.8

BOD and suspended solids loads from brewery processes
{ROBERTSON et al. (1979) [40]}

Process	Flow (hl/day)	BOD (kg/day)	Suspended solids (kg/day)
1. Mash mixer	69·4	3·6	7·3
2. Lauter tun	68·0	189·6	122·5
3. Brew kettle (copper)	97·2	27·7	3·6
4. Hot wort tank (whirlpool)	26·6	284·0	98·0
5. Wort cooler	24·1	0·5	0
6. Fermenter	290·0	240·4	155·1
7. Aging tank	407·7	93·0	137·9
8. Primary filter	81·9	46·3	231·3
9. Secondary storage	385·1	40·4	59·0
10. Secondary filter	92·6	8·2	42·2
11. Bottling tank	39·9	0·5	0
12. Filler	922·4	23·1	9·5
13. Pasteurizer	3582·6	34·5	3·2
14. Bottle washer	2933·5	68·0	29·0
15. Cooling water	4332·2	0	0
16. Miscellaneous flows	92·2	0·5	0·5

Table 7.8 indicates that the major sources of both BOD and suspended solids are the lauter tun, hot wort tank, fermenters, filters and bottling equipment. Discharge into sewers of spent grains, wort, trub, tank bottoms, yeast and beer is particularly serious. Table 7.9 gives figures for three breweries in the North East of England dating back some twelve years, while Fig. 7.3 indicates the major waste discharge from breweries.

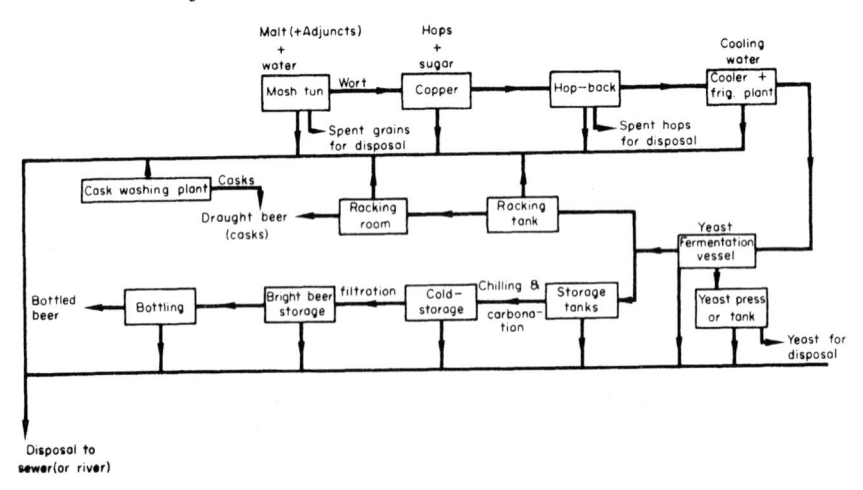

Fig. 7.3 Process flow diagram of brewing operations showing major waste discharges. {Ault (1969) [38]}.

TABLE 7.9

Analyses of malting and brewery effluents. All analyses except pH are mg/l {ISAAC (1976) [36]}

	pH	Permanganate value (4 hr)	Permanganate value (30 min)	5-day BOD	Total solids	Suspended solids (glass fibre)	Synthetic detergents	Total acidity as CaCO$_3$	Phenolphthalein alkalinity as CaCO$_3$
Brewery A									
Fermenting-vessel washings	6·50	177	109	880	2070	1080	—	47	—
Cask washing	6·40	96	38	460	1000	168	—	21	—
Bottle-washing (beer)	10·70	30	7·0	210	1930	460	Trace	—	350
Bottle-washing (mineral)	9·40	3·0	0	100	1230	450	0	—	60
Steep liquor from maltings	6·35	250	157	650	2910	218	—	87	—
Brewery B									
Steep liquor 1st drain	5·90	451	355	2350	3990	145	—	284	—
Steep liquor final drain	5·65	404	280	4050	3070	112	—	508	—
From cask-washer	6·90	32	19	110	690	96	Trace	26	—
From bottle-washer	9·00	19	9·4	105	1140	438	Trace	—	34
Brewery C									
Steep liquor composite sample from one batch of barley	6·40	30·7	16·6	170	408	27·5	—	26	—
Steep liquor 1st drain	6·65	13·6	9·5	80	341	16·0	—	14	—
Cask-washing	5·05	49·2	42·3	700	727	53·0	—	67	—
Fermenting-vessel washing	5·15	223·0	100·2	1340	1680	111·5	—	212	—
Bottle-washing (prerinse)	6·90	20·1	17·4	140	406	34·0	—	12	—
Draught beer: best bitter	3·40	2046	1020	63 000	38 585	325	—	1607	—
Draught beer: I.P.A.	3·60	1924	920	50 000	24 038	235	—	1528	—
Bottled beer: amber ale	4·05	4400	—	51 000	23 900	—	—	150*	—
Bottled beer: brown ale	4·30	8·1	—	69 000	39 600	—	—	188*	—

* CO$_2$ lost before titration.

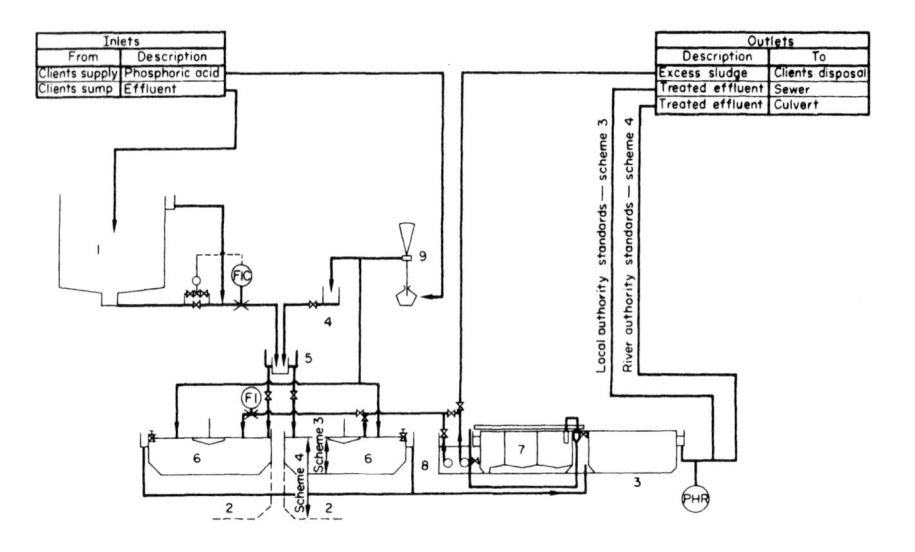

Fig. 7.4 A scheme for treating brewery effluent by activated sludge method on site. {Ault (1969) [38]}.

Item no.	Description	No. required
1	Effluent balancing tank	1
2	Aeration tank	2
3	Settling tank	1
4	Chemical nutrient tank	1
5	Splitter box	1
6	Simcar aerator	2
7	Scraper gear	1
8	Sludge recirculation pump	2
9	Hand pump	1

The brewery may decide to treat or partly treat their own effluent, and Fig. 7.4 depicts an activated sludge method for this purpose. Another approach is to use a biotower after settling and screening suspended solids. The biotower contains vertically arranged corrugated plastic sheets. The effluent is sprayed over the sheets and micro-organisms develop within the corrugations, so that they aerobically assimilate the organic material of the effluent. It is necessary to adjust pH to 6·5–8·5 and the BOD/N/P ratio to about 100:5:1. Living yeast cells in the effluent have to be killed by chlorination before the effluent can be accepted for treatment. Such a multistage plant can cope with 0·87–3·18 kg of BOD per day per m^3 of surface, provided that there is reasonably even flow. With a BOD loading of 98–353 kg/day (mean 210), the percentage BOD removed is claimed to be 70–94.

A further possibility for the large brewery is the deep-shaft system where a large bore hole is used as an aerobic treatment plant.

As mentioned, some or all of the purification may be carried out at the maltings or brewery. This may give extremely efficient use of plant because the demands can be accurately assessed, particularly if there is a 24 hr working in the brewery so that flow into the plant is maintained at a steady rate. If only partial purification is undertaken, the local authorities have the responsibility for final treatment – usually the more difficult task.

Untreated and partially treated effluents may be welcomed by those authority sewage works at which very large volumes of domestic sewage are treated. Domestic sewage has a BOD/N ratio of 17:1, which is less than the optimal for treatment, while malting effluents have a ratio of about 36:1 and brewery effluents of between 25:1 and 25 000:1. (The nitrogen is, for this purpose, the sum of half the organic nitrogen plus all the ammonia nitrogen.) From this standpoint it is desirable that breweries be sited close to large centres of population. Where breweries are rural, it is often desirable that they carry out some effluent treatment on site to reduce BOD and suspended solids.

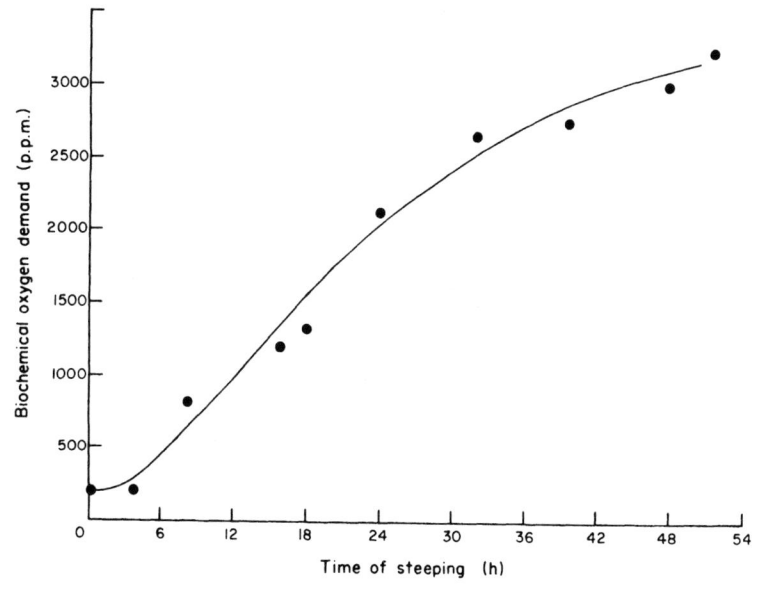

Fig. 7.5 The rise in biochemical oxygen demand with increase in the duration of the first steep in malting. [Courtesy of O. T. Griffin.]

Large volumes of water are used in maltings for steeping. Submerging barley requires 9–11 hl/tonne (200–250 gal/ton), but several changes of steep water may give rise to a total use of 21·8–27·2 hl/tonne (480–600 gal/ton). An illustration of the high BOD figures for the first steep is given in Fig. 7.5. A second steep water after a change at 24 hr has a BOD of 2050 mg/l after 48 hr of steeping, although of course steeping times far shorter than this are used in practice. A typical BOD value for second steep would be around

1000 mg/l. Contributing to the BOD of the steep liquor are soluble materials principally comprising amino, organic and phosphoric acids, simple sugars, peptides and tannins. While materials inhibitory to germination of barley may be present, barley and maize steep liquors are stimulatory to yeast in at least some brewery fermentations. The pH values of the steep liquor vary

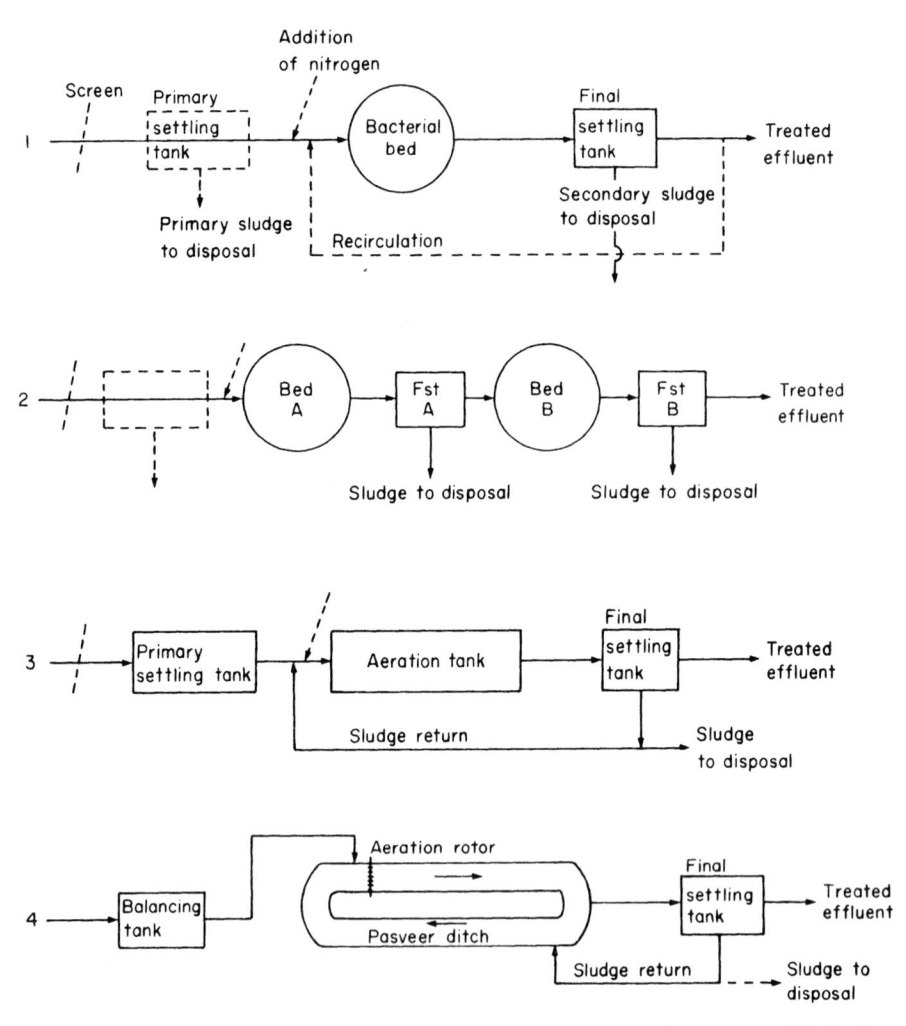

Fig. 7.6 Methods which are adopted by maltings for treating effluent (by courtesy of P. C. G. Isaac).
1. Single filtration (with recirculation)
2. Alternating double filtration. Primary settling tank, screen and nitrogen addition in broken line. Fst = Final settling tank.
3. Standard activated-sludge treatment.
4. The Pasveer ditch used for malting effluent.

from 5 to 8, depending more on aeration of the steep than the alkalinity of the water used. Organic nitrogen may range from 12 to 132 mg/l and ammonia nitrogen from 2 to 32 mg/l. Water use for maltings using four conventional steeps would be 27 hl/tonne (600 gal/ton).

In Britain, for the steep liquor to be discharged into rivers directly, the River Authority specifies that it must be diluted to give a BOD figure of 20 mg/l and suspended solids of 30 mg/l. Maltings may use trickling-filter systems, activated sludge systems or discharge on to fields at the rate of 6 in (15 cm) of water per acre (0·404 ha) per week (Fig. 7.6). The combined steep waters in maltings provide the soil with about 40 mg of nitrogen/litre, 10 mg of phosphorous/litre and 90 mg of potassium/litre. With activated sludge systems the BOD can be reduced to 30 mg/l.

7.11.2 DISCHARGE INTO WATERWAYS [5, 23, 27]
When too much crude sewage is discharged into rivers and other waterways they become polluted; much of the normal fauna and flora is poisoned or deprived of oxygen, and the waterway becomes an offensively smelling open sewer. This is because there is insufficient oxygen in the river water to support the oxidation of the sewage by micro-organisms. In Britain, there are many laws to restrict the discharge of effluents into fresh and brackish waters such as the Rivers Prevention of Pollution Acts of 1951 and 1961 for England and Wales, and corresponding acts for Scotland of 1951 and 1957.

Strict controls on the levels of suspended solids and materials capable of taking up oxygen reduce pollution of the rivers. Crude sewage may be discharged into rivers when it is not toxic, where the volumes involved are relatively small, and where legislation does not preclude it. Very roughly 1 volume of sewage to every 500 volumes or more of river water has been considered satisfactory in the past. Under these conditions, water from the river can be withdrawn and treated for public supplies.

7.12 *Water in beer*
About 92–95% of the weight of beer is water. Apart from being enjoyed for its flavour and aroma, beer therefore offers an alternative to water when water supplies are contaminated by chemical, biological or radioactive materials. The presence of various inorganic ions, arising in part from the brewing water (and in part from raw materials) may also be of considerable benefit to those drinking beer rather than soft water. Certain physical properties of water which are of importance in brewing are shown in Table 7.10.

Summarizing this chapter, water supplies have always influenced the development of breweries, originally the quality of the water with respect to mashing. With modern technology, the emphasis has shifted to the quality and quantity of water available for washing, cooling and boiler feed. The high cost of water and effluent disposal in many centres of brewing has led to better use of water. More recently the high cost of fuel has emphasized the need for better 'housekeeping' in breweries, including more attention to power saving in dewatering, water purification and recovery of solids.

TABLE 7.10

Certain physical properties of water
{MOLL (1979) [41]; HALE (1958) [42]}

Temperature (°C)	Density (kg/m³)	Viscosity η(cP)	Surface tension (dyne/cm)	Specific heat (cal/g/K) atmos. press.
0	999·84	1·787	75·6	1·00738
20	998·21	1·002	72·75	0·99883
100	958·37	0·2818	58·90	—
ice	917	—	—	—

Solubilities (mg/kg) at 10°C and atmospheric pressure in equilibrium with pure gases

N_2	23·2
O_2	54·3
CO_2	2318
H_2S	5112

Temperature (°C)	Oxygen in solution (ml/litre) in equilibrium with air
0	10·19
5	8·89
10	7·86
20	6·36
30	5·29

REFERENCES

[1] HOLMES, A. (1965). *Principles of Physical Geology*, Nelson, London.
[2] TODD, D. K. (1959). *Ground Water Hydrology*, Wiley, New York.
[3] SHOTTON, F. W. (1952). *J. Inst. Brewing*, **58**, 449.
[4] RUDIN, A. D. (1976). *Brewer's Guard.*, **105**(12), 30.
[5] TEBBUTT, T. H. Y. (1977). *Principles of Water Quality Control*, 2nd edn., Pergamon, Oxford.
[6] MORRIS, R. M. (1977). *Chem. Ind.*, 653.
[7] MATTHEWS, H. A. (1963). *J. Inst. Water Eng.*, **17**, 471.
[8] WEIR, W. K. J. (1962). *Inst. Publ. Hlth. Eng. J.*, **61**, 100.
[9] ANON (1976). *Process Biochem.*, (1), 32.
[10] GOODALL, J. B. (1976). *Chem. Ind.*, 1057.
[11] HOLLIDAY, D. C. (1972). *Chem. Ind.*, 717.
[12] COMRIE, A. A. D. (1967). *J. Inst. Brewing*, **73**, 335.
[13] KIENINGER, H. (1977). *J. Inst. Brewing*, **83**, 72.
[14] BARROW, F. D. (1979). *Brewer's Guard.*, **108**, (4), 17.

[15] FEACHAM, R., MCGARRY, M. and MARA, D. (eds.) (1977). *Water, Wastes and Health in Hot Climates*, John Wiley, London.
[16] ANON (1978). U.S. Code of Federal Regulations No. 40. Protection of the Environment (Office of the Federal Register of National Archives and Records Service).
[17] ANON (1971). *International Standards for Drinking Water*, 2nd edn., WHO, Geneva.
[18] ANON (1970). *European Standards for Drinking Water*, 2nd edn., WHO, Geneva.
[19] ANON (1969). *Bacteriological Examination of Water Supplies*, Report on Public Health No. 71, HMSO, London.
[20] ANON (1965). *Standard Methods for Examination of Water and Waste Water*, American Public Health Association, 12th edn., APHA, New York.
[21] RICHARDS, M. (1978). *Brewer's Digest*, (3), 30.
[22] RICHARDS, M. (1979). *Brewer's Digest*, (1), 36.
[23] TAYLOR, E. W. (1958). *The Examination of Waters and Water Supplies*, 7th edn. Thresh, Beal and Suckling, Churchill, London.
[24] SYKES, G. (1958). *Disinfection and Sterilisation*, Spon, London.
[25] WALKER J. F. (1976). *Chem. Ind.*, 805.
[26] SCHOLEFIELD, A. J. B. (1956). *The Treatment of Brewing Water*, published privately, printed Rockcliff, Liverpool.
[27] KLIJNHOUT, A. F. (1977). *Proc. Eur. Brewery Conv., Cong. Amsterdam*, 325.
[28] COORS, J. H. and JANGAARD, N. O. (1975). *Proc. Eur. Brewery Conv. Congr. Nice*, 311.
[29] LONES, D. P. (1977). *Brewer's Guard.*, **106**(2), 21.
[30] HERRINGTON, J. J. (1977). *The Brewer*, (1), 7.
[31] RAYNES, S. H. (1977). *Proc. Eur. Brewery Conv. Congr. Amsterdam*, 335.
[32] SASAHARA, T. (1977). *Proc. Eur. Brewery Conv. Congr. Amsterdam*, 313.
[33] CAMPBELL, W. (1977). *Process Biochem.*, **12**(6), 6.
[34] SKIPPER, A. J. (1976). *Brewer's Guard.*, **105**(11), 47.
[35] CHAPMAN, J. and O'CALLAGHAN, J. R. (1977). *The Brewer*, (6), 209.
[36] ISAAC, P. G. (1976). *Process Biochem.*, **11**(3), 17.
[37] BIDWELL, R. (1975). *Brewing Rev.*, (1), 19.
[38] AULT, R. G. (1969). *Chem. Ind.*, 87.
[39] SIMPSON, J. R. (1967). *Brewer's J.*, (8), 33 and (9), 32.
[40] ROBERTSON, J. L., BROWN, L. C. and PURPHY, K. L. (1979). *Tech. Q., Master Brewers Assoc. Am.*, **16**(1), 33.
[41] MOLL, M. (1979). *Brewing Science*, Vol. 1 (ed. POLLOCK, J. R. A.), Academic Press, London, p. 539.
[42] HALE, L. J. (1958). *Biological Laboratory Data*, Methuen, London.

Chapter 8

ADJUNCTS, SUGARS, WORT SYRUPS AND INDUSTRIAL ENZYMES

8.1 *Introduction*

This chapter outlines the preparation and use of materials, other than malt, that contribute to brewers' extract. This is taken to include materials which, in addition, alter the character of the beer. Non-malt enzymes used in producing extract or altering its nature e.g. by increasing its fermentability, are also discussed.

Colouring agents and caramels will be considered in Chapter 14, Volume 2 while the characters of special malts were indicated in Chapter 5. These distinctions are artificial to the extent that these various classes, e.g. special malts and raw or roasted cereals, are processed during mashing in similar ways. Adjuncts are classified with reference to where they are used in the brewing process, whether or not they must be 'cooked' by the brewer, and to their origin and the extent to which processing is involved in their preparation. Numerous materials were once used as 'adjuncts', including orange powder, coriander seeds, capsicum, caraway seeds, salts of tartar, oyster shells, alum, quasi-lignum, gentian or 'grains of paradise', even if some were 'of a warm nature and that they disperse wind and give warmth to the stomack' (Morrice, 1802, quoted in [1]). Adjuncts used at present can be divided into: (*i*) those used in the mash tun (which can be subdivided into those that need cooking by the brewer, and those that do not); (*ii*) those added to the copper; (*iii*) finally those added to beer either to support a secondary fermentation or (*iv*) to sweeten the beer or to add other flavours.

The mash tun adjuncts include raw whole grains, such as barley and possibly wheat, rye and *Triticale*, roasted or torrified grains, or flaked whole grains of barley, wheat or oats, partially refined grain fractions which need cooking, such as maize, sorghum or rice grits, or precooked flaked maize or flaked rice. Other materials used in the mash tun include cereal flours, refined granular starches from cereals or, less usually, from other plant materials such as potato or tapioca. Historically, enzyme-rich diastatic malt extracts were added to the mash tun. Copper adjuncts, solids or syrup materials that can be dissolved or dispersed in the wort during the hop-boil include preparations of sucrose or invert sugar, comparatively pure mixtures of saccharides based on the conversion of purified starches, and malt extracts, green-malt extracts or cereal syrups which, in addition to carbohydrates, contain significant quantities of nitrogenous substances, minor yeast nutrients and other substances and may be regarded as wort replacements. In extreme cases beer has been brewed from a hopped wort extract (the Conbrew process [2, 3]), and complete malt replacement with malt and cereal extracts, hopped or otherwise, is usual in 'home brewing.' A variation which may become widespread is the separate fermentation of concentrated syrup or sugar worts ('adjunct fermentations'). The resulting bland alcoholic liquids will be blended subsequently with 'all-malt beers', with savings *inter alia* on beer-conditioning capacity [4, 5]. The preparations used in priming or sweetening are essentially starch-derived or sucrose-derived. 'High-fructose' syrups derived from high-conversion starch syrups may come to be used to a limited extent.

In recent years the average brewer's grist in Britain has varied little, and in 1976 was 76·8% malt, 9·0% mash tun adjuncts, 14·2% copper adjuncts [6]. There has been a slow rise in the use of various sugar preparations, but where ten years ago copper adjuncts and other sugars were almost entirely sucrose or invert sugar now these materials are mainly derived from the controlled hydrolysis of starch. The rates of adjunct usage actually vary widely between breweries. EEC legislation discourages the use of more than 40% of adjuncts in the grist, but in view of the 'hydrolysis gains' that may be made, and because of the high extract yields of some adjuncts, more than 40% of the extract can be of non-malt origin. In various parts of the world higher levels of adjuncts are used, e.g. in the USA grists with 60% maize grits are known, and it is technically feasible to make an acceptable wort from 95% raw barley and 5% malt if 'industrial enzymes' are also used in the mash. Conversely elsewhere, as in the Federal Republic of Germany, the food purity laws forbid the use of adjuncts except when brewing beers destined for export.

The main reason for using adjuncts is that they provide extract less expensively than malt. The exact economics of adjunct utilization are contentious, and are partly controlled by the need for special plant (e.g. cleaning, handling, storing, milling) and by changing financial situations often based on political considerations, as well as the local availability of raw materials. Adjuncts have been discussed extensively [1, 7–14].

Because of the specialized nature of plant able to handle flours, flakes or syrups, or able to mill raw grains, it is not feasible to change easily from using one type of adjunct to using another. Consequently before a brewery commits itself to using a particular material it will assure itself of a continuing supply of it, of good quality. As well as providing cheap extract some adjuncts contribute little or no non-carbohydrate material to the worts, leading to a more bland and paler product, lower in nitrogenous and phenolic materials and so less liable to form hazes, and lower in lipids and so less liable to flavour deterioration. Thus beers brewed using such adjuncts have longer shelf-lives. The adjuncts may be used to adjust the fermentability of the final wort or improve the head retention or change the colour of the beer. The character, aroma and flavour of the beer may also be adjusted. Brewers may use syrups to increase the 'brewing capacity' of their plant, since these, being dissolved in the copper, do not require the use of a mash tun, and are convenient for obtaining concentrated worts for 'high-gravity brewing.'

Factors associated with the use of some adjuncts include the need for machinery that will mix solid adjuncts thoroughly with the malt component of the grist and lautering problems when wort separation is impeded by high wort viscosities or the formation of layers of fine particles in the mash. In addition some adjuncts have a tendency to yield poorly fermentable worts.

'Industrial (non-malt) enzymes' may be used to supplement malt enzymes in the mash tun to aid in the conversion of either a poor-quality malt or adjuncts (Chapter 9). They may also be used during adjunct cooking or to convert raw cereals or starches in the manufacture of syrups of various types. They have also been added to worts and beers to adjust fermentabilities or sweetness and, by degrading polypeptides, to reduce haze potentials. Originally enzymes were derived from malt, as diastatic malt extracts or diastase preparations, or sometimes from soya beans. In addition, considerable quantities of the plant proteases papain, ficin and bromelain are used in beer-stabilization treatments. However, most 'industrial enzymes' are derived from micro-organisms and include a wide range of amylases, dextrin-debranching enzymes, β-glucanases, proteases and possibly other hydrolases. In addition, other enzymes, such as glucose oxidase, find limited uses.

8.2 *The common cereal grains*

Most of the brewers' adjuncts are based on a limited range of cereal grains (Table 8.1). Such analyses are a useful guide to the potential value of these cereals to brewers. However, a number of points must be remembered. For instance, the composition of cereal grains varies very greatly, for example, 'crude protein' contents of barleys may be in the range 8·1–15·6% (1·3–2·5 %N), and barley β-glucan levels vary widely [17]. Further, different grades or samples of wheat, maize and rice differ markedly in their suitability for preparing adjuncts. The oil (fat) contents of cereals differ and because high oil/fat levels in adjuncts are deleterious to beer quality, oats are rarely used now in brewing, and maize, sorghum and rice grains are invariably processed to remove the oil-rich germs and bran before being used to manufacture

TABLE 8.1
Representative compositions of common cereal grains on a percentage dry-weight basis
{Calculated from DE CLERCK (1957) [15]}

Cereal	Cellulose (fibre)	Starch & carbohydrates	Lipids	Other non-nitrogenous matter	Protein	Ash
Barley	5·7	71	2·5	4·0	11·8	3·1
Wheat	2·9	76	2·0	2·8	14·5	2·2
Rye	2·4	74	2·0	5·8	13·5	2·4
Oats	12·4	61	6·1	2·4	13·4	3·5
Maize	4·2	70	5·8	7·6	11·6	1·2
Maize*	(2·0)	(69·2)	(3·9)	(—)	(8·7)	(1·2)
Sorghum*	(2·3)	(70·7)	(3·0)	(—)	(10·9)	(2·1)
Rice	2·3	81	0·5	—	9·0	0·4

* MORRISON (1953) cited in STEWART and HAHN (1965) [16].

adjuncts. Even so, lipids derived from processed adjuncts may still be important in brewing. The starches in adjuncts vary in the difficulty with which they may be degraded by amylases during mashing. Ease of degradation is associated with the ease with which starch grains swell and gelatinize as the temperature of an aqueous suspension is raised. However, swelling is a gradual time-dependent process, as is 'gelatinization' (percentage of grains gelatinized) and occurs over a range of temperatures [18]. Consequently when malt is mashed, more than 90% of the extract can be recovered eventually at temperatures *below* the notional gelatinization temperature of its starch (Chapter 9). In addition, in the case of adjuncts, starch conversion during mashing depends on the adequate disruption of investing cell walls and other cellular materials as well as the nature of the starch granules themselves. Reported gelatinization temperature ranges include: maize, 62–74°C (143·5–165°F); sorghum, 69–75°C (156–167°F); rice, 61–78°C (142–172°F); wheat, 52–64°C (125·5–147°F); barley, 60–62°C (140–143·5°F), malted barley, 64–67°C (147–152·5°F); and potato, 56–69°C (133–156°F) [10, 19]. Presumably kilning caused the gelatinization temperature of the malt starch to exceed that of the barley. These values correlate with the fact that, whereas refined starches or starch in flours of wheat, barley, malted barley, potato and *Triticale* can be converted by malt enzymes during mashing, the purified starches (or grits) of maize, sorghum, rice and oats all need to be gelatinized or precooked before adequate conversion will occur. Considerations of facts like this dictate in what way a particular raw material will be used as an adjunct. In many instances quite high levels of malt replacement with adjuncts (e.g. 40%) have been reported to have little effect on the quality of finished beers.

8.3 Analyses of adjuncts

For brewing purposes adjuncts, with the possible exception of certain sugars and syrups, are usually analysed according to one of the major analytical systems [20–22]. Attention is paid to minimizing sampling errors (which may be large with flakes, for example), and they are assessed for some or all of the following: physical characteristics and purity (odour, presence of foreign matter, moulds, insects etc.), the clarity and colour of a 10% solids solution (in the case of soluble preparations), moisture content (%), solids content (%), extract yield and sometimes extract fermentability, the spectrum of carbohydrates present, the levels of reducing sugars, iodine reaction, oil/fat content, acidity/pH of an extract or solution, nitrogen ('protein') content and/or yield of soluble nitrogen, ash content, enzyme complement/diastatic power, refractive index and specific gravity (syrups), and levels of iron, copper and arsenic. In many instances microbiological contamination is also assessed.

In many respects the analytical methods used for mash-tun adjuncts are unsatisfactory. For example, oil contents often seem surprisingly different from the levels found in the materials from which the adjunct has been prepared, when complete lipid extraction has been used in analysing the starting materials. Methods for determining extract yield are either inadequately standardized or may not reflect the recoveries obtained during brewing (which probably differ between breweries). This is because unmalted adjuncts are deficient or wholly lacking in the enzymes needed to solubilize their starch and convert it to an acceptable spectrum of carbohydrates. Usually the adjunct (milled or not and then cooked and boiled with or without an addition of enzyme(s) as specified in the chosen method) is mixed with malt to give a final 50:50 mix. The mixture is then mashed at 65°C (149°F) [21] or through a programmed temperature cycle [20, 22]. The extract of the adjunct is calculated from the gravity of the resulting wort on the basis that the malt will have contributed half of what it would do if it made up all the grist. The methods are essentially those used with coloured malts. Such techniques are incapable of a high degree of accuracy. Other proposed methods are based on converting adjuncts with additions of partially purified enzymes, microbial or otherwise, or the enzymes present in cold-water extracts of highly diastatic malts. In each case allowance is made for the 'extract' in the enzyme preparation. In any case the mashing conditions differ considerably from those found in breweries.

In the Institute of Brewing method [21] the extract of syrups is calculated as extract (litre°/kg) = $10 \times SG$ (20 ± 0.5°C), where the SG is determined on a solution of 25·00 g of syrup diluted to 250·0 ml, also to 20°C (68°F).

Control in sugar factories and refineries in which sucrose is processed in solution is based on measurement of the specific gravity and the optical rotation [23]. Balling originated the graduation of hydrometers on the basis of the percentage of weight of pure sucrose solutions, but his results were later recalculated by Brix. A more accurate set of measurements due to F. Plato and his collaborators was made in 1900 and it is this scale which is used today. Thus, °Balling, °Brix and °Plato are identical except for the fifth and sixth place of decimals but, whereas °Balling and °Plato are the currency of the brewing industry, the sugar industry usually speaks of °Brix. In continental brewing practice °Plato are normally used to express the strength of the wort obtained from the cereal mash as well as sugar solutions. In America the results are usually calculated using the Balling tables, then are expressed as '% extract'. Sugars are optically active, their solutions rotating the plane of polarized light. Polarimetry therefore provides a convenient method for assaying pure sugar solutions [23].

The specific rotation is defined as:

$$[\alpha]_D^{20} = \frac{100\alpha}{lc}$$

where α = the observed rotation in circular degrees (at 20°C; 68°F); c = the concentration of the solution (in g/100 ml); l = the length of the solution (in dm; 10 cm).

Measurements are usually made using the D-line of the sodium spectrum (589 nm) and at 20°C (68°F). For pure sucrose $[\alpha]_D^{20}$ is $+66\cdot53°$. In sugar factories and refineries special polarimeters calibrated in 'International Sugar Degrees' and known as saccharimeters are used. By definition, a solution of 26·00 g of pure sucrose dissolved in 100 cm³ of distilled water in a 2 dm (20 cm) tube rotates the plane of polarized light 100 International Sugar Degrees (which are therefore equal to 34·62 circular degrees). Using a saccharimeter a direct reading of the strength of a sucrose solution (called the 'Pol') can be obtained. The purity of a sugar solution can then be defined as:

$$\text{Purity} = \frac{\text{Pol}}{°\text{Brix}} \times 100\%$$

This expression assumes that the impurities present have no optical activity, which is generally untrue. Sucrose is hydrolysed, either by acid or the enzyme invertase into an equimolecular mixture of glucose and fructose:

$$\underset{\text{Sucrose}}{C_{12}H_{22}O_{11}} + H_2O = \underset{\text{Glucose}}{C_6H_{12}O_6} + \underset{\text{Fructose}}{C_6H_{12}O_6}$$

$[\alpha]_D^{20}$ $+66\cdot53°$ $+52\cdot7°$ $-92\cdot4°$

$-39\cdot7°$
Invert sugar

Whereas glucose (dextrose) is dextrorotatory, fructose (laevulose) is strongly laevorotatory so the equimolecular mixture has a negative rotation; it is therefore known as invert sugar. (The optical rotations given are those

TABLE 8.2

Representative analyses (IoB) of various brewing materials currently available in Britain
{After LLOYD (1978) [25]; COLLIER (1975) [24]; MAIDEN (1978) [6]; ANON (1980) [7]}

Raw material	Moisture (%)	Laboratory analysis Institute of Brewing methods			Typical brewhouse extract	
		Hot-water extract (Litre°/kg; 20°C)	(Brewer's lb per 336 lb; 1977)	Viscosity laboratory wort (1976 method cP at 20°C)	(Litre°/kg; 20°C)	(Brewer's lb per 336 lb)
Pale malt	3	296	100	1·4	287	97
Adjuncts cooked in brewery						
Maize grits	12	307	104	1·4	301	102
Uncooked mash-tun adjuncts						
Hammer-milled wheat	12	—	—	—	262	88
Hammer-milled barley (coarse)	12	—	—	High, variable	219	73
Hammer-milled barley (with enzymes)	12	—	—	Enzymes reduce	239	80
Brewer's wheat flour, pelleted wheat flour	11	325	110	1·8	301	102
Potato starch	18	319	108	1·4	313	106
Precooked mash-tun adjuncts						
Flaked maize	9	313	106	1·4	307	104
Flaked rice	10	322	109	1·4	316	107
Flaked barley	9	253	85	2–7	245	82
Flaked wheat	8	279	94	1·7	279	94
Torrified barley	5	253	85	2·3	253	85
Torrified wheat	5	273	92	1·5	273	92

Copper adjuncts	Hot-water extract		Brewhouse extract	
	(Litre°/kg; 20°C)	(Brewer's lb per 224 lb)	(Litre°/kg; 20°C)	(Brewer's lb per 224 lb)
Invert sugar liquid	319	72	319	72
Sucrose (solid)	384	87·1	384	87·1
Maize sugar syrups	306	69	306	69
Malt extract	301	68	301	68
Barley syrup	297	67	297	67
Wheat syrup	297	67	297	67

TABLE 8.3

Analyses (ASBC) of various adjuncts

{Data of CANALES and SIERRA (1976) [26]; COORS (1976) [27]; CANALES (1979) [10]}

	Moisture (%, as is)	Extract (%) (as is)	Extract (%) (on dry)	Protein (%, as is)	Fat/oil (%, as is)	Fibre (%, as is)	Ash (%, as is)	pH	Gelat.* (°C)	Gelat.* (°F)
Maize (corn) grits	11·6; 11·9	79·3; 80	90·0	8·5; 9·2	0·8	0·7	0·3	5·8	61·5–73·9	143–165
Rice grits	11·0; 12·2	81·7; 82·1	92·7	5·4; 7·5	0·2–0·6	0·6	0·8	6·4	61·1–77·8	142–172
Sorghum grits	10·8	81·3	91·1	8·7	0·5	0·8	0·3	—	—	—
Refined maize starch	11·0	92·9	103·1	0·4	0·04	—	—	5·0	61·5–73·9	143–165
Refined wheat starch	11·4	95·2	105·2	0·2	0·4	—	—	5·7	51·5–63·9	125–147
Torrified wheat	4·9	74·4	78·2	12·2	1·0	—	—	6·2	—	—
Torrified barley	6·0	67·9	72·2	13·5	1·5	—	—	5·9	—	—

* Gelat., gelatinization temperature range.

which pertain after mutarotation is complete.) Since invert sugars are common impurities in sugars and syrups, it will be seen that the single polarization measurement is often inadequate and more refined analyses are based on measurements made before and after complete inversion [23]. Representative analyses of adjuncts are given in Tables 8·2 and 8·3.

Sugars and syrups prepared from purified starches are also evaluated with reference to their density and often their sugar spectra. Higher oligosaccharides can often be partly resolved by gel exclusion chromatography, and lower saccharides are well resolved by high-pressure liquid chromatography (h.p.l.c.) or gas–liquid chromatography (g.l.c.) of volatile derivatives such as the trimethylsilyl ethers. However, routine monitoring is by the 'dextrinization equivalent' (DE) in which the reducing power of a solution expressed as glucose (dextrose) is expressed as a percentage of the value that would be obtained if the starch was wholly hydrolysed to glucose (theoretical DE = 100%). Preparations obtained by hydrolysing starch in different ways may have similar DE values but widely different carbohydrate spectra. The importance of the sugar spectrum is a consequence of the relative fermentabilities of the sugars when presented as a mixture to normal brewing yeasts. Sucrose, glucose and fructose are rapidly fermented, maltose is fermented more slowly, maltotriose is fermented very slowly and higher oligosaccharides and α-$(1 \rightarrow 6)$-linked gluco-oligosaccharides and some other di- and trisaccharides are not fermented. In the presence of high glucose concentrations maltose and maltotriose may be incompletely metabolized by brewing yeast.

8.4 Raw cereal grains

Low starch-gelatinization temperatures, low oil contents and many trials indicate that, when suitably milled, raw barley, wheat, rye and *Triticale* grains should be useful adjuncts that may be used directly in the mash tun without prior cooking [10, 12, 25, 28]. The feasibility of using raw barley in brewing has been recognized, since at least 1800 [6], and Irish whiskey distillers have used large proportions, e.g. 68% of 'low dried' barley (i.e. barley dried to 4% moisture, ground between stones or in hammer mills) in their grist [17]. Both wheat (1·8–2·1%N) and barley (1·5–1·8%N) have been used directly in infusion mashing after hammer milling [to pass about 1·6 mm (0·06 in) and 4·8 mm (0·19 in) screens respectively] in 26–35% of the grist yielding 261–268 and 208–234 litre°/kg (20°C as is; 87–90 and 68·6–78·0 lb/Qr.) of extract respectively [29, 30]. However, most interest has concentrated on the use of whole barley in breweries, although wheat flours are employed (see also Chapter 9). Whole grain is easy to handle provided that milling facilities are adequate.

The raw barley is selected for cleanliness and is sieved to obtain material with a size range suitable for the milling system in use. Viability is irrelevant. The grain contributes useful amounts of β-amylase and possibly limited amounts of other enzymes, such as protease and phosphatase, to the mash, as well as less desirable protease inhibitors and β-glucans. It is necessary therefore to ensure that for starch conversion the mash should contain adequate levels of α-amylase, and for easy run-off, β-glucanase. Further, if significant contributions of nitrogenous materials to the wort are needed, then the level of proteolytic activity in the mash is also important. Low levels of raw barley in the grist (10–20%) can usually be adequately converted by the enzymes from good-quality malt. However, for high levels of replacement, in order for the wort to be adequately fermentable and have a correct balance of nitrogenous materials, the mash may need to be supplemented by a suitable mixture of microbial enzymes, adequately standardized for amylase, protease and β-glucanase activities. Such an enzyme mixture is used in the manufacture of some barley syrups, sometimes with a small addition of highly diastatic malt [31, 32]. It is desirable to select low β-glucan barleys for use as adjuncts, and if breeders could incorporate the newly discovered 'high-residual α-amylase' characteristic into new varieties this would allow 'self liquefaction' of the barley starch during mashing [33].

If raw barley is too finely milled, it can cause problems of wort run off and it is advantageous to keep the husk as intact as possible. Besides hammer milling, barley has been successfully roller-milled dry or impact-milled to loosen the husk and then burr-milled or roller-milled, or roller-milled after steam conditioning, or wet-milled after a presoak [34–42]. There are reported to be marked advantages to washing raw barley before it is wet-milled [39, 42], and in the interests of cleanliness it has been proposed to use an alkaline wash followed by a treatment with sulphur dioxide [42] or a warm wash (70°C; 158°F/4h) [43], even if a small amount of extract (say 1%) is lost. When using high levels of malt replacement with raw barley, it is desirable to use either a decoction or a temperature-programmed mashing schedule and additional enzymes to ensure sufficient proteolysis and β-glucanolysis. In general, temperature rests at about 50°C (122°F), 63–65°C (145°–149°F) and 70–75°C (158–167°F) appear to be suitable, adjusted in duration to 'match' the proportion of raw cereal in the mash, the level of enzyme supplementation, and the nature of the wort required [34, 40, 41, 43]. Raw barley used alone as a malt replacement can give a good extract, but reduces the fermentability and level of soluble nitrogen in the wort, as well as slightly reducing the polyphenol levels (Chapter 9). However, when barley-rich grists are supplemented with a balanced mixture of hydrolytic enzymes the wort obtained can be very similar to that prepared from an all-malt grist.

8.5 Cooked intact cereal grains

Roasted barley and torrified (*syn.* torrefied) or micronized barley and wheat grains contain no active enzymes, but by virtue of the heating they have undergone their starch is partly precooked, and the grain's internal structure is partly disrupted. The products are easily handled and milled, and yield modest extracts above those obtainable from raw grains (Tables 8.2 and 8.3). Heating can partially degrade the β-glucans and other hemicelluloses so that extracts from torrified grains are less viscous than those from the raw materials. At the same time at about 140°C (284°F) starch is partly broken down to 'pyrodextrins' or 'torrifaction dextrins.' Depending on the conditions of heating, sugars may be caramelized, melanoidins may be formed, polyphenols may be oxidized to 'phlobaphenes' and many flavour substances, including sulphur compounds, aldehydes, ketones, lactones, aromatic substances and heterocyclic compounds, are formed, together with bitter diketopiperazines [17]. Cooking reduces the quantity of soluble nitrogen recovered from grains.

Traditionally barley was roasted in directly heated drums or slowly rotating iron cylinders. Overheating could result in grain carbonization. Now indirect hot-air cooking may be used (Chapter 6). Roasted barley has a characteristic sharp dry flavour and, in addition to providing extract, it is used, for example, to give colour or character to stouts. It is prepared from a clean high-nitrogen barley, and has a hot-water extract of about 267–277 litre°/kg (90–93 lb/Qr.), a moisture content of 1–2% and a colour of 1000–1550 EBC units.

Historically barley was cooked or 'torrified' by passing it through a stream of red-hot sand, when it popped and gained a nutty flavour, and experimentally it has been cooked using microwaves to cause dielectric heating. At present barley may be pre-wet [at 65°C (149°F)] to a moisture content of 14–18% and cooked in a stream of hot air at about 220°C (500°F), when it swells and nearly 'pops' [27, 44]. Alternatively it may be carried on vibrating wedge-wire belts below ceramic tiles, heated to red heat with gas flames, when it is cooked, or 'micronized' by the infrared radiation. Its temperature rises to about 140°C (284°F), and the rise in internal water-vapour pressure swells the grain and disrupts its internal structure. After cooling it has a moisture content of 4–6% [24, 45]. These lightly cooked cereals have low colours (e.g. about 2°EBC), and bland flavours. Micronized grains may be flaked before cooling (see below).

8.6 Grits

Grits consist of uncooked nearly pure fragments of starchy endosperm derived from cereal grains. They may be used directly in the brewery, where they must be cooked, or they may be flaked (see below). The removal of the surface layers of the grain reduces the lipid, ash, and fibre contents of the material, and results in an enrichment of the starch.

Experimentally pearled barley [17] or debranned barley [46] has been used in brewing. The debranned material, prepared in 75% yield, had an extract of 86% (compared with 75% for the original barley). Its lipid, fibre, ash and polyphenol contents were reduced, but, because of an enrichment in β-glucans, it yielded a more viscous wort. This viscosity could be reduced by additions of β-glucanase [46].

Rice grits or broken rice are usually fragments of endosperm produced during the removal of hulls, bran, embryos and aleurone layers while debranning and milling grain intended for human consumption. As the efficiency of milling has improved, so less of this by-product is produced [10, 27]. The product is low in lipid and fibre, and it must be cooked well before it will yield its extract (Table 8.3). Different types of rice vary widely in their suitability for use in brewing. In general, the flavour imparted to beer is 'neutral', so this adjunct is popular.

Maize grits, prepared from yellow dent corn, are the most widely used adjuncts in the USA [10]. The grits are prepared by one of several possible dry-milling processes [47, 48]. Selected maize grain is screened and washed, then conditioned in steam to soften the surface layers of the endosperm, the skin and the germ. The grain is decorticated and degermed by abrasive milling and is successively passed through break-rolls and screens to prepare a pure grits fraction. The grits are rich in starch and contain much less oil and fat than the original grain. After cooking, they yield a good extract during mashing (Tables 8·1, 8·2 and 8·3).

Originally, experiments designed to use sorghum as a brewing adjunct had only limited success, partly because of contamination by the surface layers, which led to poor-quality bitter-tasting beers. However, by selecting improved varieties and by more efficient milling the original difficulties were overcome [10, 26, 48]. The grits may be processed like maize grits, and appear equal to them in quality (Tables 8·1 and 8·3). Both give rise to extracts with bland flavours.

For use in the brewery all grits must be cooked to disrupt them and gelatinize their starch (Chapter 9). Afterwards they are usually mixed into a 'malt mash' to complete the starch conversion. The grits are mixed with a small proportion of highly diastatic malt (e.g. 5%) or with microbial amylase, and are heated. Normally temperatures have been raised to boiling or the grits may even be cooked under pressure. However, at least in some trials, grits have been cooked at 85°C (185°F) with bacterial α-amylase. This mash may have its pH adjusted and a calcium salt added to protect the α-amylase from heat inactivation. The starch swells and is gelatinized, and it is partially degraded and 'liquefied' before the heat inactivates the α-amylase. Consequently the viscosity of the mixture does not rise too high, nor does it set to a gel on cooling, so it does not 'cling' and burn on to the heating surfaces, and it can be pumped to, and mixed with, the malt mash.

8.7 Flaked cereals

Among a wide range of possibilities the flaked materials actually in use appear to be prepared from whole grains of barley, wheat or possibly oats, or from grits of yellow or white maize or sometimes rice grits. For a while, pearl barley was also flaked [24, 45]. Flakes are manufactured by two different processes. In the older process the moisture content of the raw material is adjusted to about 18-20%, then cooked with steam under slight pressure to soften the internal structures and begin starch gelatinization. The cooked material is delivered through feed rolls to flaking rolls held at about 85°C (185°F). The squashed flaked product is dried in a counter-current of warm air, usually to 8-10% moisture, and cooled. The product, which has a low bulk density (i.e. it is light in weight for the volume it occupies), may be slightly crushed to make it more compact [19]. In the newer process the raw materials are micronized and are passed immediately between flaking rolls and are cooled. An advantage is that the final product has a moisture content of 7-8%, and so does not need to be dried [24].

Flaked cereals have been used for many years in the UK, as they are easily used in infusion mashing. They are readily handled and mixed with ground malt. Their thinness and the disruption of their cellular structures allows the ready penetration of enzymes and leaching of products of hydrolysis. Furthermore no cooker is needed at the brewery, yet because their starch is gelatinized, it is readily converted. They produce relatively little soluble nitrogen, so their extract serves as a 'wort nitrogen-diluent.' They usually make little contribution to flavour. In the past, flaked oats, which appeared rather soft and grey, were used in the preparation of some stouts, to which they were thought to add a distinctive character. On a dry-weight basis their extract yield was low, about 254-282 litre°/kg (20°C; 85-95 lb/Qr.), their protein content was about 15%, and they contained a high level of oil and lipid, about 9% [49]. Flaked rice, which is normally white, is not used at present for economic reasons. However, it is a well-liked adjunct with a good extract yield (Table 8.2) and neutral flavour characteristics. Flaked maize grits and flaked whole wheat grains and barley grains are in use, and flaked pearl barley was available. Representative analyses of some of these are shown in Table 8.2. In some pilot brewhouse trials, flaked micronized barley yielded higher extracts than some traditionally prepared flakes (73·5% against 72·0% dry-weight basis), but micronized maize flakes gave a lower extract than the more traditional products (76·5% as against 78·0% dry-weight basis) [45].

On occasion, traditional barley flakes give rise to highly viscous extracts, due possibly, in part, to uneven steam cooking before flaking. It is less in micronized barley flakes. By spraying flakes with a dissolved preparation of bacterial α-amylase, which contains β-glucanase, the problem is overcome. The enzymes are destroyed during drying, but the flakes, which otherwise appear normal, have a high cold-water extract and yield wort with a low viscosity [45].

8.8 Flours

Raw wheat and barley may be hammer-milled in breweries to produce coarse flours suitable for use in the mash tun (see above). However, flour milling in the usual sense is a complex process carried out in specialized mills, and the finished products are purchased by the brewers. In principle, it is possible to produce flours from all cereals, and it is feasible to fractionate the mill streams in various ways to modify their compositions [48]. 'Flours' are formed during the production of maize and sorghum grits, and barley can be converted to a flour. All these products may be air-classified (see Fig. 8.1) to yield for example a barley starch flour fraction of 72% starch and 10·2% protein as against values of 57% and 15·2% respectively for the original barley (10, 17, 48, 50]. However, it seems that these materials are not used regularly in brewing. Wheat flour preparations, however, are used widely (Table 8.2 and Chapter 9).

Wheat, in contrast with most varieties of barley, is a naked grain with a deep crease running along its ventral side, which complicates the cleaning and milling processes. Wheat selected for brewing is usually a 'soft' variety, having a 'mealy' or 'floury' endosperm and a nitrogen content of up to 1·8%. Wheat milling is a complex process, which is described in outline [48]. The prime objective in milling is to collect the starchy endosperm, as flour, separated as cleanly as possible both from the bran (pericarp, testa and aleurone layer) and the germ. The grain is cleaned by various screening, dry-scouring and agitated-washing steps and other impurities are removed by disc separators for example. The grain is conditioned to regulate the distribution of its moisture, and to increase or reduce the moisture level to an optimum value, usually in the range 15·0–17·5%. The wheat is fed into the 'break system' of a mill in which the grains are split open and endosperm is scraped from the bran. Between each stage, fractions are separated by sieving ('scalping' in the early stages) before being processed further. Endosperm-rich streams are roller-milled and reduced to flour, while germ and bran are collected elsewhere. Care is taken to maximize the yield of the 'best size' range of broken endosperm fragments which make up the flour. In the final flour, starch and protein fragments are never entirely enclosed in unbroken endosperm cell walls. This is very important when brewing flour is being produced, since unduly small flour particles slow wort separation.

Such wheat flour may be used directly in the mash tun. However, by using a series of air-classification stages it is possible to collect a starch-enriched low-protein flour fraction (for example, having 7% nitrogen as against 9·5% of the original flour), with particles 17–35 μm in diameter, which has a better extract and fewer fines than the original flour [48, 51, 52] (Fig. 8.1). The other grades are used for example to enrich bread-flour (< 17 μm) or in biscuit manufacture (> 35 μm).

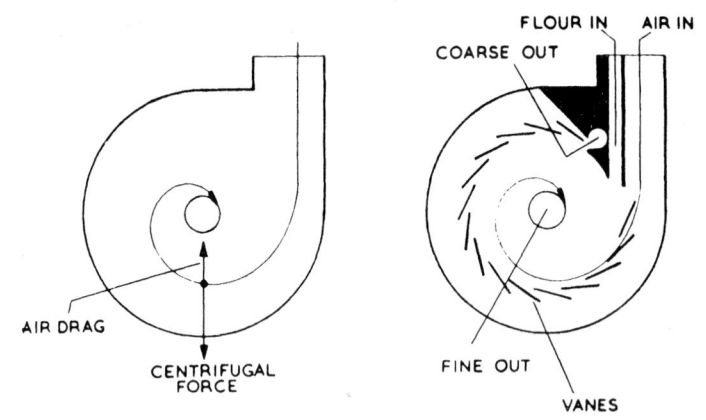

Fig. 8.1 The principle of one type of air classifier. Air and flour are introduced into an air vortex in a modified cyclone. The particles sort into streams regulated by their size and densities such that the centrifugal outward force balances the drag drawing them into the air vortex. By altering the positions of the vanes it is possible to regulate the size at which the 'cut' between the two fractions occurs. {Rumpf and Kaiser (1952) [51]}.

An alternative to air classification for producing brewing flours is to adopt a milling procedure that produces flours in which some 45% of the particles consist of 'clusters' of starch granules and protein of greater than 100 μm average diameter and in which particles of less than 20 μm make up only 3–4% of the whole [53].

Flours provide unfamiliar handling problems for brewers who may buy them in sacks or in bulk. Although they have a higher bulk density than malt [13], they need to be held in steep-sided containers often fitted with vibrators to ensure that they flow freely. They need to be moved horizontally either on vibrating conveyors or pneumatically. To overcome problems caused by static electricity and to minimize risks of electrical sparks triggering explosions in dusty flour–air mixtures all metal containers for flour (or refined starch), or handling equipment, must be grounded or earthed. Plastic chutes and tubes are avoided because of the ease with which they become electrically charged. It requires skill to mix the flour evenly with the ground malt in the grist. Large-particle flours flow more freely and are more readily handled than fine flours [53]. 'Agglomerated' wheat flours, held in pellets by binding agents, are still more easily handled, although the run-off characteristics they contribute to a grist are comparable with those of a normal flour [10, 54].

It is not essential to gelatinize wheat flour starch before adding it to the mash tun, although this additional treatment may result in a slightly more complete conversion of starch to sugars. However, it has been suggested that flour can usefully be presoaked before use in mashing. In N. American practice it may be slurried and cooked, much like grits, except that it is recommended that the slurry be made at less than 52°C (approx. 125°F) to minimize clumping and that cooking be at less than 98°C (approx. 208°F) to avoid excessive frothing [10]. In infusion mashing a 10% replacement of malt by wheat flour is easily tolerated. Replacement levels of 25% and even 36% have been used with good results where care has been taken with the mixing and the malt is sufficiently rich enzymically [29, 30] (Chapter 9). The flour itself contributes limited quantities of β-amylase to the grist. Run-off problems with wheat starch appear to be caused by the formation of impermeable layers of 'fines', largely hemicellulosic in nature, in the grist. It seems possible that additions of β-glucanase and low-temperature rests in the mashing cycle would overcome this problem. As well as reducing the total soluble nitrogen levels and α-amino nitrogen levels of worts, wheat flours alter the spectrum of nitrogenous substances significantly. They add a high-molecular-weight fraction, which probably contains glycoproteins, and give improved head retention (foam stability) to the finished beer [55, 56]. Probably because of the overall reduction in particular soluble nitrogenous fractions and polyphenols, beer brewed with wheat flour has a reduced tendency to form haze, i.e. an enhanced shelf-life.

8.9 Refined starches

Refined starches can be prepared from many cereal grains and other vegetable sources such as tapioca (manioc, cassava) and potatoes. Experimentally, starch has been prepared from barleys, for example, using an alkali-wash procedure [57]. In commercial practice refined wheat starch, potato starch and maize starches have been used in breweries and maize starches in particular are used in the preparation of glucose syrups (see below). Potato starch or farina is readily converted in infusion or congress mashes at a 20% replacement rate [58]. However, brewers are generally unwilling to use it because of consumer resistance to the idea of 'potato beer'. Wheat starch has been employed in breweries in Australia and Canada where local conditions make it economic to use. Methods are known for preparing it by dry milling or wet processing procedures [10, 27]. However, the most important source of refined starch is maize.

Maize starch (corn flour; refined maize grits; wet-milled pearl starch) is prepared from selected grain by a continuous wet-milling procedure [10, 14,

48, 59]. The clean grain, from which broken grains have been removed, is steeped in warm water (48–52°C; about 118–126°F) – below the gelatinization temperature of the starch for about 2 days. The water contains sulphur dioxide (0·1–0·2%) which limits microbial growth but allows desirable lactobacilli to multiply, and weakens the proteins of the maize endosperm by chemically reducing and so breaking the interchain disulphide bridges. The corn-steep liquor (6–8% solids) is concentrated to about 50% solids and is sold as a component of animal feeds, or as a nutrient for the industrial culture of micro-organisms. The softened grain is coarsely ground and the released embryos, which are rich in oil and have a low density, are separated by floatation in hydrocyclones. After drying they are pressed or extracted with solvents to yield maize oil (about 50%) and a residue for cattlefeed. The degermed material is ground more finely, then the hulls and fibre are collected on sieves while the starch and gluten pass through. The starch slurry (about 40% solids) is purified by counter-current washing. The slurry may be used directly in the manufacture of glucose syrups (see below). Alternatively the granular starch may be collected in continuous centrifuges, then dried to 1–10% moisture. The gluten, which is mainly in suspension, is recovered by centrifugation and, together with the hulls and fibre, goes for cattlefood.

Refined starches are the 'purest' mash-tun adjuncts available to the brewer. When dry they must be handled with all the precautions used with flours and, as with flours, some handling problems may be reduced by 'granulation' or 'pelletizing' [10]. Potato and wheat starches, having low gelatinization temperatures, can be added directly to the grist. The starches may be handled as a slurry, mixed with water, but on no account must the mixture be allowed to set or settle. However, wheat starch may be cooked and liquefied before use, and maize starch must be cooked. Cooking is carried out as for grits, and starch may be mixed with them, as well as with a source of α-amylase to ensure liquefaction. Refined starches contain very little nitrogenous material, and on a dry-weight basis have extracts of 103–105% because of hydrolysis gains (Table 8.3) or 350–380 litre°/kg; 20°C (120–130 lb/Qr.). Since starches are wholly converted to soluble materials, they do not cause run-off problems. Furthermore pure starches make no contribution towards beer flavour. However, some maize and wheat starch preparations contain appreciable quantities of lipids (e.g. 0·6%).

8.10 *Sucrose and invert sugars*
Sucrose (α-D-glucopyranosyl-(1→2)-β-D-fructofuranoside) and its hydrolysis product 'invert sugar', are widely used in the brewing industry. Historically lactose (β-D-galactopyranosyl-(1→4)-D-glucopyranose) derived from milk whey was also used for example in the manufacture of milk stouts. Probably over 80 million tonnes of sucrose are produced annually of which about

20 million tonnes enters international commerce (Sturrock in reference [60]). The relative costs are much influenced by trading agreements and import controls. Sucrose is prepared from sugar cane, a giant tropical grass (*Saccharum officinarum* L., and related spp.), and the swollen root of the sugar beet (*Beta vulgaris* L.), an annual which grows in temperate climates. Fully refined white sugar or invert sugar from either source may be used in brewing. However, impure grades of beet sugar have unpleasant flavour characteristics and are not used, whereas incompletely refined preparations of cane sugar have flavours which may actually be desirable.

Sugar cane is a perennial crop which is propagated by cuttings. When fully grown it is harvested, by machine or by hand, and is taken to be processed immediately to minimize sucrose inversion and sugar losses [23, 60]. The cane occurs in long jointed lengths 2–5 cm (1–2 in) in diameter. Processes for preparing raw sugar differ significantly. In outline the cane is successively crushed between rollers and shredded and the expressed juice is collected. The solid residue, bagasse, is washed and pressed to collect the last sugar. It is then burned as fuel, or is used to make hard-board or insulating board or cattle bedding for example. About 10–16% of the cane is soluble solids, and the cane contains 73–76% juice. Of the solids 75–92% are total sugars. Sucrose makes up 70–88% of the total solids [23]. The dark-green acidic (pH 5–5·5) and muddy-looking juice is screened and passed through a cyclone to separate suspended solids, then is mixed with a calculated amount of a lime suspension (calcium hydroxide), and possibly other substances, raising the pH (7·5–8·0), and is heated under pressure. Coagulation of protein and precipitation of calcium phosphates occurs, and the solids separate together with many impurities, including wax. The 'mud' is separated by settling and filtration and is returned to the cane fields. The clear yellow juice contains sucrose, with lesser amounts of invert sugar, minerals and organic substances. It is concentrated from about 12–15% solids to 60–65% solids by heating in multiple-effect vacuum evaporators. The syrup is carefully concentrated further in vacuum pans until the sugar starts to crystallize. Because of the impurities present, crystallization is incomplete. The mixture of crystals suspended in the syrup (molasses) is called massecuite. Solid sugar is collected by centrifugation and after washing is ready to dispatch to the refiners. Residual syrup is concentrated and the sugar is separated twice more, the successive massecuites being designated A, B and C. The later crops of sugar are progressively less fine, and the final crop is redissolved and returned to the start of the process stream. The final C molasses, from which no more sucrose can be crystallized, may be used in the manufacture of rum or as a medium for growing bakers' yeast. The raw sugar leaving the factory is usually subjected to microbiological control. It is about 95–98% sucrose and contains minerals, some insoluble impurities, coloured materials and residual syrup.

It is the usual article of international trade, but is generally refined further before use.

Beet roots are harvested, and the green leaves are removed mechanically. The leaves may be crushed and fed to cattle. The detopped beets are moved quickly to the factory [60, 61]. After washing away adherent soil and removing stones, the beet is cut up into long thin strips (cossettes) about 7·8 × 0·6 × 0·3 cm (approx. 3 × 0·2 × 0·1 in). In highly automated plant the sugar, with many other substances, is extracted by diffusion. The extracting liquid, which may be warmed, and the cossettes move 'counter-current' to each other so that fresh cossettes are extracted by the strongest solution, while those that are nearly exhausted are washed by clean water. The spent sugar beet pulp may be dried, pelleted and used in animal feed. The pH and microbial purity of the juice is controlled. Initially its sugar content is 10–15%. It is finely screened, then is treated first with lime, followed by carbon dioxide. The calcium carbonate formed, together with associated impurities are separated as a sludge. This treatment is repeated several times, sometimes using other coagulating agents. The product is a less-coloured clear juice which is concentrated to 50–65% solids. This in turn is boiled in vacuum pans until crystallization occurs. The solid sugar is collected in centrifuges. The crude beet sugar is too impure to be used directly, being contaminated with nitrogenous and other materials, some of which have unpleasant flavours and foam-destroying properties. It is fully refined by further processing before it leaves the factory, and may be used by brewers.

Brewers rarely use raw cane sugar as such, because of its variable composition. More usually they use partially refined preparations or the fully refined product. Probably residual impurities add to the 'character' as well as the colour of some brewing sugar preparations. Sugar refining takes place in large highly efficient units [23]. The first step is affination, when the molasses clinging to the raw sugar crystals are removed by washing successively with warm syrup and water, the liquids and solids being separated in centrifugal machines. The impure affination syrups are processed separately. The washed sugar is dissolved or 'melted' in water. Insoluble impurities are removed and lime is added to the solution followed by carbon dioxide. The precipitate of calcium carbonate carries down many impurities. Alternatively in 'phosphatation' or 'sulphitation' processes liming and phosphoric acid or sulphur dioxide are used to form adsorbent inorganic precipitates. The amber coloured solutions may be decolourized by treatment with animal charcoal (bone char), and perhaps other adsorbents and mixed-bed ion-exchange resins, before being concentrated under vacuum until sugar crystallization occurs. The sugar is separated in centrifugal machines.

Partially purified raw cane sugar that has been treated with active carbon may be purchased as a brewing syrup. A representative syrup has an extract of 258 litre°/kg (20°C), a colour of 30 EBC units and a nitrogen content of about 0·01%. Although the main sugar is sucrose, the syrup is still 'impure' and only about 95% fermentable. It will not keep because of its low sugar concentration, often about 62% w/w [7]. This cannot be increased further or crystallization will occur at a concentration of about 67% w/w. Solid preparations of sucrose, of varying degrees of purity, may be dissolved in the copper, but with care to ensure that it does not sink to the bottom and burn on to the heating surfaces. It may more safely be dissolved in water in a separate stirred container, before being added to the copper. Granulated sugar, the fully refined material, is better than 99·9% pure, colourless and wholly fermentable. However, it may generate colour during boiling owing to interactions with other wort constituents. Its hot-water extract is high and completely recovered in the brewery (Table 8.2). Some brewers once used candy sugar, crystals of sucrose formed into strings or rods – others preferred coloured less-refined brown preparations.

Invert sugars are prepared by the hydrolysis of sucrose, often raw sugar, and, although this can be done with the enzyme invertase, it is usually carried out under controlled conditions with dilute sulphuric acid or hydrochloric acid. The product is neutralized with calcium carbonate or soda ash, and, after partial refining by a charcoal treatment, is concentrated to a syrup under vacuum. It may be supplied as a syrup of up to 83% w/w solids, which must be stored warm at 40–50°C (104–122°F) to prevent crystallization [6]. Alternatively it may be seeded with solid invert sugar, and cooled, and the solidified material, which still contains water (about 17%), supplied to the brewer as blocks (50 kg; approx. 1 cwt). Extract yields are good (Table 8.2). Invert sugars may contain 0·01–0·04% nitrogen, and have colours in the range 30–500 EBC units (in 25 mm cells, 100% sample). The more deeply coloured samples have the higher nitrogen contents and are less fermentable, e.g. 91% against 95% for the paler types [7]. As the main sugars are all wholly fermentable, it must be supposed that the reason fermentabilities are not 100% is that impurities and non-fermentable products formed during acid inversion are responsible. The colour is partly derived from the original raw sucrose and partly from side reactions which occur during acid inversion, especially involving fructose, which produce such compounds as hydroxymethylfurfural, laevulinic acid and melanoidin-like condensation products between the sugars and nitrogenous materials. It is these by-products which give 'characters' to invert sugars. Pure sucrose and pure invert sugar are equally well utilized by yeasts with no differential effects on beer flavour. Thus sucrose, being cheaper, is the preferred source of extract, but invert sugars may confer particular flavours and characters as well as colour to the final product. Typical sugar primings are a blend of sucrose with invert sugar (or partially inverted sucrose), and sometimes some starch-derived sugars (see below). A ratio of about 55 invert sugar: 45 sucrose is usual. Sugars do not crystallize readily from this mixture, which is supplied as a concentrated syrup with an extended storage life, even at room temperature, and extracts of 290–340 litre°/kg (20°C). Primings can be provided with a variety of colours, nitrogen contents and fermentabilities resembling those of invert sugars. In addition, they are chosen for their special flavours, 'bland', sweet (various intensities), 'burnt note', 'raw sugar', and so on.

8.11 Sugars and syrups from purified starch

Syrups and sugars from the partial or essentially complete hydrolysis of refined maize starch contain no significant quantities of nitrogenous materials or other yeast nutrients. Thus, like sucrose-based preparations, they are a source of extract, but are wort extenders rather than wort replacements. During their preparation contaminating lipids may have to be removed [9]. In general, the starch, prepared in a wet-milling process (see above), is used as a slurry (about 40% solids) in water, so the drying stages are omitted. Hydrolysis of the starch is carried out in two stages. In the first stage the starch slurry is 'solubilized', 'pasted,' 'cooked,' 'liquefied' or 'dextrinized.' The granules are wholly degraded, and the products of partial hydrolysis are soluble, have a low viscosity and do not set to a gel or retrograde (come out of solution) when the mixture is cooled. In the second stage the dextrins are converted to the desired spectrum of sugars. The two hydrolytic stages may be achieved using dilute mineral acid throughout (acid/acid procedures), or an acid-catalysed liquefaction stage followed by an enzymic saccharification stage (acid/enzyme processes), or both stages may be carried out using enzymes (enzyme/enzyme processes). A wide range of products may be prepared, and their compositions accurately controlled.

TABLE 8.4

Composition of glucose syrups derived from starches by various routes
{Data from HOWLING (1979) [9]}

Syrup type	Acid conversion				Acid/enzyme				Enzyme/enzyme	
					High maltose		High conversion			
Dextrose equivalent ...	30	34–36	42–43	55	42	48	63	70	42	65
Dextrose (glucose)	10	13·5	19	31	6	9	37	43	2·5	34
Maltose	9	11·5	14	18	45	52	32	30	56	47
Maltotriose	10	10	12	13	12	15	11	7	16	3
Maltotetraose	8	9	10	10	3	2	4	5	0·7	2
Maltopentaose	7	8	8	7	2	2	4	3	0·4	1·5
Maltohexaose	6	6	7	5	2	2	3	2	0·7	1
Maltoheptaose	5	5·5	5	4	} 30	} 18	} 9	} 10	} 23·7	} 11·5
Higher sugars	45	36·5	25	12						

Wholly acid conversions may be used to prepare 'confectioners' glucoses' with a range of compositions and fermentabilities (Table 8.4), e.g. with hydrochloric acid subsequently neutralized with soda ash filtered, and treated with carbon or ion exchange resins [7, 9, 59]. These have good extracts (Table 8.2), are colourless and may have fermentabilities of 30–50%. If acid hydrolysis is extended, coloured by-products, bitter-flavoured substances and reversion products, such as polymeric materials and gentiobiose, may be formed [1]. The product, termed 'chip-sugar' may be supplied as blocks, or chips, or as a syrup blended with other materials. Extract may be 319 litre°/kg (20°C), and the material is about 82% fermentable, and has a colour of 200–500 EBC units (25 mm cell, 100% sample). The bitter and coloured components may be partly removed by charcoal treatments. Chip-sugars confer a characteristic 'dry' flavour to beer.

Products prepared by acid/enzyme and enzyme/enzyme treatments may be provided as solids or syrups and can have their compositions regulated over a wider range [7] (Table 8.4). Some syrups must be held in the brewery at about 50°C (122°F) to prevent solidification and may be supplied at 60–71°C (140–160°F). In addition, enzyme-converted materials are less rich in inorganic salts. Fermentabilities of 75–85% are common, but lower and higher values can be achieved. The products are usually colourless, but their colours can be adjusted. In the UK these materials are usually made from maize starch, but wheat-starch-derived materials (acid/enzyme conversion) are also available.

Enzymic liquefaction of a starch slurry is normally achieved using a preparation of bacterial α-amylase stabilized by additions of calcium salts, and progressively heating the mixture up to 90°C (194°F). Alternatively the starch in a slurry can be liquefied continuously in the tube converter [working at about 140°C (294°F)], using an exceptionally heat-stable α-amylase from a thermophilic bacterium. Oil and coagulated materials may be removed at this stage, and carbon adsorption treatments may also be used. Further conversion may be obtained, after cooling and pH adjustment, by the use of fungal amylases, highly diastatic malt, or β-amylases from wheat or soya (Table 8.4). Highly fermentable preparations of nearly pure glucose (95% 'dextrose') are obtained using amyloglucosidase (glucamylase) with or without α-amylase. Reaction may proceed for 2–3 days at 55–60°C (131–140°F). A product containing 92–93% glucose is obtained if an acid/enzyme process is used. High-glucose syrups must be kept below about 35% solids, or crystallization will occur. However, with at least some varieties of brewing yeasts, high levels of glucose in the wort suppress the fermentation of higher saccharides, causing 'hanging fermentations' and incomplete attenuation [62]. To reduce this problem 'high-maltose' syrups (Table 8.4) may be prepared using plant β-amylases (malt, soya flour), fungal enzymes or *Bacillus polymyxa* β-amylase to convert the starch to about 50% maltose. If, in addition, a debranching enzyme, such as pullulanase, capable of breaking the α-(1→6) linkages in the dextrins, is added with the β-amylase, a syrup with a DE value of 70–80% may be obtained [9].

Before use or during manufacture, syrups must be refined to remove residual lipids (as much as 0·5–0·6% in maize and wheat starches), mineral salts and other unwanted materials. They may be centrifuged or filtered, treated with active carbon and perhaps mixed-bed ion-exchange resins. If desired, anti-foam agents may be added or sulphur dioxide may be used as a preservative [9]. Sulphur dioxide is commonly present in many syrups (2–40 p.p.m.), but excessive levels may cause problems, such as reduced head-stability in beer, so brewers specify the maximum levels allowed.

Recently starch-derived 'high-fructose' syrups have become available [9, 14, 63]. Using α-amylase and amyloglucosidase, starch is converted to a syrup having a DE of 98%, about 96% glucose. The syrup is purified and passed through columns of insolubilized microbial 'ketol isomerase', which catalyses the reversible isomerization of glucose to fructose. After carbon treatment, the syrup is concentrated to 70–72% solids, and is 99% fermentable. Typically its composition (percentage dry-weight basis) would be: fructose, 42; glucose, 54; maltose, 2; maltotriose, 1; higher saccharides, 1. The potential priming value of this material to brewers resides in its sweetness. Fructose is sweeter than sucrose, which in turn is sweeter than glucose. The syrup is about as sweet as sucrose, but is cheaper, and could find a use in some special beers. However, in the EEC production is limited by legislation.

Other sorts of syrups have been prepared experimentally, e.g. by using transfructosylases, dextrins with terminal fructose residues have been prepared. If they have useful properties, some use may be found for them in brewing.

8.12 *Malt extracts and wort replacement syrups*

Malt extracts are prepared by mashing malt, collecting the sweet wort in an essentially normal fashion, then concentrating it to a syrup by evaporation under reduced pressure. In the past, some 'diastatic extracts', rich in malt enzymes, were added to the mash tun at the rate of about 1 part to 36 parts of grist (w/w) to supplement the enzymes there, or were added to cooled wort (despite the risk of microbiological contamination) to adjust its fermentability. Non-diastatic malt extracts, which are dissolved in the copper on boiling, represent a convenient source of extract that is independent of the mashing process. Diastatic malt extracts are rarely, if ever, used in brewing in the UK at present, but limited quantities of other malt extracts are used (despite being an expensive source of extract), and concentrated syrups prepared from various grists and converted in a variety of ways are also used (see below). In addition, hopped wort extracts, as well as being used by some 'home brewers', have been produced by the 'Conbrew system' in which the syrup (minimum of 75% solids) was produced in Newfoundland and shipped to the Bahamas or other small isolated markets in 19 hl (500 US gal) collapsible rubber containers lined with food-grade rubber. On arrival it was diluted and fermented [2, 3]. For diastatic malt extracts, processing is carried out to maximize enzyme survival [17, 64]. A high-nitrogen (1·7–2·0%) highly diastatic malt is ground and mashed at less than 49°C (120°F). After a stand the first worts are collected. The residual solids are remashed at about 68°C (154°F) to complete starch conversion and recover the extract. Spargings and last runnings may be used in the mashing-in liquor of the next mash to maintain high wort concentration, and so minimize the amount of water to be evaporated. The combined worts are concentrated at the lowest possible temperatures in multiple-effect vacuum evaporators. Initial temperatures are 32–35°C (90–95°F), but the temperature rises to 43–46°C (109–115°F) as the extract concentration rises towards 80% solids.

Malt extracts and syrups represent compact forms of extract, 3·7 units (volume) being approximately equivalent to 10 units (volume) of malt. An extract of 302 litre°/kg (20°C) would be usual. They contain 75–82% solids with an SG of 1400–1450. Higher concentrations are used in preparations intended for the tropics. The high concentrations are necessary to prevent microbial growth. The contaminants most likely to occur are osmophilic yeasts. The extracts must be handled warm, e.g. 49°C (120°F), to reduce their viscosities, and this can cause changes in flavour and increases in colour. At ambient temperatures they are stable indefinitely provided that local dilution does not occur, e.g. as the result of condensate from the structures above the syrup dripping or running back. To overcome condensation problems the surface of a syrup and the upper parts of its container may be ventilated with warm filtered air. Malt extracts and syrups may be delivered to breweries in bulk in tankers, or in smaller quantities in metal drums. Malt extracts may be prepared to meet a wide range of specifications: colour, 3–520 units EBC (3–50 EBC units with diastatic malt extracts); enzyme activity, zero to high levels of amylases and proteases; DP 0–400°L. Solids may have maltose equivalents of 58–96%, glucose equivalents of 45–62%, and 'crude protein' contents of 4–9%. Fermentabilities of about 65–93% can be produced. Pale extracts have comparatively bland flavours, whereas some dark melanoidin-rich preparations are fine, luscious and rich to taste. It is technically feasible to dry the syrups to flaky powders on heated belts moving *in vacuo*, but such products are not used in commercial brewing.

Malt extracts are expensive both because malt is expensive and so are the operations for concentrating the wort to a syrup. Numerous syrupy products have now been produced from raw cereal grains converted with enzymes

TABLE 8.5

Compositions of some barley syrups compared with a conventional wort
{From ROBERTS and RAINBOW (1971) [39]}

	Carbohydrate components (expressed as % of total carbohydrate)						Nitrogen fractions* (mg/litre at SG 1038·8)		
	Fructose	Glucose	Sucrose	Maltose	Maltotriose	Dextrins†	Total N	α-Amino N	HMW N
Pale ale wort	1–4	8–11	2–5	45–49	13–15	21–25	650–750	140–190	155–225
Standard syrup (A)	1–5	7–19	1–7	41–51	7–15	19–27	618–965	123–232	145–328
Low-dextrin syrup (A)	2–3	9	2–4	54–56	13–14	17–18	913–916	222–232	266–270
Standard syrup (B)	2–5	9–14	1–2	36–58	7–17	17–31	523–805	126–190	156–242
Low-dextrin syrup (B)	2–3	9–10	2–6	47–57	10–15	16–22	623–745	112–179	201–236

* Nitrogen fractions concentrations in solution in extracts diluted to an SG of 1038·8, approximately 10% solids; HMW N, high-molecular-weight (> 5000) nitrogen containing fractions.
† Dextrins, the unfermentable carbohydrate fraction.

from microbes, malt or green malt or, at least experimentally, with enzymes after a preliminary acid hydrolysis. These approaches can produce the wide range of sugar spectra that are encountered in products made from refined starches. However, these cereal-based products are intended for use as wort replacements rather than merely carbohydrate extenders and so their nitrogen composition is also of importance. Extracts can be prepared that closely match malt worts in composition (Table 8.5) [1, 39, 65]. Compared with normal worts from malt mashes, proline levels are low, and the amino acid spectra of cereal syrups are not exactly the same [39]. It seems that the syrups do not lack essential yeast nutrients, but they may be sufficiently poor in zinc for this to limit fermentations. Syrups are used as copper adjuncts to increase the brewing capacity of plants, produce worts for high-gravity brewing, and adjust wort composition. Probably whole wort replacement is technically feasible. Brewers buy syrups from outside suppliers, and demand a high degree of consistency in their compositions [39]. In principle, syrups could be made in the brewery, and the technology used to convert the cereals is largely applicable to adjunct conversions. 'Liquid malt', legally a malt extract (and so exempted from the restrictions applied to the use of raw-grain adjuncts in some countries) was a product produced exclusively from green unkilned malt [45, 66]. The difficult milling and wort-separation problems were overcome, in the latter case using bowl centrifuges. The syrup was low in anthocyanogens (see Chapter 9). The unpleasant 'green grain' flavour was removed during the evaporative concentration steps. The process was attractive in that, as the green malt was unkilned, it saved the cost of this expensive step, which tended to offset the cost of syrup concentration. Barley may be ground wet or dry in various ways for making 'barley syrups,' as is the case when the grain is used as a mash-tun adjunct [1, 13, 39–41, 45, 65–67]. In addition the grain has been 'dehusked' by impacting in a pin-mill and the dehusked portion coarsely ground between fluted rolls before being recombined with the husks [67]. A coarse grind is needed if wort is to be separated in a lauter tun, but finer grinds can be used if bowl centrifuges, filter presses or continuous-belt vacuum filters are to be used. By using weak last-runnings as the foundation water for the following mash, high-gravity worts are produced. Typically a grist containing 90–95% ground barley, 5–10% highly diastatic malt and a mixture of bacterial and perhaps fungal enzymes would be mashed in and subjected to a programmed rising temperature cycle [38, 40, 41]. The enzymes from the malt are complemented with the β-amylase of the raw barley and extra β-glucanase, protease and α-amylase (and perhaps other starch-degrading enzymes) in the mixture of microbial enzymes. Stands occur at 50–55°C (122–131°F) to allow glucanolysis and proteolysis, at 63–65°C (145–149°F) for the major part of amylolysis and saccharification and at 75°C (169°F) to ensure complete starch solubilization. Alternatively an all-barley mash, supplemented by microbial enzymes including an active α-amylase, is subjected to a temperature-programmed mash rising to 85°C (185°F). This temperature is held for a period until all the starch is liquefied. At this stage adequate glucanolysis and proteolysis has occurred, and the starch is liquefied, but the mixture of starch-breakdown products is incorrect, and the fermentability is too low. Then the mixture is cooled to 62–65°C (144–149°F) and is mixed with malt and/or fungal enzymes to complete the saccharification stages, which may last for several hours. To achieve adequate conversions, mashing cycles may last for 6–12 hr. The worts are vacuum-concentrated, and the syrups have solids contents of 78–83% and an SG of 1400–1430. 'Crude protein' contents may be 3–5%, and about 85% of the solids are carbohydrates (compared with about 98% for syrups from maize starch). Colours may be adjusted to a wide range of values. Fermentabilities are usually arranged to be 65–75% although these too can be adjusted. Clearly the potential range of cereal-based adjuncts and syrups is very large, and only a proportion of those possible are commercially available.

8.13 *Industrial enzymes*

'Industrial enzymes' is here taken to mean those enzymes not derived from malt or raw cereals which are used in brewing. They are used to make good enzyme deficiencies in grists or to adjust wort or beet compositions, or in the manufacture of brewing adjuncts. Limited quantities of enzyme preparations derived from plants have been or are used, such as the proteolytic enzymes papain (papaya latex), bromelain (pineapple) and ficin (fig) used to 'chill-proof' beer, by degrading high-molecular-weight haze precursors, or the β-amylase from soya beans used to adjust fermentabilities by increasing the maltose contents of syrups and worts. Rarely small quantities of enzymes from animal sources, such as the proteolytic enzyme pepsin from pig gastric juice, have also been employed. However, most industrial enzymes are prepared from specially cultured micro-organisms, usually fungi or bacteria. Historically these were grown on solid media, cooked cereals or bran, as *Aspergillus oryzae* is grown on steamed rice to make the traditional 'koji' in Japan. Most industrial enzymes are now made in deep stirred liquid cultures. They may be supplied in 'liquid form' in solution, or in the 'solid form' as dry powders, often mixed with 'extenders' or carrier substances.

The evaluation of industrial enzymes is hampered by a lack of generally agreed standards of purity and methods for assessing their activities. However, to be acceptable the enzymes must be prepared under hygienic conditions from microbes that do not produce mycotoxins or other noxious materials. It is thought that most fungal enzymes are derived from *Aspergillus* spp., *Asp. flavus*, *Asp. oryzae* and *Asp. niger*. However, enzymes from *Trichoderma viride* and *Rhizopus* spp. have been used, at least experimentally. Bacterial enzymes probably come mainly from *Bacillus* spp. such as *B. subtilis* and

B. stearothermophilus. The preparations must be adequately stable and consistent, i.e. activities of all the significant components must be the same in successive batches. Furthermore the enzymes are not pure, although they may be partially purified. They may contain residual materials from the culture medium (which must be unobjectionable). Also present may be a range of other enzymes, whose presence may be advantageous (β-glucanase in bacterial α-amylase) or potentially disadvantageous (e.g. protease). The latter can lead to excessive levels of wort nitrogen, or to too extreme degradation of the high-molecular-weight proteinaceous materials in the wort resulting in a loss of head retention or other alterations in beer character. By selecting microbial strains and culturing under different conditions, the proportions of enzymes occurring in the cultures can be varied widely. Furthermore the preparations must not represent a microbiological hazard to the brewery, and must not contain viable organisms or spores. On the whole, brewers prefer to buy standardized mixtures of enzymes from specialist suppliers, who, in many instances, do not state their origins, but which often come from *Bacillus* spp. The preparations of most interest to brewers are those containing hydrolytic enzymes which degrade starch and dextrins, β-glucans, proteins and possibly phosphates including phytate. However, lipases may be important in some instances (the levels of free fatty acids in worts should not be excessive), while nucleic acids should be degraded. Furthermore other enzymes may be used to regulate beer properties, e.g. glucose oxidase may be used to 'scavenge' oxygen and diacetyl reductase may be used to change vicinal diketones into diols in beer, with an improvement in flavour. The wide and increasing range of enzymes known to be produced by microbes and improved ways of manufacturing them in increased yields will mean that enzymes with new and unfamiliar properties will become available [35, 38, 43, 68–72]. Some properties of currently available enzymes are summarized below.

The starch- and dextrin-degrading enzymes of bacteria and fungi have been extensively studied. In general, fungal α-amylases have convenient pH optima, around 5·5, but are inconveniently heat-labile [e.g. they are inactivated at 60°C (140°F)] for use in mashing, although a 'high-maltose'-producing fungal α-amylase has been used to 'prime' beer and its heat-lability is an advantage, since it is readily inactivated during pasteurization. A common type of bacterial α-amylase has a high pH optimum of 6–7, but it has a useful level of activity of pH 5–6 and, in the presence of calcium and chloride ions, it remains active for a period at 75–80°C (167–176°F), but is destroyed on boiling. It is much used in cereal cooking and starch liquefaction. Other heat-stable α-amylases will, under correct conditions, remain active during boiling and are efficient in liquefying starch slurries for example. However, their resistance to boiling means that they could survive into finished beer.

Commercial preparations of bacterial α-amylases are commonly accompanied by β-glucanase and proteases. Selected mixed preparations may be used to convert barley grists in syrup manufacture, supplement mashes made with undermodified malts and malt–adjunct mashes [40, 41]. Debranching enzymes such as isoamylases and pullulanase, which are able to hydrolyse α-(1→6)-interchain linkages in branched starch-derived dextrins are potentially of value in mashing and particularly in adjusting the fermentability of syrups or worts when used in conjunction with β-amylase or amyloglucosidase [9, 69, 73, 74]. Pullulanase, probably prepared from *Klebsiella* (*Aerobacter*) *aerogenes*, is commercially available and approved for use. Because of its thermolability it must be used at 45–50°C (113–122°F). Wort of normal fermentability has been prepared from a mash of 90% raw barley and 10% malt using an unusual mashing cycle in which the temperature was reduced after liquefaction, and pullulanase was added and allowed to act for 2 hr at 45°C (113°F) [73]. This enzyme may be destroyed at will by pasteurization and may be used to degrade limit dextrins in beers to fermentable sugars and so replace the use of priming sugars. Pullulanase seems preferable to amyloglucosidase (which is too heat-stable, and continues to act after some pasteurization treatments, causing the beer to become progressively sweeter) and fungal α-amylase. Some fungal preparations have 'saccharifying' activity that is distinct from α-amylase or amyloglucosidase, and preferentially produces maltose. Some bacteria (e.g. *Bacillus polymyxa*) produce an enzyme which, like plant β-amylase, converts starch to maltose. Highly attenuated light (lite; dietetic; diabetic; low-carbohydrate) beers are produced from worts treated with carbohydrases to degrade residual dextrins to fermentable sugars. Amyloglucosidase or a mixture of β-amylase with pullulanase can be used successfully [74].

Amyloglucosidase (AMG; glucamylase) preparations are usually prepared from *Rhizopus* spp. and *Aspergillus niger*, and, although they are optimally active at pH 4·4, they are usefully active in brewing pH values, e.g. 5·5, and at 60–65°C (140–149°F). The enzyme preparations, which often contain α-amylase, will eventually degrade liquefied starch nearly completely to glucose. However, the enzymes are slow to cleave α-[1→6] links and so degradation is accelerated by the addition of debranching enzymes. Some preparations contain a transglucosylase which is disadvantageous, as it produces isomaltose, panose and perhaps other unwanted and unfermentable disaccharides. In addition, some preparations contain protease, lipase, β-glucanase and cellulase activities. Amyloglucosidase has limited value when added to the mash tun, but it may be used to obtain highly attenuated beers (see above), and has been tested as a replacement for priming sugars (see above). It is indispensible for the production of high-glucose and high-fructose syrups. It seems attractive to use columns of immobilized amyloglucosidase (and some other enzymes) to treat beers, but in fact the need for extended residence times and problems of maintaining sterility have restricted the acceptance of this technique.

β-Glucanases may be prepared from bacteria or fungi. The bacterial enzyme is more heat-stable and is the more useful [43, 75–77]. Even so when raw barley is the substrate the enzyme may need 60 min at 45–50°C (113–122°F) to reduce wort viscosity and facilitate its recovery. Indeed the enzyme may also degrade insoluble barley β-glucans, which contribute to 'bed-stickiness' in mashes and retard wort separation [54].

Proteases from bacteria and fungi are available with a range of properties. *Bacillus* spp. tend to produce proteinases with neutral and alkaline pH optima of which the former is successful in solubilizing barley proteins during mashing, and in making barley syrups. Sufficient α-amino nitrogen, including compounds assimilable by yeast, are produced to support yeast growth. Bacterial neutral proteinase is more heat-labile than α-amylase, so that it requires a temperature stand at about 50°C (122°F) in the mashing programme.

8.14 *Future possibilities*

Adjuncts and industrial enzymes will undoubtedly continue to play an important role in brewing practice, despite the fact that the quantities of adjuncts employed are often limited by legislation. The materials chosen for use as adjuncts will depend on local supplies and economics, yet the preparation of adjuncts may be refined by careful selection of cereal varieties, improvements in the details of milling and other processing, and perhaps by the addition of microbial or other enzymes the activities of which will be utilized later, in the brewery. It is also likely that a wider range of industrial enzymes will become available, and the composition of the preparations supplied will be more closely defined. Among the factors limiting the use of several known enzymes of potential value are patent restrictions, food laws, high costs, poor stability, the need for toxic metallic ions as co-factors and inconvenient pH optima. In addition, rather than relying on outside suppliers, brewers may in future alter the pattern of adjunct production and utilization. For example, by deciding to make 'barley mashes' on site, and then blending the 'worts' with their all-malt worts the need for the expensive concentration steps, used in the production of syrups, would be avoided.

REFERENCES

[1] IMRIE, F. K. and MARTIN, F. J. (1969). *Brewer's Guild J.*, **55**, 252.
[2] PAPADOPOULOS, A. (1964). *Brewer's J.*, **100**, 617.
[3] RAMSAY, G. (1964). *Brewer's J.*, **100**, 786.
[4] POLLOCK, J. R. A. and WEIR, M. J. (1976). *MBAA Tech. Q.*, **13**, 22.

[5] MOLL, M. and DUTEURTRE, B. (1976). *MBAA Tech. Q.*, **13**, 26.
[6] MAIDEN, A. M. (1978). *Pauls and Whites Brewing Room Book*, Pauls and Whites, Ipswich, p. 64.
[7] ANON (1980). *Pauls and Whites Brewing Room Book, 1980 and 1981*, Pauls and Whites, Ipswich.
[8] HARRIS, J. O. (1976). *J. Inst. Brewing*, **82**, 27.
[9] HOWLING, D. (1979). *Sugar: Science and Technology* (eds. G. G. BIRCH, and K. J. PARKER), Applied Science Publishers, London, p. 259.
[10] CANALES, A. M. (1979). *Brewing Science*, Vol. 1 (ed. J. R. A. POLLOCK), Academic Press, London, p. 233.
[11] MAIDEN, A. M. (1975). *Brewing Distilling Int.*, **5**(7), 36.
[12] MARTIN, P. A. (1978-Aug.). *Brewer's Guard.*, **107**, 29.
[13] MAULE, A. P. and GREENSHIELDS, R. N. (1970). *Process Biochem.*, **5**(2), 39.
[14] SMITH, J. BARRY, (1977-Feb.), *The Brewer*, **63**(2), 50.
[15] DE CLERCK, J. (1957). *A Textbook of Brewing*, Vol. 1 (translated by K. BARTON-WRIGHT), Chapman & Hall, London.
[16] STEWART, E. D. and HAHN, R. H. (1965-July). *Am. Brewer*, p. 21.
[17] BRIGGS, D. E. (1978). *Barley*, Chapman & Hall, London, 612pp.
[18] RADLEY, J. A. (ed.) (1968). *Starch and its Derivatives*, Chapman & Hall, London.
[19] MAIDEN, A. M. (1971). *The Brewer*, **57**, 76.
[20] Analysis Committee of the European Brewery Convention (1975). *Analytica – EBC*, Schweizer Brauerei-Rundschau, Zurich.
[21] Institute of Brewing Analysis Committee (1977). *Recommended Methods of Analysis*, The Institute of Brewing, London.
[22] Technical and Editorial Committees of the American Society of Brewing Chemists (1976). *Methods of Analysis of the American Society of Brewing Chemists* (7th edn., revised), A.S.B.C., St Paul, Minnesota.
[23] MEADE, G. P. and CHEN, J. C. P. (1977). *Sugar Cane Handbook* (10th edn.), John Wiley and Sons, New York, 947pp.
[24] COLLIER, J. A. (1975-Aug.), *The Brewer*, **61**, 350.
[25] LLOYD, W. J. W. (1978-Mar.). *The Brewer*, **64**, 84.
[26] CANALES, A. M. and SIERRA, J. A. (1976). *MBAA Tech. Q.* **13**, 114.
[27] COORS, J. (1976). *MBAA Tech. Q.*, **13**, 117.
[28] KOSZYK, P. F. and LEWIS, M. J. (1977). *J. Am. Soc. Brewing Chem.*, **35**(2), 77.
[29] SCULLY, P. A. S. and LLOYD, M. J. (1965). *J. Inst. Brewing*, **71**, 156.
[30] SCULLY, P. A. S. and LLOYD, M. J. (1967). *Brewer's Guild J.*, **53**, 29; 559.
[31] SCHERRER, A., PFENNINGER, H. B. and WIEG, A. J. (1973). *Brauwissenschaft*, **26**(4), 101.
[32] SCHERRER, A., PFENNINGER, H. B., KORNICKER, W. and MULLER, P. W. (1973). *Brauwissenschaft*, **26**(5), 129.
[33] GOERING, K. J. and ESLICK, R. F. (1976). *Cereal Chem.* **53**, 174.
[34] BUTTON, A. H. and PALMER, J. R. (1974). *J. Inst. Brewing*, **80**, 206.
[35] IMRIE, F. K. (1968-Apr.). *Process Biochem.*, **3**(4), 21.
[36] MARTIN, F. J. (1967). *Brewer's Guild J.*, **53**, 354.
[37] PFENNINGER, H. B., SCHUR, F. and WIEG, A. J. (1971). *Proc. Eur. Brewery Conv. Congr. 13th, Estoril*, p. 171.
[38] RAO, B. A. S. and NARASIMHAM, V. V. L. (1976). *J. Food Sci. Technol.*, **13**, 119.
[39] ROBERTS, R. H. and RAINBOW, C. (1971). *MBAA Tech. Q.* **8**, 1.

[40] WEIG, A. J. (1970–Aug.). *Process Biochem.*, **5**(8), 46.
[41] WIEG, A. J. (1973), *MBAA Tech. Q.*, **10**, 79.
[42] WITT, P. R. (1970–Feb.). *Brewer's Digest*, **45**(2), 48.
[43] SORENSEN, S. A. (1970). *Process Biochem.*, **5**(4), 60.
[44] BRITNELL, J. (1973). *MBAA Tech. Q.*, **10**, 176.
[45] COLLIER, J. A. (1970–May). *Brewer's Guild J.*, **56**, 242.
[46] MUNCK, L. and LORENZEN, K. (1977). *Proc. Eur. Brewery Conv. Congr. 16th Amsterdam*, p. 369.
[47] MAIDEN, A. M. (1966–Oct.), *Brewer's Guard.*, **95**, (10, Suppl.), 59.
[48] KENT, N. L. (1978). *Technology of Cereals* (2nd edn.), Pergamon Press, Oxford.
[49] HIND, H. L. (1950). *Brewing Science and Practice* Vol. 1, Chapman & Hall, London.
[50] VOSE, J. R. and YOUNGS, C. G. (1978). *Cereal Chem.*, **55**, 280.
[51] RUMPF, H. and KAISER, F. (1952). *Chemie-Ingenieur-Technik.*, **24**(3), 129.
[52] EGGITT, P. W. R. (1964–Nov.). *Brewer's Guild J.*, **50**, 533.
[53] HOUGH, J. S., WADESON, A. and DANIELS, N. W. R. (1976-Mar.). *Brewer's Guard.*, **105**(3), 38.
[54] WEBSTER, R. (1978–July). *Brewer's Guard.*, **107**(7), 51.
[55] ANDERSON, F. B. (1966). *J. Inst. Brewing*, **72**, 384.
[56] LEACH, A. A. (1968). *J. Inst. Brewing*, **74**, 183.
[57] GOERING, K. J. and IMSANDE, J. D. (1960). *Agric. Food Chem.*, **8**(5), 368.
[58] KLOPPER, W. J. (1973). *The Brewer*, **59**, 395.
[59] SWAIN, E. F. (1976). *MBAA Tech. Q.*, **13**, 108.
[60] BIRCH, G. G. and PARKER, K. J. (eds.) (1979). *Sugar: Science and Technology*, Applied Science Publishers, London, 475pp.
[61] MCGINNIS, R. A. (1971). *Sugar Beet Technology* (2nd edn.). Sugar Beet Development Foundation, Fort Collins, Colorado.
[62] PFISTERER, E. A., GARRISON, I. F. and MCKEE, R. A. (1978). *MBAA Tech. Q.*, **15**, 59.
[63] CASEY, J. P. (1977). *Cereal Foods World*, **22**(2), 48.
[64] HALL, J. A. D. (1978–July). *Process Biochem.*, **13**(7), 20.
[65] MACWILLIAM, I. C. (1971). *J. Inst. Brewing*, **77**, 295.
[66] GRIFFIN, O. T., COLLIER, J. A. and SHIELDS, P. D. (1968). *J. Inst. Brewing*, **74**, 154.
[67] WICKENS, D. (1973–Nov.). *Brewer's Guard.*, **102**(11), 47.
[68] ASCHENGREEN, N. H. (1969). *Process Biochem.*, **4**(8), 23.
[69] ENEVOLDSEN, B. S. (1975). *Proc. Eur. Brewery Conv. Congr. 15th, Nice*, p. 683.
[70] FOGARTY, W. M., GRIFFIN, P. J. and JOYCE, A. M. (1974). *Process Biochem.*, **9**(6), 11; (7), 29.
[71] MARSHALL, J. J. (1975). *Stärke*, **27**(11), 377.
[72] NIELSEN, E. BJERL, (1971). *Proc. Eur. Brewery Conv. Congr. 13th, Estoril*, p. 149.
[73] ENEVOLDSEN, B. S. (1970). *J. Inst. Brewing*, **76**, 546.
[74] WILLOX, I. C., RADER, S. R., RIOLO, J. M. and STERN, H. (1977). *MBAA Tech. Q.*, **14**, 105.
[75] ERDAL, K. and GJERTSEN, P. (1971). *Proc. Eur. Brewery Conv. Congr. 13th, Estoril*, p. 49.
[76] ENARI, T.-M. and MARKKANEN, P. H. (1975). *Proc. Am. Soc. Brewing Chem.*, **33**(1), 13.
[77] ENKENLUND, J. (1972). *Process Biochem.*, **7**(8), 27.

Chapter 9

THE CHEMISTRY AND BIOCHEMISTRY OF MASHING

9.1 Introduction

Brewers are concerned to obtain raw materials that are as uniform as possible, and of the best quality for producing their beer at the lowest cost. To some extent the need to produce an identical beer from each brew has resulted in an unwillingness to test new raw materials in the grist and also a tendency to set unrealistically inflexible specifications for malt. Malt specifications only rarely take into account the uncertainties involved in the analytical methods used to characterize each sample. Indeed it is not usually known whether or not indistinguishable beers may be produced from grists of slightly different compositions. Malt analysis has been reviewed in detail [1, 2], so only limited aspects of this topic will be discussed here. Some characteristics of adjuncts are noted in Chapter 8.

Brewers now buy malt by weight, although in the past it was sometimes purchased by volume. The malt they buy must meet specification, that is, it must meet or exceed certain analytical limits such as: (*i*) a stated hot-water extract, often of 300 litre°/kg (about 100 brewers' lb/Qr.) of dry malt; (*ii*) an acceptable moisture content; (*iii*) it must give wort of a certain colour; (*iv*) have a certain diastatic power; (*v*) it must not have more than a stated maximum total nitrogen content and yield not more than a certain amount of soluble nitrogen; (*vi*) it sometimes must have an acceptable cold-water extract; (*vii*) it must have a certain minimum β-glucanase activity. Further specifications are (*viii*) limited wort-filtration times and (*ix*) maximum differences of extract between finely and coarsely ground samples. In the UK this measurement may be standardized in the near-future. Further special analyses may be required by particular brewers, for example (*i*) wort fermentability, (*ii*) wort viscosity, (*iii*) wort β-glucan, (*iv*) wort total polyphenols or anthocyanogens, (*v*) malt hardness, and (*vi*) the proportions of different nitrogenous fractions present in a standard wort. Many of the values specified will be traditional for a particular brewery. Others, such as colour, must obviously be correct or at least not too high. In the UK, breweries may or may not use adjuncts in their grists, but in some countries the use of adjuncts is forbidden. Where it is forbidden to use additives in their preparation, malts may be analysed for residues, e.g. unduly elevated levels of bromide would indicate that bromates had been used.

Malt analysis serves two related functions. It provides (*i*) an estimate of the brewing value of the malt and (*ii*) a basis for the evaluation of the malt in relative terms that will serve to guide commercial transactions. Since breweries use different mashing and brewing schedules, no single analytical scheme is likely to predict the value of a malt in all cases. Consequently, alternative methods of malt analysis are frequently proposed. At one time many individual methods of analysis were in use and each was thought to be particularly useful for an individual brewery to measure the extract obtainable from a malt in that brewery. Gradually, and against substantial opposition, 'standard' or 'recommended' methods of analysis came to be employed. The three main systems of analysis that are in use at present, and which are frequently revised and up-dated, are:

(1) *The Recommended Methods of Analysis of Barley, Malt and Adjuncts* The Institute of Brewing (IoB), as used in Britain [3].
(2) *Analytica-EBC* (*European Brewery Convention*) used in Continental Europe and many British lager breweries [4], also called *Congress Methods of Analysis*.
(3) *Methods of Analysis – ASBC* (*American Society of Brewing Chemists*), chiefly used in North America [5].

9.2 Hot-water extract

Extract' means the dissolved materials present in wort, derived from the grist. In addition, salts initially dissolved in the mashing liquor may also be present, but some components, e.g. calcium ions, may be partly retained in the grist, or are precipitated in the hop-boil (see below). The hot-water extract (HWE) is an analytical measure of the quantity of dissolved solids in a sweet wort prepared from malt or other materials by a small-scale mashing process. Small samples of malt or other grist materials are ground in a specified way, mashed with distilled water under rigidly standardized conditions of agitation, time and temperature, made up to a fixed volume or weight and filtered. The specific gravity of the liquid wort so obtained is measured at a fixed temperature, and from it the extract derived from the malt or other grist sample is calculated. Such analytical mashes differ from brewery mashes in a number of important ways. In particular, (*i*) the mashes used in analyses are very dilute compared with those used in brewing practice, so, for example, the 'wort' may have a specific gravity (SG) of about 1029 compared with an SG of 1085 for first brewery runnings. (*ii*) In contrast with brewhouse practice, the mashing liquor is devoid of salts, being distilled water. (*iii*) No attempt is made to adjust the pH of the mash to a favourable value. (*iv*) There is no 'sparging' of the spent grains. The analytical figures do not indicate how the malt will behave in a mixed grist containing various malts and adjuncts nor in important ways do the tests indicate the 'quality' of the derived wort, aspects of which must be checked separately. For example, the quantity of nitrogenous materials and the fermentability of the carbohydrates that are present may be estimated. To minimize some of these problems, certain breweries are equipped with 'micro-breweries' which process several kilograms of malt to produce worts more closely comparable with those obtained from the production plant. The worts can then be characterized as extensively as desired. The proportion of the total nitrogen extracted into the test 'wort' in a laboratory mash is usually less than that extracted into a brewery wort prepared from the same malt.

The extract from a given weight of malt weighs more on a dry-weight basis than the decline in weight of the malt solids, because water is bound during the hydrolytic processes that occur during mashing. On the other hand, the extract obtained is inversely related to the quantity of 'draff' or 'spent grains', the insoluble residue from the malt remaining at the end of mashing. The quantity of material in the wort is generally calculated from the specific gravity of the wort measured at a particular temperature. The specific gravity is usually determined directly by weighing, using a specific gravity bottle or a pycnometer. Rapid less-accurate determinations may be made by weighing objects suspended in air, in water and in wort, or by using hydrometers, or by floating drops of wort in density gradients made with organic liquids which are immiscible with water. Specific gravity can also be determined indirectly from refractive index measurements, provided that the refractometer used is very precise, its temperature is controlled to better than 0.1°C, and that it has been 'calibrated' against a series of worts of known specific gravity, made from closely similar grists [6].

In the pre-1977 British analytical system, extract was expressed in terms of brewers' pounds per quarter of malt [lb/Qr., when Qr. (malt) = 336 lb]. This unit originated from the custom of determining the weight of the contents of a barrel (36 imperial gallons) of wort, and subtracting the weight of one barrel of water (360 lb). Thus the extract, in brewers' pounds, was the excess weight of a barrel of wort above 360 lb at 15·5°C (60°F), and:

$$\text{the specific gravity of a wort} = \frac{\text{Excess brewers' lb} + 360}{360}$$

In brewing usage the specific gravity (sg) is multiplied by 1000 so that the SG of water is 1000·0 in contrast with the more familiar value of (sg) = 1·0000. Therefore:

Extract, brewers' lb = [Specific gravity (SG) − 1000] × 0·36
= Excess specific gravity × 0.36

In the standard mash (IoB), ground malt (50 g) is mixed with water at 65°C (approx. 149°F) and is shaken every 10 min. After 1 hr the mixture is quickly cooled to 20°C (68°F), and the volume is made to 515 ml. The mixture is then filtered. Originally, when mashing flasks calibrated to 515 cc. were introduced, it was assumed that the volume of the draff from 50 g of malt was 15 cm³, so that 50 g of malt gave rise to 500 ml of wort, that is a '10% wort'. Thus the hot-water extract of the malt, in brewers' lb/Qr. (336 lb) of malt was taken as:

Apparent HWE (lb/Qr.)
$$= \frac{\text{Excess SG of 10\% wort (at 15·5°C)} \times 10 \times 0·36 \times 336}{360}$$
$$= (SG - 1000.0) \times 3·36$$

Even when this formula was in use, it was realized that the draff occupied a varying volume that was inversely proportional to the quantity of solids dissolved as extract. Thus 50 g of a well-modified malt yields 8–9 ml of draff, while 50 g of oat adjuncts yield about 16 ml of draff. Because the specific gravity of the wort altered (*i*) with the materials extracted and (*ii*) with the volume occupied by the draff, the oversimplified formula was corrected [7, 8]. Allowance was made for the fact that greater extracts would be reflected in smaller quantities of draff and hence a larger volume of wort. A decision was made to retain the 515 cm³ mashing flasks as a practical convenience. It was shown by trial that:

True HWE (lb/Qr. 'as is') $= (1·061 \times 3·36 \times \text{Excess SG}) - 4·8$
$$= (3·565 \times \text{Excess SG}) - 4·8$$

However, malts contain varying amounts of moisture, and so for absolute comparisons, the HWE 'as is' (that is fresh weight, which is of most value to brewers) is best converted to HWE 'on dry' (that is to the dry-weight basis.) The results of an analysis may be calculated using the formula or from a table. The specific gravity of the wort must be measured to five significant figures, because it is the excess specific gravity (i.e. wort SG − 1000.0) that is used to calculate the results, to the required accuracy of 0·1%.

In 1977, and subsequently, the IoB method was altered in two significant ways. (*i*) The malt must now be ground in a Bühler–Miag disc mill (instead of the older two-roll Boby mill), which will enable controlled coarse and fine grind mashes to be introduced when the mill settings for these determinations have been agreed. (*ii*) The units of extract have been altered to litre°/kg, i.e. extract is expressed as excess SG × the wort volume in litres at 20°C, derived from 1 kg of malt. The formula used to calculate the results is E (extract, litre°/kg) $= 10·13\, G$, where G is the excess SG at 20°C. The new calculation assumes an average value for the volume of draff, namely 8·5 ml, hence a fixed volume of wort, and an 'average' factor to multiply the excess SG of the wort. The volume of the draff (cc) is taken as the washed, dried weight (g), divided by 1·48. For the interconversion of the old and new sets of units see also the Appendix. These units cannot be interconverted by the use of a single factor because of the different assumptions made in the calculations and the differences in the experimental conditions [9]. From 1st December 1979 the gravities of the hot-water extracts have been measured at 20°C (68°F) instead of 15·5°C (60°F) as used previously, so the reported extracts are less by 1·6 litre°/kg compared with the 1977 method. Brewers' lb/Qr. are now officially obsolete but remain in limited use.

In continental European and North American practice different mashing procedures and different units of extract are used. Because the mashing methods are fundamentally different there are no wholly valid factors for converting hot-water extract from one unit into another, although *very approximate* equivalents may be established (see the Appendix). In the European 'Congress' EBC analytical mash, water (200 ml) at 45°C (113°F) is added to grist (50 g) at 45°C (113°F), and the mixture is stirred mechanically and held at this temperature for 30 min. Then the temperature is raised at the rate of 1°C/min for 25 min. At 70°C (158°F) more water (100 ml) at the same temperature is added, and the temperature is maintained for 1 hr. The mash is then cooled, the stirrers are washed off, and the weight of the mash is made up to 450 g [4]. The North American 'ASBC' mash is very similar [5]. These mashing sequences, which use increasing mashing temperatures, resemble temperature-programmed brewery mashes or, less closely, the decoction mashes of traditional lager brewing practice. The ASBC and EBC practice is to express the extract as a percentage of the malt using tables relating the specific gravities of sucrose solutions to their solid contents (Appendix). As already noted, the solids in solution weigh more than the difference between the dry weights of the malt and the draff because of the water that is bound during the hydrolyses which take place during mashing. Also expressing the weight of material in solution as the weight of sucrose giving a solution of equal SG gives slightly different results from those actually obtained with starch-conversion products and, in any case, takes no account of the non-carbohydrate components of the extract. Thus in spite of the logical attraction of the 'percentage extract' units they are not an exact measure of the proportion of the malt that has come into solution.

The foregoing discussion on determining malt extracts relates to 'normal' malts with adequate enzyme complements. When the extracts of highly coloured malts, roasted barleys or mash-tun adjuncts deficient in hydrolytic enzymes are to be determined they must be mashed mixed with a known proportion of a highly enzymic malt, or some other source of hydrolytic enzymes [3] (Chapter 8).

Mathematical analysis has shown that, of all the chemical analyses made on malt, the hot-water extract gave most data on malt quality, and correlated well with a range of other analyses such as wort run-off time, and the volume and gravity of the collected wort [10].

9.3 Other analyses

The quality of the wort produced in the chosen 'hot-water' extraction procedure may be evaluated further. For example, the total soluble nitrogen (TSN), the formol nitrogen [11] or the amino acid nitrogen may be determined directly. These values are usually below those obtained in the thicker mashes of a brewery. If the wort is collected and boiled for a standard time after mashing, to inactivate carbohydrases, then the carbohydrate composition of the wort is fixed. Subsequently valid estimates of fermentability may be made either by a standardized forced fermentation procedure [3, 12], or the carbohydrates present in the wort may be investigated by chromatography. Apparently no relationship has been established between the fermentabilities of hot-water extract worts and brewery worts prepared from the same malt. Various indirect measures of wort fermentability have been proposed, such as those based on the reducing power of the mixture of carbohydrates. Such techniques may serve a limited purpose, for quality control for example, but they will fail if the wort has an unusual composition, such as being unusually rich in glucose or sucrose. However, a method based on a wort carbohydrate balance works well, and, provided that allowance is made for non-carbohydrate wort solids, the results agree with those obtained by forced fermentation tests [13].

'Diastase' is a collective name for all the malt enzymes that are involved in the breakdown of starch in mashing. Although 'diastatic power' has been estimated in extracts from malt for many years, the significance of the results is doubtful because (*i*) there has been a failure to use reliable quantitative extraction procedures for the enzymes, (*ii*) allowance has not been made for the interfering effects of amino acids and other materials from the malt on the estimation of reducing sugars in the diastase extract–soluble starch digests, and (*iii*) no account is taken of the variable composition of the enzymic mixture. Estimations of α-amylase have come into use in breweries. A widely used method, based on the findings of Sandstedt, Kneen and Blish, follows the degradation of β-limit dextrin in the presence of an excess of β-amylase. It is not specific for α-amylase, however, since debranching enzymes, including pullulanase and α-glucosidase, contribute to the values reported as α-amylase activity. Recently a standardized method for determining β-glucanase in malt has been proposed [14]. This analysis is probably of importance when the malt is to be used in mashes made with adjuncts rich in β-glucans, such as flaked barley [10].

Maltsters and brewers have long been preoccupied with obtaining 'well-modified malts', and in the course of time many chemical 'indices of modification' have been suggested. These various indices do not correlate closely when malts prepared in different ways are considered, and by using different malting techniques it is possible, within limits, to vary the indices independently (Chapter 5). However, for controlling one particular malting process such estimates of 'modification' are valuable. For example, the cold-water extract (CWE) of malt is often determined in British practice. In this extract only the preformed soluble materials, sometimes called 'matters soluble' or 'preformed sugars' are leached into solution from the ground malt under conditions (ice-cold or alkaline) which preclude the intervention of enzymes. The values are calculated from the specific gravity of the extract using a conventional formula [3]. Malts with cold-water extracts of less than 18% of dry matter are regarded as undermodified, 18–20% are well-modified, and over 22% they are overmodified or 'forced'. In contrast, malt hot-water extracts are normally in the range 75–82%. The difference between these values is a measure of the materials solubilized during mashing.

The nitrogenous materials in wort are generally determined as total soluble nitrogen (TSN). More rarely they are estimated in terms of permanently soluble nitrogen (PSN), that is the nitrogen remaining in solution after the wort has been boiled under standard conditions, and the coagulable nitrogen removed by filtration. The amount of 'coagulable nitrogen' found in sweet wort is strongly influenced by the pH, the time of boiling and other factors [15]. It has been a convention (IoB) to multiply TSN by a factor of 0·94 to obtain the estimated PSN for use in computing one nitrogen index of modification. Although empirical methods of wort fractionation by selective precipitation techniques, such as that due to Lundin [1], by gel-exclusion chromatography, or the estimation of individual amino acids are available, they are not carried out on a routine basis. Estimations of total α-amino nitrogen are now widely carried out. Tests of 'assimilable nitrogen', analogous to the percentage fermentability tests on carbohydrates, are difficult to perform in a reliable way, since the removal of nitrogenous substances from wort by yeast depends very greatly on the yeast strain used, the degree of aeration of the fermentation, and on the pitching rate [16]. Even so, a method designed to overcome these difficulties and others has been proposed [17]. The protein indices of modification, in which the total soluble nitrogen, or in other cases permanently soluble nitrogen, of the worts are expressed as percentages of the total barley or malt nitrogen contents are also much used. The ratio (100 × total wort soluble nitrogen)/total malt nitrogen is commonly called the soluble nitrogen ratio, SNR, when the IoB mashing procedure is used, and the Kolbach index when the EBC mash is employed. In Britain the nitrogen index of modification was taken as (100 × wort permanently soluble nitrogen)/total malt nitrogen, using IoB analyses. Values in the range 30–33% are said to represent undermodification and 37–40% overmodification. However, by using bromate, malts may be produced having a high carbohydrate extract, but a low nitrogen index of modification (Chapter 5).

Modification has also been evaluated by mashing several samples of a malt at different temperatures, and judging the results on the extracts obtained. In the best known of these techniques extracts are found from samples mashed at 25, 45, 65 and 85°C (77, 113, 149 and 185°F) and by the EBC procedure. The 'Hartong number' is calculated by expressing the extracts as percentages of the congress extract, averaging the results, and subtracting 60. 'Ideal' malts should yield a Hartong number of 5; lower and higher values indicate under- and over-modification respectively. It is claimed that much other information can be gained from this laborious technique, or its variants [1].

A method of assessing modification that is of considerable interest involves mashing samples of a malt that have been (*i*) coarsely and (*ii*) finely ground using agreed settings on the mill [18, 19], and noting the difference between the hot-water extracts so obtained. The more finely ground a malt is, the more readily it will yield its extract when mashed because the protein and cell walls investing the starch grains are largely disrupted by fine grinding, allowing hydrolytic enzymes to gain access to them. In addition, the diffusion of substances into and out of fine particles ceases to be rate-limiting. Therefore the fine-grind mash may be expected to estimate the total recoverable extract in the malt sample limited by the quantities of sugars, starch and enzymes that are present. The proportion of this extract yielded by the coarsely ground sample will be regulated by the 'degree of modification'. Thus with increasing grinding a good-quality malt will eventually yield a high maximum extract, and a poor-quality malt a lower extract than that given by the good malt, whether they are well or poorly modified [1, 18, 20, 21] (Fig. 9.1). Only well-modified malts will yield near maximal extracts when coarsely ground. This method is evidently soundly based, although one might expect, in some instances, that the extract obtainable from the finely ground sample would be enhanced by supplementing the starch-degrading enzymes available, thereby giving a better estimate of the total potential extract. Recently the Institute of Brewing adopted the Bühler-Miag disc mill, in the hope that it will prove suitable for preparing 'fine' and 'coarse' as well as 'standard' grists from malts [13] (Fig. 9.1). In practice it is difficult to mill successive samples so that the same proportions of different particle sizes (flour, fine grits, coarse grits) are always obtained. Further, the hot-water extracts are numerically large compared with the differences between the 'coarse' and 'fine' extracts that are obtained with most malts. To obtain the necessary degree of reliability, each determination must therefore be replicated and the significance of the differences should be evaluated by statistical tests. It seems that the 'quality' of newer sorts of malt and unmalted adjuncts and the nature of the wort and beer that may be obtained from them can only be fully evaluated by brewing trials on a scale that is sufficiently large to reflect performance on the production scale. The mashing of as much as 150 kg (approx. 1 Qr.) of malt may be necessary to ensure that the evaluation is adequate. When flavour trials are to be made, it seems that test brewing must be on an even larger semi-production scale.

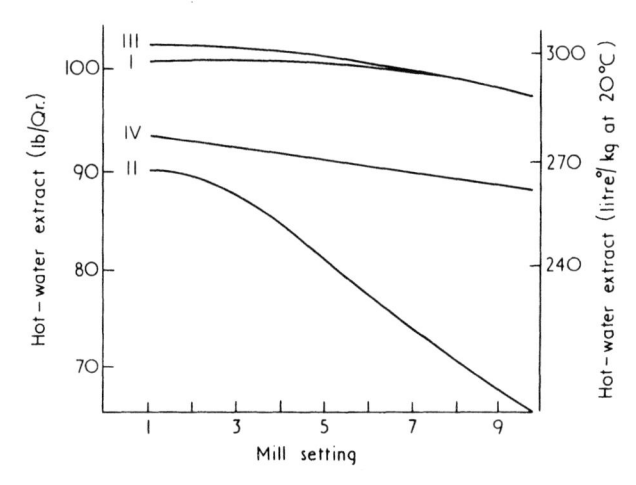

Fig. 9.1 Hot-water extracts obtained from four different malts ground to different degrees of fineness, and mashed by the IoB method. The setting numbers refer to the Bühler-Miag disc mill {Buckee *et al.* (1976) [18]}. Malt I, a well-modified ale malt. Malt II, an under-modified malt, Proctor barley germinated for 1·5 days. Malt III, a malt made by germinating Proctor barley for 3 days (medium modification). Malt IV, an undermodified commercial malt. Lower mill settings give more finely ground grists.

9.4 *Mashing – some general considerations* [22, 23]

In mashing, when ground malt is mixed with warm water usually containing salts, many interrelated chemical and physical changes begin and proceed simultaneously. The grist particles begin to hydrate and swell; at higher temperatures starch granules begin to gelatinize; simple substances having low molecular weights (and represented at least in part by the cold-water extract) are rapidly dissolved and may be further degraded by enzymes. Enzymes progressively catalyse the hydrolysis of insoluble polymers, such as proteins, nucleic acids and carbohydrates, to give soluble products. Further, some of these enzymes may solubilize, that is 'convert', the starch of unmalted adjuncts when these form part of the grist. Enzymes may take an appreciable time to reach their substrates by diffusing through the mash, and the soluble products of hydrolysis must diffuse from within the mash particles into the interstitial fluid, which will be withdrawn as wort. Until the grist particles are hydrated enzymes will not be free to diffuse or to attack the insoluble substrates. At the temperatures at which infusion mashing is carried out, for example beginning at about 63–67°C (approx. 145–153°F) and being raised to 68–70°C (approx. 154–158°F) by an underlet, then being sparged at 75–77°C (approx. 167–170°F), enzymes are inactivated at various rates and some may be entirely destroyed early in the process. Thus variations in (*i*) the enzyme complement contributed to the mash by the malt, or microbial preparations, (*ii*) the mashing temperature regime and (*iii*) the duration of mashing all alter the composition of the final wort. Fortunately, mashes show many self-regulating properties, and worts produced by the infusion-mashing process vary less with alterations in mashing conditions than might be expected. Presumably this is because starch conversion, for example, tends to go to completion during mashing, and so the initial rate of conversion is less critical than in an enzyme estimation test. However, alterations in the mashing conditions do change wort composition to a significant extent, as will be discussed below. Wort contains simple sugars, more-complex polysaccharides, amino acids, peptides, proteins, other nitrogenous materials, vitamins, organic and inorganic phosphates, mineral salts, polyphenols (including tannin precursors and tannins), small quantities of lipids and many other minor components, many of which have not yet been identified [24].

Infusion mashing at about 65°C (149°F) was developed using thick mashes (0·021–0·027 hl of liquor/kg of grist; 2·0–2·5 brl/Qr., barrel/quarter of 336 lb) made with well-modified malts. However, when less-well-modified malts are used, or when high levels of adjuncts are used in mashes (whether these are converted by enzymes from malt or microbes), it is often desirable to begin mashing at a lower temperature to allow β-glucanases, proteases and other heat-labile enzymes to act, before increasing the temperature of the mash to allow starch conversion, and later enzyme inactivation and reduced wort viscosity.

In temperature-programmed mashing (rising-temperature infusion mashing), which has only recently come into widespread use, the temperature of a stirred mash is progressively raised, then is held until the wort is separated and the residual goods are sparged. For an undermodified malt, or a mash containing adjuncts, a mashing temperature programme might be (*i*) mashing in at 35°C (95°F) and holding for 30 min, then raising the temperature, over 15 min, at 1°C/min, to 50°C (122°F). After a further 'hold' of 30 min the temperature would be raised to 65°C (149°F). After a further 30 min the mash may be raised to 70°C (158°F) for 30 min, then 75°C (167°F) for 30 min [25–27]. For better-modified malts a shorter mashing programme with rests at 50, 65 and 75°C (122, 149 and 167°F) might be typical. Clearly such programmes may be varied in many ways, altering the number, temperatures and durations of the rests, or temperature-holding periods, and the rates of heating used during temperature increases.

If raw grits (maize, sorghum, rice) are used, these must be cooked in a special vessel before being mixed in with the malt mash (Chapter 8). However, flaked cereals are used directly, mixed with the grist, as are wheat or barley flours, or potato starch.

Decoction processes were devised for irregularly or poorly modified malts, but a host of variations allow well-modified malts or mixed grists of malt and starchy adjuncts to be mashed successfully. If adjuncts with starch having a high gelatinization temperature, such as rice or maize grits, are to be used, they are precooked in a mash kettle, or other cereal cooker, mixed with 5–10% of a highly enzymic malt, or microbial enzymes, often before mashing begins. This brings about gelatinization and initial liquefaction of the raw starch and some breakdown of the cell walls of the adjuncts so that subsequently, in the main mash, enzymic breakdown of the liquefied starch can occur readily. Similarly, the starch of undermodified malt can be gelatinized during the boiling stage of a decoction.

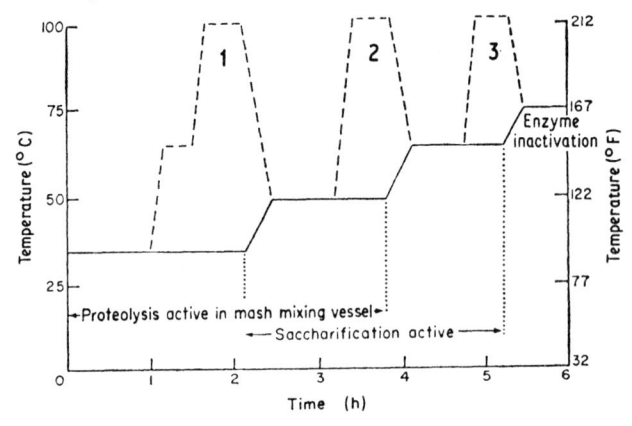

Fig. 9.2 A typical triple-decoction mashing process {Hind (1950) [28]}. ———, Temperatures in the mash vessel; – – – –, temperatures in the mash copper during the first, second and third decoctions. About one-third of the mash is used in each decoction. At 75°C most enzymes in the mash are inactivated.

In the classical triple-decoction process, mashing-in is carried out at ambient temperature and solution of the soluble components of the mash occurs [28, 29] (Fig. 9.2). With light beers, the liquor/grist ratio is about 4·8–5·4 hl/100 kg (4·5–5 brl/Qr.), whereas with dark beers it is 3·0–4·0 hl/100 kg (2·75–3·75 brl/Qr.). About one-third of the liquor is initially withheld in the mash copper and is heated. The hot liquor is then added to the cold mash, with mixing, to bring its temperature to about 35–40°C (95–105°F). One-third of the mash is then pumped to the mash copper, heated to about 65°C (149°F) and held at this temperature for about 20 min to allow starch conversion. It is then brought to boiling point over some 40 min and held for 15 min (pale beers) to 45 min (dark beers) before being pumped back to the mash-mixing vessel. During the boil enzymes are inactivated, but residual starch is fully gelatinized and cellular structures are broken apart. Mixed with the rest of the mash, a temperature of about 52°C (125°F) is achieved, permitting proteolytic enzymes to operate. Again, one-third of the mash is pumped to the mash copper, brought to boiling in 30 min (with or without a hold at 65°C), held at this temperature for 30 min, and returned to the main mash, with mixing. This time the mash temperature rises to about 65°C (149°F), which permits the enzymic breakdown of starch. After the goods have settled, a third portion is treated similarly to the second and when mixed with the main mash the temperature reaches about 76°C (170°F), and the mash enzymes are inactivated. The mash is then transferred to the lauter tun. The overall processing time is about 6 hr from mashing-in to lautering. There are many minor variations on the classical triple-decoction mash, especially with regard to speeds of heating, the rate at which the hot 'decoction' part of the mash is mixed with the main mash, and the lengths of the stands. The overall time for mashing is too long by modern standards and unnecessarily complicated for malts which are reasonably modified.

For better-modified pale malts with an adequate enzyme complement, a more-rapid double-decoction process, or variations of it are often adopted (Fig. 9.3). Mashing-in is either with cold liquor, when the temperature of the

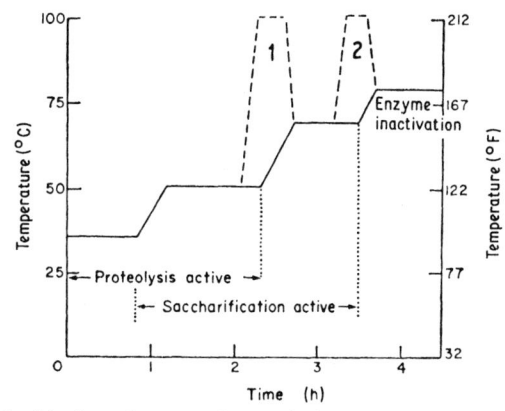

Fig. 9.3 A typical double-decoction procedure {Hind (1950) [18]}. The key is as for Fig. 9.2. In this instance about one-quarter of the mash is used in each decoction.

mash is subsequently raised by steam-heating or the mash is made with warm liquor. The protein rest or stand at about 52°C (125°F) may be substantially reduced in duration for fully modified malts and the overall mash temperature raised to 65–70°C (150–158°F) by taking, as soon as the temperature reaches 52°C (125°F), half of the mash into the mash copper, boiling it for about 10 min and returning it to the rest of the mash for mixing. After about 30 min saccharification stand, one-quarter of the mash is withdrawn to the copper, boiled for about 10 min and returned to the main mash to raise the temperature to about 76°C (170°F). The total time from mixing the mash to lautering is of the order of 3–4·5 hr.

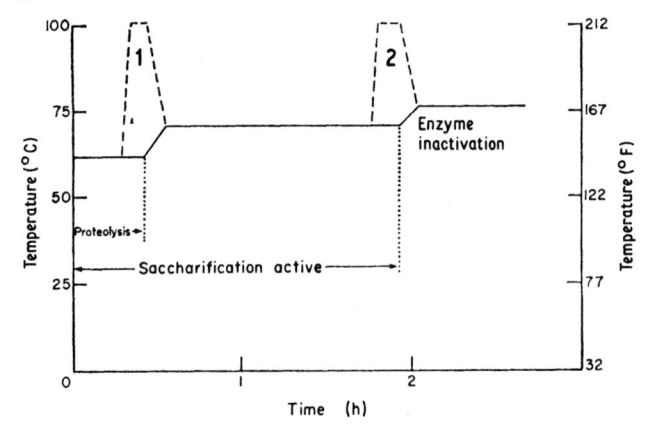

Fig. 9.4 A short-time double-decoction procedure {Hind (1950) [18]}. For key, see Fig. 9.2. About one-quarter of the mash is used in each decoction.

A number of short double-decoction mashing processes have been devised. For instance (Fig. 9.4), mashing-in may take place at 63°C (145°F), one-quarter of the mash is immediately withdrawn for boiling and is returned to the main mash 10–30 min later to raise the overall temperature to 70°C (158°F). After a 45–60 min saccharification rest, one-quarter of the mash is withdrawn, boiled and returned to raise the main mash temperature to about 77°C (170°F). This mashing process takes 2–3 hr up to the lautering stage and is successful for well-modified malts requiring no protein rest.

Single-decoction processes offer a further simplification and speedy processing, but for dark beers they are sometimes preceded by long cold liquor rests ('vormaischverfahren'). In the Schmitz process, the initial mash temperature is 50°C (122°F) and the whole mash is heated to 65°C (150°F), while being stirred. The mash is allowed to settle, then the upper layers containing dissolved sugars and enzymes are pumped off. The thick mash remaining is boiled to gelatinize any residual starch and break up intact grist particles, and then is cooled. The thin mash is returned and mixed with the thick mash for a saccharification rest. Finally, the temperature is raised to 77°C (170°F). Under some conditions decoction mashes yield 1·5–2·0% more extract than is recovered by infusion mashing, but this is by no means always the case, particularly if the malt is well-modified and the infusion mash is temperature-programmed.

The use of high proportions of starchy adjuncts, such as rice, maize or sorghum grits, together with well-modified highly diastatic malts has led to the development, particularly in North America, of double-mash systems

CHEMISTRY AND BIOCHEMISTRY OF MASHING

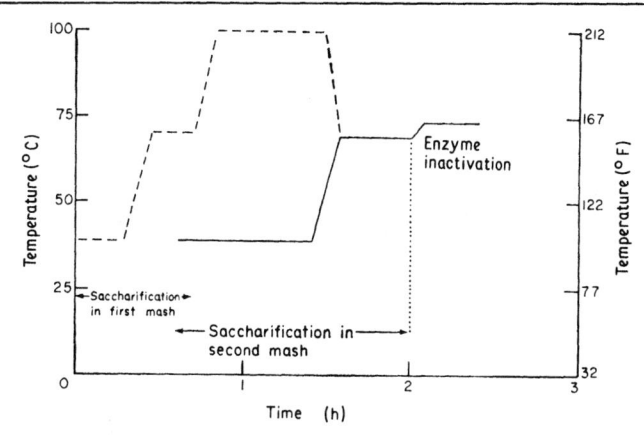

Fig. 9.5 A double-mash procedure {Hind (1950) [18]}. -----, Temperature of the adjunct mash in the cereal cooker; ———, temperature of the malt mash, and the combined mash during and after mixing.

in which two separate mashes are prepared (Fig. 9.5). The malt mash is made from either an all-malt grist or a grist having a large proportion of malt compared with starchy adjunct. The second adjunct mash comprises the starchy adjunct with a small proportion of malt. In American brewing practice, the adjuncts may represent 25–60% of the combined grists. The malt has a high proteinase activity and similarly the amylolytic activity (diastatic power) is as high as 80–200°Lintner. Both mashes are mashed-in ('doughed-in') at a low temperature about 35°C (95°F), and are permitted to stand for 30–60 min in order to dissolve mash constituents and allow the grist to hydrate. The adjunct mash is prepared first in the mash copper or 'cereal cooker'. The temperature is raised to about 70°C (158°F) for a saccharification rest of 20 min (Fig. 9.5). The malt and the grits are softened during this period. The adjunct mash is then heated to boiling while the malt mash is being mixed. Where bacterial α-amylase is used to liquefy the starch of maize grits, processing temperatures below boiling, e.g. of 85°C (185°F), may be appropriate, with a consequent saving of heat. After the adjunct mash has been boiled for about 45 min, possibly under a slight positive pressure, it is mixed with the malt mash to achieve an overall mash temperature of about 68°C (155°F). The mixing of the two mashes commonly takes place in a third vessel. After some 15 min, during which the starch of the grits is saccharified, the temperature of the whole mash is raised to about 73°C (163°F) by additions of hot liquor or direct steam injection. The use of starchy adjuncts, often as a high percentage of the grist, is now common throughout the world (Chapter 8). Choice of the adjunct is mainly determined by the cost of each unit weight of extract, but attention is also paid to the effects of the adjuncts or beer properties, such as flavour and foam stability. Normally, the level of soluble nitrogen extracted from the adjuncts is very low, so that they are used as a diluent of soluble nitrogen in the wort, an important consideration when malts of high nitrogen content are used in the grist.

9.5 Carbohydrates in the wort [22, 25]

The details of the mixture of carbohydrates found in wort is dependent on the nature of the grist, and the mashing conditions used. However, except when mashes have been made with additions of microbial enzymes such as pullulanase or amyloglucosidase (Chapter 8), the proportions of the carbohydrates found in different worts are generally similar. A comparison of the total carbohydrates found in wort with the 'potentially extractable carbohydrates' of malt illustrates very clearly the overall hydrolytic changes that occur in mashing (Table 9.1). This type of balance, which represents the broad fractions of carbohydrates rather than a detailed analysis of the sugars present, indicates that most of the extract obtained from malt comes from

TABLE 9.1

A comparison of the major fractions of the wort carbohydrates with the 'potentially extractable' carbohydrates of malt. Extract of malt, laboratory 104·6 lb/Qr.; Brewery 102·2 lb/Qr. Recovery of potentially extractable hexose apparently 99·0%
{HALL et al. (1956) [30]}

Malt carbohydrates (monosaccharide equivalents as % of wort solids)		Carbohydrates of wort (% of total wort carbohydrate)	
Starch	85·8	Dextrins*	22·2
Glucosan	2·5	Maltotetraose	6·1
Fructosan	1·4		
Maltotriose	0·58	Maltotriose	14·0
Maltose	0·99	Maltose†	41·1
Sucrose	5·1	Sucrose	5·5
Glucose	1·7	Glucose ⎱	8·9
Fructose	0·71	Fructose ⎰	
Total	98·78	Total	97·8

* Includes pentosans.
† Includes trace of iso-maltose.

starch and is additional to the 'preformed soluble sugars' of the cold-water extract. The major simple sugars of worts are the monosaccharides glucose and fructose, the disaccharides sucrose and maltose, and the trisaccharide maltotriose. The amylolysis of starch will be discussed later. It seems that in infusion worts only about 2% of wort carbohydrates comes from insoluble non-starchy polysaccharides such as hemicelluloses, but in decoction worts this value may reach 6% [31]. Appreciable quantities of pentoses may be found in decoction worts. At various times, worts have been reported to contain, in addition to the major carbohydrates, arabinose, xylose, ribose, pentosans, β-glucans, fructosans, maltopentaose and other α-(1→4)-gluco-oligosaccharides, isomaltose, panose, isopanose, maltulose, maltulotriose, nigerose and 'gluco-difructose'. Pentoses, and some other minor sugars, are not always present. During mashing some fructose arises from the limited hydrolysis of sucrose and fructosans. Some of the gluco-oligosaccharides contain α-(1→6) bonds. Of these dextrins about 80% contain 4–20 glucose residues and 20% contain more than 20 glucose residues per molecule. The frequency of distribution of chain lengths and degrees of branching is complex [32, 33].

In forced fermentation tests on boiled wort most brewing yeasts ferment all the major mono, di- and tri-saccharides. Estimates of the percentage fermentability of the carbohydrates, usually about 75%, may thus be made on the chromatographic analysis of wort carbohydrates by expressing the quantities of the mono-, di- and tri-saccharides as a percentage of the total. Some 92% of the wort solids are carbohydrates. In fermentation the carbohydrates are the major energy source of the yeast, and are degraded to the alcohol and carbon dioxide found in the beer. The calorific food value of beer is principally due to the ethanol and unfermented carbohydrates, but it is doubtful whether the latter contribute significantly to any other beer character, although glycoproteins may act as foam stabilizers [34].

9.6 Nitrogenous materials in sweet wort [17, 18]
Nitrogenous materials account for 5–6% of the wort solids from an all-malt mash, and they are equivalent to some 30–40% of the total nitrogenous materials found in the malt. In IoB hot-water extracts 94% of the total soluble nitrogen (TSN) has, by convention, been taken to be permanently soluble (PSN), that is it is not coagulated and precipitated when the wort is boiled. In fact the PSN/TSN ratio varies according to the nature of the mash, and so this assumption is wrong. About 40% of the PSN titrates as formol nitrogen, which is supposed to be mainly the nitrogen of amino acids. When a range of comparable malts, differing in nitrogen content, are mashed, it is found that, although the level of soluble nitrogen increases with malt nitrogen content, it does not increase in simple proportion, so the Kolbach index of nitrogen modification is lower in the 'high-nitrogen' malts, and these contribute less soluble nitrogen to the wort than might have been predicted [25].

The total soluble nitrogen fraction of wort contains the basic materials choline, betaine and ammonium ions, together with amino acids, proline, peptides of varying complexity, proteins, vitamins, purine and pyrimidine bases and their nucleosides and deoxynucleosides. The latter are derived from the nucleic acids of the malt. In decoction mashing the low initial mashing temperatures favour proteolysis, and it was thought that about half of the soluble nitrogen found in the wort is formed in the mash, the other half being preformed in the malt. In contrast, it was believed that in infusion mashing only about one-third of the soluble nitrogen found in the wort is formed in the mash. More recent experiments indicate that previously the extent of proteolysis during mashing had been underestimated [35]. Revised estimates indicate that in infusion mashes proteolysis accounts for the formation of about half the TSN and half the α-amino nitrogen found in the wort [35]. The continued formation of soluble nitrogenous substrates in mashing is due to the comparatively heat-stable endopeptidases acting in conjunction with heat-stable carboxypeptidases [36, 37]. The complexity of the nitrogenous fractions of malts, worts and beers has prevented their complete analyses. However, numerous partial analyses have been made. The high-molecular-weight fractions, proteins and polypeptides, have been variously separated by fractional precipitations with tannin and inorganic salts or by gel-exclusion chromatography and have been characterized by electrophoresis and isoelectric focusing. The presence of large quantities of high-molecular-weight material, reflecting inadequate proteolysis in the mash, may contribute to haze formation in the beer. In this connection, proteins and peptides, together with polyphenols are the major constituents of the most important sorts of beer haze. Peptides and proteins have been implicated in contributing to 'palate fullness' and improving head retention. They probably bring about these effects by lowering surface tension, increasing viscosity and forming films on the surfaces of gas bubbles, thus stabilizing foam.

The amino acids of wort and malt and the alterations that occur during mashing have been investigated by ion-exchange chromatography (Table 9.2). The imino acid proline is the major constituent, and, as it is hardly metabolized by fermenting yeast, it appears in the finished beer. In contrast, the amino acids present in wort are metabolized during fermentation and the spectrum of amino acids available to the yeast plays an important part in determining the volatile constituents of the final beer. For example, it is known that in large quantities valine tends to suppress the formation of diacetyl during fermentation. Currently it is thought desirable to keep a uniform spectrum of amino acids in successive worts [25]. Estimates vary, but to achieve good yeast growth and rapid fermentation it seems that wort should contain about 150 mg of α-amino nitrogen/litre, and certainly not less than 100 mg/litre. Brewers' worts have been found to contain 155–225 mg of protein nitrogen/litre in a total of 650–750 mg of soluble nitrogen/litre [25].

The nucleic acids make up about 0·18% of the dry weight of malt. They are at least 95% degraded and solubilized in mashing. Nucleotides, the

TABLE 9.2

Amino acids of malt and derived worts during infusion mashing at 65°C (149°F), carried out for various periods of time
{JONES and PIERCE (1963) [38]}

Mash time ...	Amino (or imino) N (μg/g of dry malt)					
	0 (malt)	20 min	40 min	1 hr	3 hr	5 hr
Aspartic acid	35·6	37·3	36·0	39·4	43·0	45·6
Threonine	29·1	34·7	28·9	37·2	42·6	46·6
Serine and amides	114·4	140·8	143·7	151·4	158·3	158·0
Glutamic acid	54·0	36·7	30·0	40·4	53·6	60·6
Proline (imino acid)	408·9	452·9	455·2	436·2	455·6	468·7
Glycine	15·6	20·3	23·0	24·6	28·3	32·1
Alanine	36·5	46·1	51·5	54·3	62·3	72·4
Valine	61·9	80·6	89·5	87·0	99·2	105·2
Methionine	8·9	12·5	15·9	16·3	19·7	22·4
Isoleucine	34·8	39·8	42·1	43·5	49·1	52·9
Leucine	52·5	75·2	85·7	88·3	101·4	108·9
Tyrosine	36·9	45·0	46·9	48·4	52·4	55·5
Phenylalanine	53·1	68·3	71·0	71·9	76·8	79·7
Tryptophan	18·7	23·4	24·9	24·9	29·5	29·8
Lysine	39·5	55·9	66·2	67·5	80·3	86·9
Histidine	25·1	32·0	34·4	34·6	38·1	40·1
Ammonia	77·3	63·1	71·3	72·9	93·6	119·4
Arginine	54·0	66·2	70·1	70·6	80·5	86·5

phosphorylated products of partial hydrolytic degradation, are not found in worts because they are hydrolysed further by the very active phosphatases of malt to the nucleosides, e.g. adenosine and guanosine which, in turn, may be hydrolysed further to the free purine bases, adenine and guanine. Some of the adenine and guanine may be deaminated, and the products hypoxanthine and xanthine are found in wort. The deoxyribosides of guanine, uracil and thymine also occur in wort [39–43].

Amines derived from malt have been found in sweet wort and beer, for example, methylamine, dimethylamine, ethylamine, butylamine, amylamine and pyrrolidine. Also hordenine (2 mg/litre), tyramine (150 μg/litre) and choline (200–250 mg/litre) have been detected. Probably other basic materials, including trimethylamine, are present. Little is known of the extraction of these materials during mashing [31, 44].

9.7 Vitamins

Vitamin C is destroyed in kilning and so does not pass into the wort. Some other 'vitamins' and growth factors for yeast that are found in worts are summarized in Table 9.3. Other estimates differ appreciably from these. In general, growth factors are detected and assayed by notoriously difficult biological techniques. However, it appears that the quantities of these vitamins that occur in sweet-wort may be very variable. Experience suggests that rarely, if ever, are they present in such small amounts that they limit yeast performance in brewing. Some vitamins are known to be present in malt in bound forms, from which they may be liberated by enzyme action during mashing.

TABLE 9.3

Vitamins in sweet wort

Vitamin	Quantity and Comments	Reference
Choline	200–250 mg/l of decoction wort	[44]
myo-Inositol (*i*)	930 μg of free and 189 μg of total/g dry weight of malt extract (therefore approx. 9·3 mg of free and 18·9 mg of total/100 ml of wort, 1040 SG)	[45]
(*ii*)	1·5–3·5 mg of free inositol/g of dry extract in decoction wort	[46]
(*iii*)	5·5 mg/100 ml of decoction wort	[47]
Thiamin (Aneurine, vitamin B$_1$)	60 μg/100 ml of infusion wort (SG 1060)	[48]
Riboflavin	33–46 μg/100 ml of infusion wort (SG 1038) (very variable)	[48–50]
Folic acid, folinic acid and unidentified compounds	0·1–1 μg/g of dry extract decoction wort (very variable)	[42]
Nicotinic acid	1000–1200 μg/100 ml of infusion wort (SG 1040)	[51]
Calcium pantothenate	45–65 μg/100 ml of infusion wort (SG 1040)	[52]
Pyridoxin, pyridoxal, pyridoxamine	85 μg/100 ml of infusion wort (SG 1040)	[53]
Biotin	0·65 μg/100 ml of infusion wort (SG 1040)	[54]

9.8 *Polyphenols*

Polyphenols present in malt are leached into wort. They contribute astringency to beers, and some polymeric phenols called phlobaphenes may contribute to colour. The quantities extracted from the grist increase with increasing wort pH and sparging temperatures [55]. Thus the proportion of polyphenols in the wort solids increases in the last runnings to the detriment of the flavour and of the stability of the derived beer (see below). The polyphenols may, in part, be adsorbed on proteins or be present as covalently bound complexes with them. During mashing and wort run-off different phenolic fractions are extracted in the worts at different rates (Fig. 9.6) [56, 57]. Anthocyanogens, as reactive 'pro-tannins', have been extensively studied. Oxidized polyphenols are likely to be tannins. Oxidizable polyphenols can pass through the brewing process into beer, where they may become converted into tannins and then interact with proteins to give rise to hazes. Decoction worts contain more oxidizable polyphenols and fewer oxidized polyphenols than infusion worts. The last mash-tun runnings are rich in the oxidizable polyphenols, but most oxidized polyphenols emerge in earlier wort factions [58].

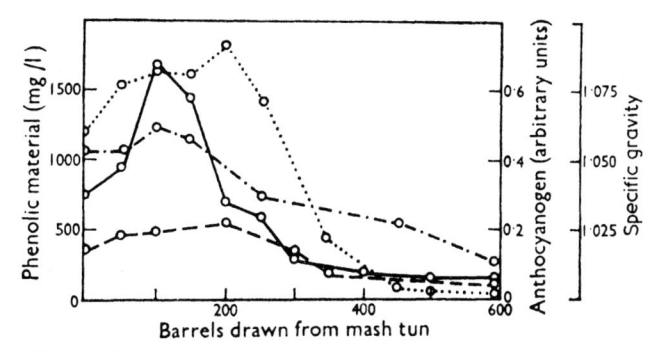

Fig. 9.6 Phenolic materials in successive sweet-wort fractions during run-off from the mash tun {Woof and Pierce (1966) [56]}. o——o, total phenols; o- - -o, dialysable phenols; o· · ·o, specific gravity (SG); o–·–·o, anthocyanogens.

9.9 Organic acids and lipids

Various other acidic materials of non-phenolic nature are also leached into wort. For example, citric, fumaric, malic, lactic, succinic, pyruvic, laevulinic (*syn.* levulinic), mesaconic, oxalic and α-oxoglutaric acids together with an unindentified acid have been reported [59]. Various free fatty acids also occur (see below). Calcium added to the mash tends to remove oxalic acid as the insoluble calcium oxalate. Later in the brewing process calcium oxalate may be deposited in the fermenters as beer stone.

A minor proportion of the malt lipids, usually less than 2% of the total originally present in the grist, is dispersed into the wort [60]. The remainder stays with the draff. Although some of the dispersed lipid is lost during further processing, being carried down with the trub for example, a little reaches the final beer. The actual quantities extracted vary with the means used to filter and sparge the goods. Rapid lautering techniques yield worts with enhanced lipid contents. Filtering through kieselguhr reduces the levels of fats [61]. Centrifugation of sweet wort has been used experimentally to separate malt fat, probably mostly triacylglycerols (about 0·7 mg/litre). The beer brewed from the 'defatted' wort had improved head characteristics. Conversely, the addition of extra malt lipids to a wort reduced the foaming quality of beer [62].

Lipids noted in unhopped worts (in mg/litre) were: triacylglycerols, 5–8; diacylglycerols, 0·2–0·5; monoacylglycerols, 1·6–1·8; sterol esters, 0·1–0·2; and free sterols, 0·2–0·4 [60]. All malt wort contains free fatty acids (18–25 mg/ litre at 12° Plato), mainly palmitic acid, but also the unsaturated acids, oleic, linoleic and a little linolenic [59]. Grouped according to their chain lengths, the quantities of fatty acids were C_0–C_{14}, 0–1 mg/litre C_{12}–C_{18}, 18–26 mg/litre. Improved methods of analysis have greatly increased interest in the fatty acids of wort [25]. True phospholipids may be absent from sweet worts, although a β-lysolecithin with foam-stabilizing properties may be present. Some of these compounds may have organoleptic properties of importance to beer character. Successful yeast multiplication under anaerobic conditions and its continued viability during storage is dependent on an adequate supply of particular unsaturated fatty acids and sterols in the wort. The levels in brewery worts may be critical, because supplementing the wort with lecithin or other sources of unsaturated fatty acids improved yeast viability during subsequent storage [63]. Unsaturated fatty acids and products of their oxidation catalysed by lipoxygenase may decompose in staling beer, giving rise to numerous aldehydes such as *trans*-2-nonenal, hexenal and hexanal, which contribute to the development of 'off flavours' [64]. It has been proposed that the liquor pressed from spent grains be used as an anti-foaming agent during fermentations [65]. Presumably the active agents are the lipids which are present in the press liquor.

9.10 *Mineral salts in mashing*
The mineral salts present in mashing and sparging liquors have been mentioned already (Chapter 7). Some of these will be retained in the spent grains at the end of mashing, and the sweet wort will contain the residuum together with mineral substances leached from the goods. Thus the sweet wort may be saturated with silica derived from the malt husk. Considerable amounts of inorganic phosphate and lesser amounts of organic phosphate, the latter being mainly phytate, and lower phosphate esters of inositol are present in sweet wort [24]. The interaction of these with salts present in the mashing liquor have substantial effects on the pH of the mash, the changes of pH occurring in the hop-boil, and ultimately on the character of the beer. Darker malts give rise to more acidic worts and beers [70].

Analyses of the salts normally present in sweet wort are available [24, 25]. The mixture does not normally appear to be deficient in its ability to support yeast growth, although in some instances the trace component zinc may be limiting. However, the situation is complicated, since what constitutes a limiting (or an excessive) dose of zinc is influenced by the quantities of manganese present [66–68]. The mineral salts present in the mash, both those provided by the mashing and sparge liquors and those leached from the malt, have a profound influence on the composition of the wort. The ions into which these salts are dissociated (*i*) influence the pH of the mash, (*ii*) act as buffers, (*iii*) influence the extraction of proteins, vitamins and other materials and (*iv*) also affect the stability and activities of some enzymes with resultant changes in the extract recovered, in the fermentability of the wort, the efficiency of hop extraction, and the nature of the finished beer. The flavour of the beer is altered by the quantities of chloride (full flavour), sulphate (dry flavour) and lactate (softer flavour) ions that are present (Chapter 7). Contamination with ions such as copper or iron, picked up from plant, may inactivate enzymes in the mash. Thus they alter the quality of the extract and subsequently are toxic to yeast and may potentiate haze formation in beer. Calcium ions help to precipitate 'nucleoprotein', phytate, inorganic phosphate, oxalate and perhaps other trub constituents in the hop-boil. Their presence in the mash is necessary to stabilize the enzyme α-amylase [69, 70]. Calcium exerts a most useful effect in helping to regulate the pH of the mash.

9.11 Water hardness and mash pH [23, 28]

The processes for reducing temporary hardness, due to carbonates and bicarbonates, and for augmenting permanent hardness with calcium sulphate (Burtonization) and other salts in the brewing liquor have been outlined (Chapter 7). Excessive amounts of temporary hardness due to the presence of carbonates and bicarbonates of calcium and magnesium are harmful in mashing and sparging liquor. This is because these substances act as weak bases and raise the pH of mashes and sparges to an undesirable extent (Table 9.4). When such water is heated and particularly if it is boiled or aerated, carbon dioxide leaves the system, hydroxyl ions are produced and the liquor becomes alkaline:

$$Ca(HCO_3)_2 \rightleftharpoons Ca^{2+} + 2\,HCO_3^-$$
$$HCO_3^- + H_2O \rightleftharpoons H_2CO_3 + OH^-$$
$$H_2CO_3 \rightleftharpoons H_2O + CO_2 \uparrow$$

TABLE 9.4
The effect of water hardness on wort pH after decoction mashing
{After HOPKINS and KRAUSE (1947) [23]}

Preparation and nature of water	pH of mash-tun wort at room temperature
Cold-water extract, distilled water	6·2–6·3
Wort pH, water with temporary hardness (about 15 grains* of CaCO₃/gal†)	5·89
Wort pH, distilled water	5·76
Wort pH, permanent hardness (4 grains of CaSO₄/gal)	5·65

* 1 grain is 0·0648 g.
† Imperial gallon is 4·546 litre.

When wort or decoction mashes are boiled, the adverse effects on pH of 'temporary hardness' are particularly noticeable. The effect is less apparent in infusion mashes, but is very important in sparging when the buffering capacity of the mash is reduced by the elution of the buffering substances initially present, and the water is hotter. As the pH of the last runnings increases, particularly over pH 6.0, silicates, tannins and harshly flavoured materials are increasingly extracted into the wort (see below; Figs. 9.14 and 9.15). The pH of bicarbonate waters is buffered to about 7·2–7·5 regardless of the quantity of bicarbonate present. Carbonates/bicarbonates may be removed before mashing, or their effects may be countered by the addition of gypsum and/or of materials such as sulphuric acid or lactic acid to the mashing liquors. The choice of acid will depend in part upon considerations of flavour. A moderately temporarily hard water may require 0·5 ml of 1N acid/100 ml water to achieve a desirable mash pH of about 5·3.

The addition of calcium sulphate, and to a lesser extent magnesium sulphate, to a mash results in a desirable reduction in pH. Some pH alteration occurs instantaneously in the cold, and is due to the interaction of the calcium and magnesium ions with secondary inorganic phosphates, phytate and possibly other substances such as peptides and proteins, resulting in the formation of undissociated calcium complexes. These may or may not remain in solution. Such interactions allow the release of equivalent quantities of hydrogen ions. For example at wort pH values:

$$3\,Ca^{2+} + 2\,HPO_4^{2-} \rightleftharpoons 2H^+ + Ca_3(PO_4)_2 \downarrow$$

Up to four calcium ions react in a similar manner with one molecule of phytic acid (inositol hexaphosphate), which has an extremely high affinity for calcium. Inorganic secondary phosphates reacting alone with calcium could not induce some of the observed falls in pH. To a limited extent the observed falls in pH may be due to interactions between calcium ions and proteins. The calcium salts are more soluble in wort than in pure water, but they tend to precipitate on boiling, e.g. in a decoction mashing boil or in a hop-boil. This precipitation results in a displacement of the equilibria involved, and a further reduction in pH. The magnesium salts are more soluble than the calcium salts and so they exert a lesser effect on pH.

In decoction mashing, 60–65% of the phytate present in the malt may be hydrolysed at pH 5·8 and 90% at pH 5·3. The hydrolysis may be incomplete giving rise to lower phosphates of inositol as well as the free vitamin [71]. In infusion mashing one-third to one-half of the phytate phosphorus of the malt is extracted into wort in about 30 min and little phytase activity can be detected. Phytate levels of 5–30 mg (as P_2O_5)/100 ml are recorded in infusion worts [72, 73].

After infusion mashing much of the phosphate present, mainly phytate, is precipitated in the copper boil. Within limits the more calcium that is present in the liquor, the greater the desired reduction in pH, but at the same time, more phosphate is removed and so becomes unavailable as a yeast nutrient (Table 9.5). When mashing with distilled water, about 30% of the total phosphate is left in the spent grains, whereas with gypseous or carbonated waters values of 59% and 49% respectively have been observed. In the case of carbonate liquors the acidifying action of the calcium is more than counterbalanced by the bicarbonate effect. The addition of bisulphites, such as calcium bisulphite or potassium metabisulphite, to the mash tun or copper results in increased break formation during the boil [74], but this additional source of sulphur may ultimately be deleterious to the flavour of the beer.

TABLE 9.5
The effects of adding gypsum to the liquor when mashing malt
{Data of HIND (1950) [28]}

$CaSO_4$ added (mg/ 100 ml of liquor)	Extract (litre°/kg at 20°C) apparent	corrected for ash content	Unboiled wort Ash (mg/100 ml)	Phosphates as P_2O_5 (mg/100 ml)	Boiled wort phosphate as P_2O_5 (mg/100 ml)	Phosphate precipitated after boiling (%)
0	296·7	286·4	138	70	68	2·9
38	300·5	289·3	148	63	59	15·7
76	302·9	290·4	167	56	54	22·9
114	305·7	290·6	197	54	50	28·6

9.12 Other substances added to mashes

Occasionally formaldehyde is added to the mash tun, when it reduces the levels of anthocyanogens found in the derived worts and beers [75], and the beers are slow to become hazy. However, this treatment may lead to a wort of low fermentability. A similar result may be achieved by using malt prepared using steeps or resteeps in solutions of formaldehyde [76]. Additions of hydrogen peroxide or charcoal to the mash or charcoal to the sweet wort also reduce polyphenol levels [76]. Oxygenating mashes, made with uncured or lightly cured malts, also reduces the wort anthocyanogen levels, probably through oxidations catalysed by polyphenol oxidase, and can result in beers with enhanced shelf lives [77]. The low levels of anthocyanogens found in worts from mashes of green malt may also be due to their destruction by some enzyme-catalysed oxidative mechanism possibly using oxygen from air entrained in the mash. Formaldehyde, hydrogen peroxide (helped by peroxidase) and oxygen (helped by polyphenol oxidase) polymerize polyphenols and tannins, and perhaps alter them and link them to proteins so that they are retained in the mash. Charcoal is thought to adsorb haze precursors and so remove them from the wort.

In mashes made with high levels of certain adjuncts it may be necessary to add supplementary enzymes, such as a mixture of bacterial α-amylase, protease and β-glucanase, to obtain an adequate yield of acceptable extract (Chapter 8).

9.13 *Some properties of enzymes relevant to mashing*

The conditions in a mash tun differ in several important respects from those normally used in studies on enzymes, consequently a knowledge of enzyme kinetics is not immediately helpful. Thus (*i*) the substrates on which the mixtures of enzymes are to act, such as starch or protein, are heterogeneous and chemically impure, being mixed with large quantities of other substances, and in part they are boxed in by unbroken cell walls. (*ii*) Moreover, they are largely insoluble at the beginning of the mashing process. In the case of starch the native granules are almost resistant to enzymolysis and must be swollen or partly gelatinized before enzymes will degrade them at a convenient rate; nevertheless the starch is virtually all converted into soluble substances before the end of the mashing period. (*iii*) Mashing is carried out at relatively high temperatures, about 65°C (149°F) in the infusion process, and at these temperatures the enzymes are unstable, some being rapidly inactivated while others survive in nearly unaltered amounts at the end of the mash. (*iv*) Not only the enzymic complement and chemical composition but also the pH and temperature of the mash alters with time. (*v*) The brewer is concerned with the state of the liquid part of the reaction mixture, the wort, at the end of the mashing period, while, in contrast, the enzyme chemist is usually concerned with the initial reaction rate in an incubation carried out at one temperature, with little or no enzyme inactivation and preferably with all the reactants in homogeneous solution.

As the substrate levels fall, together with the levels of 'protective colloids', enzymes tend to be destroyed more rapidly by heat inactivation and the action of any surviving proteolytic enzymes. Thus (*i*) the composition of the mashing liquor, (*ii*) the nature of the grist, (*iii*) the thickness of the mash, (*iv*) the temperature(s) used and (*v*) the duration of mashing, including the times taken to reach different temperatures and for which the temperatures are held, all alter the composition of the mash and the wort. All these variables exert their effects in inseparable ways, but the effects of changes in mashing conditions are, of necessity, considered separately.

The close interaction of these different factors is seen when discussing 'optima' for enzyme action. Optima, the most favourable conditions of pH or temperature etc., are not true invariable constants, but are the best for the set of conditions under test. Thus at a given pH and initial substrate concentration, enzyme activity measured after a fixed period will increase with increasing temperature and then decline. This phenomenon is due to the interaction of two opposing effects caused by the increasing temperature, that is, increased rates of enzyme-catalysed reaction and increased rates of enzyme inactivation. The temperature optimum adduced will be lower the longer the reaction period that is studied, and will vary with the availability of substrate, the pH, and so on. Such effects are not observed with studies of true initial reaction rates. Similar considerations apply to all 'optima', and, in mashing, each set of 'optima' will only apply exactly to one mashing schedule applied to one grist.

9.14 Mashing and pH [22, 23, 28]

The addition of some salts to the mash alters the pH as outlined above. In addition the pH of a mash or wort alters with the temperature. At 65°C (149°F) the pH of the mash will be about 0·35 unit less than at 18°C (65°F), owing to the greater dissociation of the acidic buffer substances present. Therefore enzymes whose pH optima are known from determinations at 20°C (68°F) appear to have higher pH optima in the mash if this is cooled, as is usual, before the pH is determined. An infusion mash is best carried out at pH 5·2–5·4. Consequently the pH in the cooled wort will be 5·5–5·8. Alteration of the mash pH towards this range will (*i*) allow the amylases to degrade starch more rapidly, (*ii*) enhance the activities of many other carbohydrases and proteases, (*iii*) alter the solubility and coagulability of proteins and (*iv*) minimize the extraction of tannins. With increasing acidity more soluble nitrogen appears in the wort, but the proteins are more degraded by the proteolytic enzymes, and so the wort gives beer that is less prone to nonbiological haze.

Mashes made with malt and distilled water have a pH of about 5·8 and the buffers in the grist tend to hold this value. Mash pH may be adjusted downwards by liquor treatments such as removing temporary hardness, retaining permanent hardness, and even adding gypsum to the mashing liquor (Burtonization) (Table 9.4). Also the mash may be acidified by the direct addition of sulphuric, phosphoric or lactic acids, the choice being made on a basis of safety, cost and the flavour of the final beer. In countries where the additions of 'chemicals' to foodstuffs are forbidden, the sugars in a separate mash or wort may be converted to dilute lactic acid by the thermophilic *Lactobacillus delbrückii*, and this may be added to the bulk mash to reduce the pH. More conveniently, acid malts prepared by steeping the green malt in biologically prepared lactic acid may be added at the rate of 1–5% to the grist. Dark malts usually yield a lower pH in the mash than pale ale malts which, in turn, yield a lower pH than undermodified lager malts or raw grain adjuncts. The interactions between pH and mashing temperature are complex so that, for example, it is reported that starch extraction is optimal at pH 4·9–5·3 at 50°C (122°F), pH 5·1–5·5 at 60°C (140°F) and pH 5·5–5·9 at 65°C (149°F) [78]. The most advantageous pH values for various processes occurring in mashes are summarized in Table 9.6 which shows that it is best to mash at

TABLE 9.6

'Optimal' pH values for a 'standard' infusion mash of 'usual' thickness made at 65·5°C (150°F) with a pale-ale malt (data from various sources)

Characteristic	pH
Shortest 'saccharification time'*	5·3–5·6 (wort)
Greatest extract in an infusion mash	5·2–5·4 (mash)
Greatest extract in a decoction mash	5·3–5·9
Mash pH giving most fermentable wort	5·3–5·4 (approx. 0·2 above pH for max. extract)
Mash impossible to filter	< 4·7
α-Amylase activity maximal (presence of Ca^{2+})	5·3–5·7 (wort)
β-Amylase activity maximal	5·1–5·3
Permanently soluble nitrogen maximal	about 4·6 (mash), 4·9–5·1 (wort)
Formol nitrogen maximal	about 4·6 (mash), 4·9–5·1 (wort)
Proteinase activity maximal	4·6–5·0 (mash)
Phytase activity maximal	about 5·2 (mash)
Carboxypeptidases	4·8; 5·2; 5·7

* Time for starch dextrins capable of giving colour with I_2 to disappear.

about pH 5·3. When this pH adjustment is made by the addition of calcium sulphate to the liquor, phosphates are lost from the final wort. With any favourable adjustment in pH there are (*i*) gains in extract and permanently soluble nitrogen, (*ii*) wort colour is reduced, (*iii*) mash filtration and wort run-off is improved, (*iv*) a better break is achieved on boiling with the hops and (*v*) the final beer clarifies better and is less prone to form haze. The latter is probably due to the greater degree of protein breakdown and the smaller quantities of polyphenols extracted at reduced pH levels, both from the malt and from the hops. The smaller levels of polyphenols with astringent flavours give a beer with a clearer 'bitter' taste. The presence of calcium carried through to beer helps yeast flocculation and causes the separation of oxalate as crystals of its calcium salt, which contribute to the formation of beer stone. This precipitation may necessitate special steps to remove any 'oxalate haze'. However, at the reduced pH, smaller amounts of hop resins are extracted, so reducing the 'hop utilization rate'. Permanently hard waters are typically used to produce pale ale, 'Burton'-type beers.

9.15 *Temperature, time and mashing* [22, 23, 28, 79]

The effects of altering mashing temperatures and times have to be considered together. In brewing practice, in contrast with laboratory hot-water extract determinations, the thickness of the mash allows enzymes to survive at unexpectedly high temperatures. Also the temperature of the mash is usually not held constant but is increased in steps by (*i*) decoctions or sparging in decoction mashing or (*ii*) underletting and sparging in traditional infusion mashing, or (*iii*) by the external application of heat to the vessel or an injection of steam in temperature-programmed infusion mashing.

In general, increasing the mash temperature (*i*) reduces the viscosity of the wort, (*ii*) accelerates the rates of solution, (*iii*) accelerates diffusion rates, e.g. of substances into and out of the grist particles, (*iv*) accelerates mixing, (*v*) speeds enzyme action and (*vi*) enzyme inactivation, (*vii*) accelerates the rates at which grist solids hydrate and swell, and – (*viii*) in the case of starch – gelatinize. The net result is that temperature 'optima' are higher for shorter mashing times and also vary with mash thickness, pH, grist composition and so on. Approximate temperature optima for mashes made with distilled water under simulated brewery conditions of time and mash thicknesses are shown in Table 9.7. Some variations in temperature optima with alterations in mash pH are shown in Table 9.8. Several points deserve comment. At normal mashing temperatures protease activity and survival are very dependent on mash thickness, and α-amylase is less stable and β-amylase is more stable in mashes than would be expected from their behaviour in solution.

TABLE 9.7

Temperature optima for various infusion mash-tun processes
{Most of data from HOPKINS and KRAUSE (1947) [23]}

	°C	°F
Highest extract (starch conversion)	65–68	149–155
Highest yield of reducing sugars	60–62	140–144
Highest yield of fermentable extract	65	approx. 149
Highest yield of permanently soluble nitrogen (PSN) in a 1–3 hr mash (higher temperature optima for more concentrated mashes)	50–55	122–131
Highest yield of formol N	50–55	122–131
Highest yield of (PSN)–(formol N)	55–60	131–140
Highest yield of 'acid buffers'	50–55	122–131
Maximal activity of α-amylase	70	approx. 158
Maximal activity of β-amylase	60–65	140–149

TABLE 9.8

Temperature optima for two processes at different levels of mash pH
{KOLBACH and HAASE (1939) [78]}

pH	Extraction of starch		'Maltose' production	
	°C	°F	°C	°F
6·6	55–60	131–140	55–60	131–140
6·0	60–65	140–149	55–60	131–140
5·1	55–60	131–140	50–55	122–131
4·5	50–55	122–131	45–50	113–122
4·2	45–50	113–122	35–40	95–104

It has long been known that extended mashing at 55°C (130°F) eventually produces about 90% of the possible extract [80], although this is well below the recognized gelatinization temperature of barley starch, and even at lower temperature extensive starch degradation occurs (Figs. 9.7, 9.8 and 9.9;

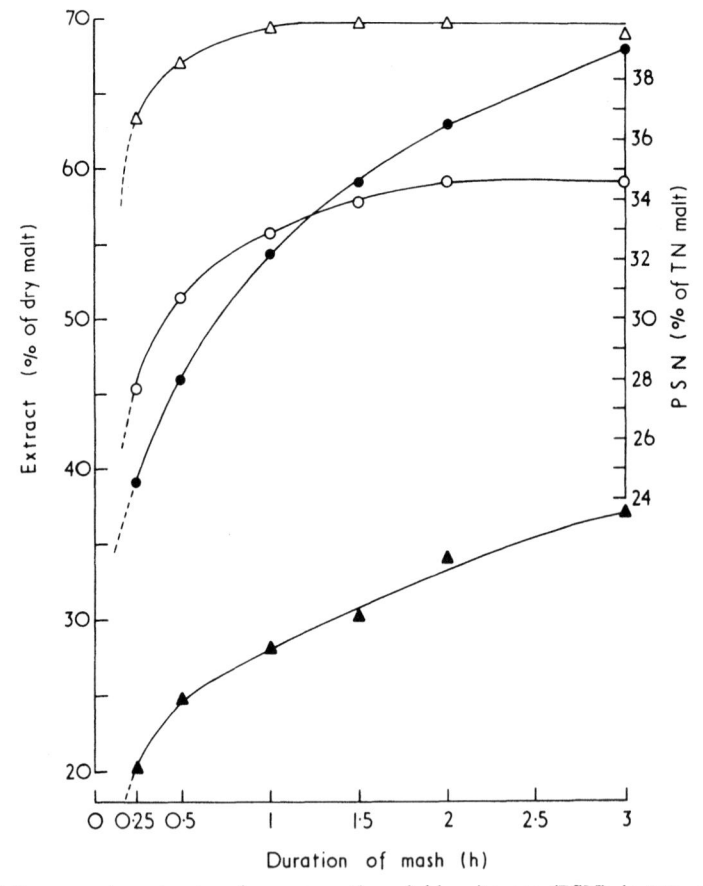

Fig. 9.7 Increases in extract and permanently soluble nitrogen (PSN) in two mashes, made at constant temperatures of 49°C (120°F) and 65·5°C (150°F) {Hind (1950) [28]}. Hot-water extract (% of dry malt) at 49°C (120°F) (▲) and 65·5°C (150°F) (△). Permanently soluble nitrogen (% of malt N) at 49°C (120°F) (●) and 65·5°C (150°F) (○).

Chapter 8). The enzymes are therefore slowly able to degrade malt starch granules and not merely a paste or gelatinized starch as is present in pre-cooked adjuncts. α-Amylase survives higher temperatures than does β-amylase, and so higher temperatures [about 70°C (158°F)] produce dextrinous extracts of lower fermentability compared with those made at 65°C (149°F), although total extract recoveries are the same [31]. The fermentability of worts and

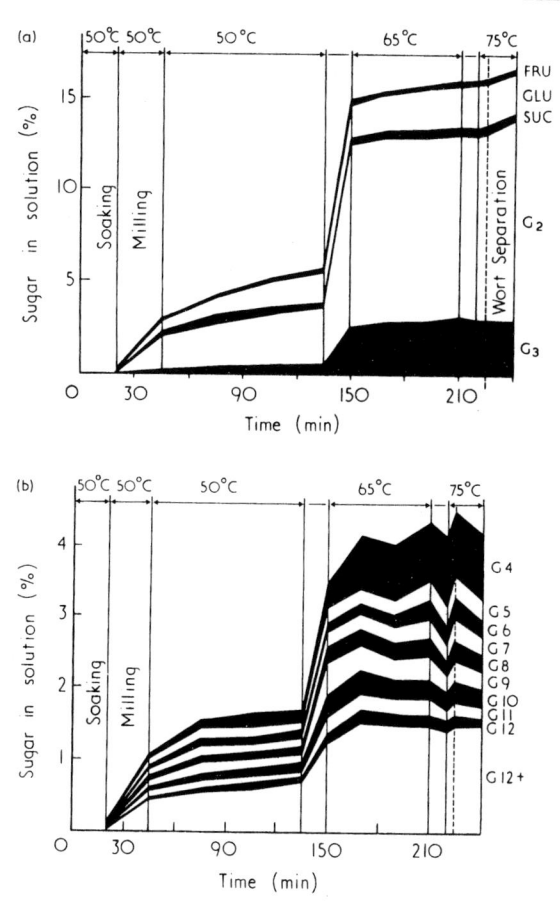

Fig. 9.8 The increase in carbohydrate fractions during a temperature-programmed mash {Schur *et al.* (1973) [81]}. (a) Fermentable carbohydrates. (b) Non-fermentable carbohydrates. Fru, fructose; Glu, glucose; Suc, sucrose; G_2, glucodisaccharides (mainly maltose); G_3, glucotrisaccharides (mainly maltotriose); G_4 etc., higher gluco-oligosaccharides.

hence the alcohol content of the final beers may be regulated in part by adjusting mashing temperatures. The generation of the major groups of soluble carbohydrates during a temperature-programmed mash is shown in Fig. 9.8, and the influence of temperature and time on extract yield in a series of fixed-temperature mashes is shown in Fig. 9.9.

If the course of starch degradation is sensitive to alterations in mashing temperature, this is even more true of the degradation of the nitrogenous fractions in the malt. Nucleic acids are almost totally solubilized in mashes; higher temperatures weaken the nucleosidases but not the nucleases or the nucleotidases. Consequently nucleotides that are formed initially are all

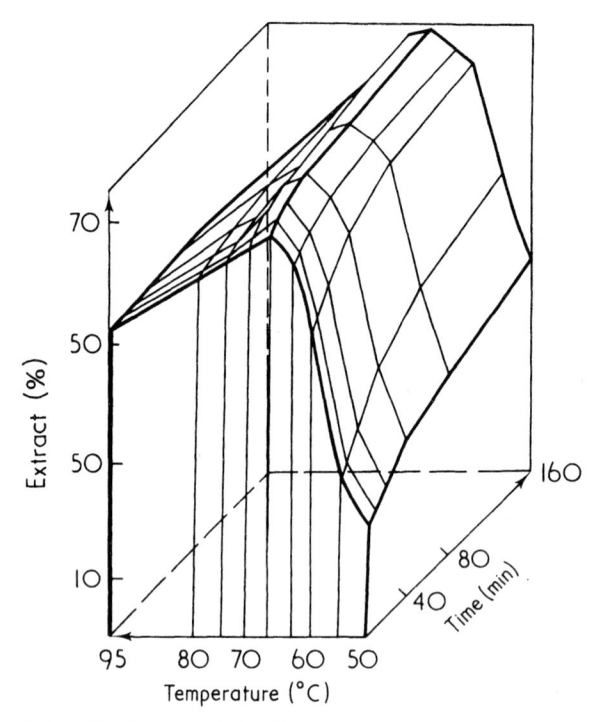

Fig. 9.9 Interrelationships between yield of extract, mashing temperature, and the duration of mashing for a series of single-temperature mashes {Schur *et al.* (1975) [82]}.

dephosphorylated, but the ratio of free bases to nucleosides falls with increasing mash temperatures. Yeasts take up the free bases, but not the nucleosides. Mashing for long periods at low temperatures favours proteolysis (Fig. 9.7) and produces worts with levels of formol nitrogen in excess of those required for adequate yeast growth. Consequently either an overgrowth of yeast may occur or a tendency to contamination by other micro-organisms in the beer may result. Also beer flavour may be altered. Levels of coagulable nitrogen in the wort decrease with increasing mash temperatures – as would be expected from the coagulation of this material in the mash. However, the formation of permanently soluble nitrogen has a higher temperature optimum than formol nitrogen formation, because at these higher temperatures a higher proportion of more complex nitrogenous materials will be present in the wort. The type of relationship between mashing temperature, time and nitrogen in the wort is illustrated in Fig. 9.10. The effects on two wort nitrogen fractions of varying the starting temperatures of temperature – programmed mashes are illustrated in Fig. 9.11.

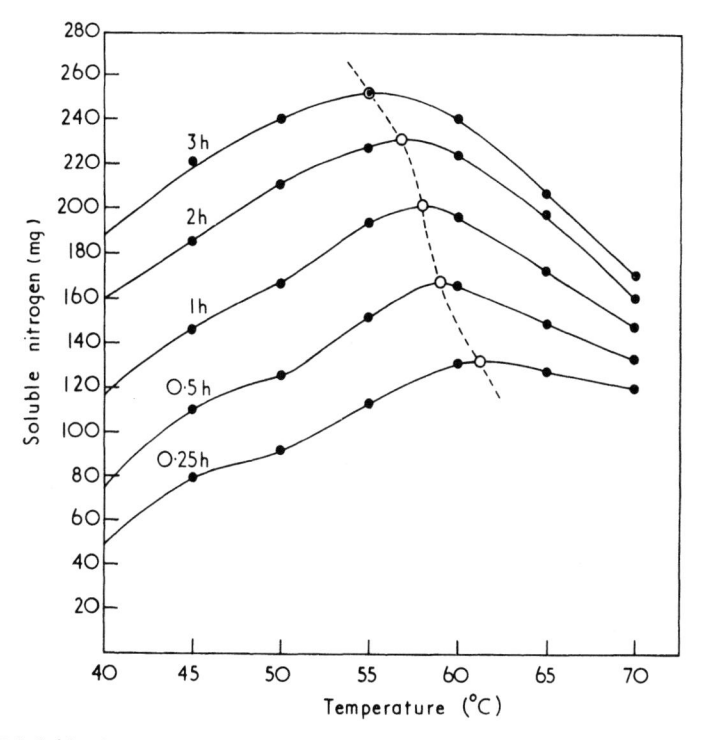

Fig. 9.10 Soluble nitrogen formed by mashing for different times at different fixed temperatures. ●, soluble nitrogen formed in mashes allowed to stand for the times shown; ○, apparent temperature optima for different mashing times. Entommen aus F. Schönfeld, *Handbuch der Brauerei und Malzerei*, Dritter Band, 1935, Verlag Paul Parey, Berlin.

The temperature-sensitivity of some of the dissolution processes involving the starch and nitrogenous materials of malt helps to explain how under-modified malts are successfully mashed by mashing techniques in which the overall temperature of the mash is gradually increased from low to high values, thus giving temperature-sensitive enzymes a chance to act. For example, the temperature of about 50°C (122°F) held in decoction mashing is termed the 'protein rest' and is included to allow proteases, peptidases and other heat-labile enzymes like β-glucanase to act before the mash temperature is increased to levels at which those enzymes are rapidly destroyed, but at which starch conversion proceeds rapidly. At the same time, in lightly kilned malts, it is probable that phytase will also act at this mash temperature and release inorganic phosphate. Table 9.9 illustrates some of these points. The results of a temperature-programmed infusion-mashing schedule are quite different from those expected from a British infusion mash, but the 'high-quick-mash' process, no doubt due to the high mashing-in temperature,

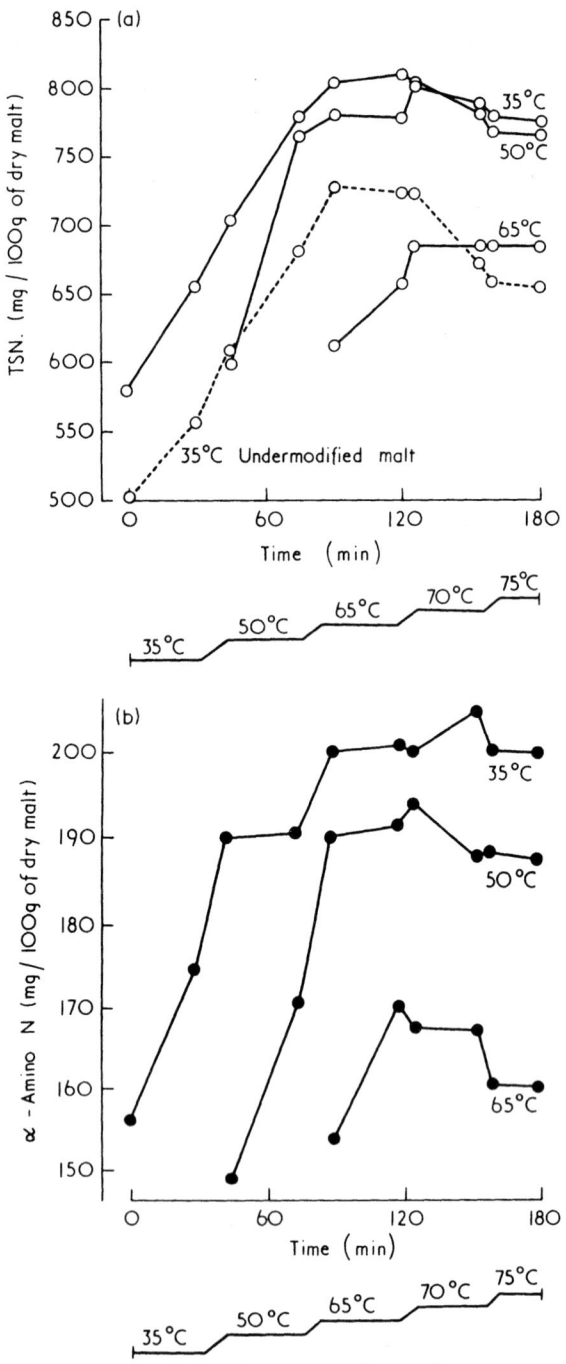

Fig. 9.11 Increases in (a) total soluble nitrogen and (b) α-amino nitrogen in temperature-programmed mashes started at the temperatures shown. ———, normal malt. - - - -, undermodified malt. {Narziss (1977) [83]}.

TABLE 9.9

Comparison of worts made with an undermodified malt (*u*) and a well-modified malt (*m*) using different mashing processes {LÜERS et al. (1934) [84]}

Degree of modification of malt ...	Brewery extract (% of dry matter) u	m	Wort attenuation (%) u	m	Relative viscosity of sweet wort u	m	Total soluble nitrogen (% of dry extract) u	m	Amino nitrogen (% of dry extract) u	m	Pi m only	Sludge on boiling (kg of dry matter) u	m
Two-mash process: Mash in at 50°C (122°F) Decoctions to 67°C (152¾°F) and then 76°C (169°F)	77·5	78·8	66·2	65·6	1·786	1·803	0·87	0·85	0·29	0·26	83·5	11·5	13·6
Three-mash process: Mash in 35·5°C (96°F) Decoctions to 50°C (122°F), 65°C (149°F), and 76°C (169°F)	78·3	79·3	63·4	64·5	1·714	1·785	0·93	0·89	0·31	0·29	86·4	13·6	9·9
'High-quick-mash process' Mash in 68°C (154¾°F), then raise to 76°C (169°F)	76·2	78·4	67·4	67·5	1·758	1·793	0·92	0·78	0·25	0·21	79·4	26·7	17·5
Temperature-programmed mash Mashed in 35°C (95°F) temperature increased with steam coils – held at 50°C (122°F), 67°C (152¾°F) and 76°C (169°F)	78·2	78·8	66·8	66·5	1·675	1·700	0·94	0·86	0·28	0·27	86·4	26·9	22·8

approaches a traditional British infusion system in its results (Table 9.9).

Investigations of the extracts obtained from infusion mashes made at different constant temperatures for different times (Figs. 9.7 and 9.9) have shown that infusion mashes carried out for shorter periods at lower temperatures can, with adequately modified malts, yield acceptable worts (Table 9.10) [85–88]. In the cases shown, recovery of the starch, as extract, was about 98%. At the lower temperatures the alteration in the Lundin fractions showed that proteolysis was more complete. This was also reflected by a higher hop utilization, probably due to the reduced break caused by the presence of lesser quantities of complex nitrogenous substances to bring down the bittering substances.

TABLE 9.10

Effects of 'standing' infusion mashes for different times at different temperatures on the composition of the worts and the derived beers
{HALL (1957) [85]}

Temperature/rest time of the mashes ...	57°C (135°F)/ 15 min	63°C (145°F)/ 30 min	65·5°C (150°F)/ 120 min
Carbohydrates (% of wort solids)			
Fermentable ⎧ Hexose	6·9	8·0	7·9
Sucrose	4·1	4·5	5·2
Maltose	44·9	41·0	41·0
⎩ Trisaccharide	13·5	15·0	14·2
Unfermentable dextrins	21·2	21·5	22·2
General wort analysis			
Colour, EBC units	9·0	7·0	6·5
PSN (mg/100 ml at SG 1040)	76·4	73·0	72·0
pH	5·52	5·52	5·46
Fermentability (%) (forced fermentation)	69·5	69·0	68·7
Beer			
Bitterness, as *iso*-compounds (mg/l)	30	27	22
Hop utilization (%)	18·2	16·4	13·3

9.16 Mash thickness [22, 23, 25, 28, 61]

Reduction in the grist/liquor ratio in the mash (*i*) reduces the temperature stability of some of the enzymes present, including proteases and disaccharidases (*ii*) dilutes the enzymes, (*iii*) dilutes the substrates on which the enzymes act, (*iv*) dilutes the products of enzyme action and (*v*) reduces the viscosity of the wort. Hydrolytic reactions may proceed at a greater rate in more

TABLE 9.11

Influence of mash temperature and concentration on the composition of sweet wort
{Data of HALL quoted by HARRIS (1962) [22]}

Mashing temperature....	60°C (140°F)			65·6°C (150°F)			68·3°C (155°F)		
Mash thickness (%)...*	67	39	29	67	39	29	67	39	29
Wort analyses†									
Hexose	12·3	10·1	9·5	11·9	9·5	8·1	11·0	10·2	8·0
Sucrose	2·8	3·4	3·4	4·1	4·2	3·8	3·7	5·0	4·0
Maltose	43·9	48·3	49·5	38·8	43·9	42·8	36·9	37·0	39·0
Trisaccharide	14·3	14·3	13·8	12·6	13·6	15·0	12·8	12·7	14·3
Dextrin	17·5	15·5	14·6	24·2	21·2	22·3	27·6	26·2	26·9
Fermentability (%)	73·3	76·1	76·2	67·4	71·2	69·7	64·4	65·0	65·3
Extract (%)	55–63	76·2	75·6	73·4	75·3	74·2	73·3	74·6	74·0
Soluble N (% of wort solids)	6·2–6·6	5·34	5·50	5·58	5·22	5·03	4·90	4·77	4·85
pH	5·46	5·40	5·50	5·31	5·33	5·38	5·31	5·35	5·30

* Parts of grist/100 parts of water.
† Carbohydrates expressed as % of wort solids.

dilute mashes, because the products of reaction are less concentrated and so inhibit enzyme activity less. More concentrated mashes yield more soluble nitrogen and more hexose sugars in the wort (Table 9.11). In British breweries the nitrogen levels in brewery worts are 0–40% above those found in small-scale analytical hot-water extracts [89]. Proteases are not stable at 60°C (140°F) in a dilute mash, and at 65°C (149°F) this instability is accentuated. At 60°C (140°F) the fermentability of the extract is usually greater in worts from mashes of lower concentrations because the more dilute solutions of sugars produced are less inhibitory to the amylases. In general, fermentability is maximal in worts prepared from mashes of 16–32% solids concentration, while maximal extracts are obtained from 39% mashes [22]. Clearly results depend on the quantities of enzymes initially present in the mash. In thick mashes the rate of saccharification is retarded, probably because the accumulating sugars competitively inhibit the hydrolytic enzymes, and so wort fermentabilities are reduced when short mashing times are used.

9.17 *Mashing and the dissolution of starch* [22, 25, 79]

The most notable change brought about in mashing is the dissolution of starch to yield the greater part of the wort carbohydrates. As indicated different mashing conditions may yield worts having different degrees of fermentability. Consideration of the relative levels of the hexoses, maltose and sucrose in Table 9.11 indicates that in thick mashes, or mashes made at low temperatures, invertase and maltase are active. The structure of starch, and the activities of the enzymes that degrade it were described in Chapter 4. The major facts of starch dissolution as it occurs in an infusion mash may be explained by the joint action of the α- and β-amylases with small contributions from debranching enzymes and α-glucosidase ('maltase'). It is probable that model experiments made with α- and β-amylase have usually been made with enzyme concentrates containing small amounts of other carbohydrases [90, 91]. α-Amylase acting on starch alone will ultimately produce a dextrinous wort containing a mixture of sugars, of which only 16–20% are fermentable. In contrast, β-amylase will not attack starch granules, but when it attacks soluble starch it produces only maltose and a β-limit dextrin still sufficiently large to give a blue colour with iodine. If the mash is made with β-amylase and pullulanase a debranching enzyme, and only small amounts of α-amylase, a highly fermentable wort, rich in maltose and maltotriose, can be obtained (Chapter 8). Acting in concert, an excess of a mixture of pure amylases (α and β) will theoretically break down starch to yield a wort that is about 70% fermentable. This figure increases to about 80% if limit dextrinase or other debranching enzyme is present. It should be recalled that α-glucosidase is able to hydrolyse terminal glucoside-α-(1→6) linkages. The carbohydrates in boiled sweet wort are usually about 75% fermentable and in unboiled wort, prepared from low-kilned malts, about 82% fermentable. Thus enzymes other than the α- and β- amylases do participate in the mashing process and may continue to act in the sweet wort. Normally their activity is terminated by the hop-boil. The fermentability values quoted here refer to the carbohydrate components of the extract. They do not take account of other wort constituents or accumulations of dissolved ethanol or carbon dioxide. Consequently they are not directly reflected in changes in the excess gravity which occur when the wort is fermented.

In mashes made at 65°C (149°F) limit dextrinase, β-amylase and no doubt α-glucosidase are progressively destroyed, although less rapidly than in solution, at least in the case of β-amylase. α-Amylase activity also declines, although in the presence of an excess of calcium ions and at pH 6·0 the impure (but not the pure) enzyme is reported to be stable in solution at 70°C (158°F), supposedly because some of the impurities exert a protective action. Possibly a low level of free calcium ions and an unfavourable pH in mashes encourage instability in this enzyme, allowing proteolytic and heat inactivation. At increasing mash temperatures smaller quantities of diastatic enzymes survive to come into solution, and at higher temperatures the enzyme activity declines more rapidly (Fig. 9.12). β-Amylase is destroyed in mashes in 40–60 min at 65°C (149°F). The α-amylase activity declines at 67°C (153°F), the last activity disappearing in about 2 hr. Thus the mashing temperature should be achieved and maintained as precisely as possible, since it has a large effect on the survival of the amylases and other starch-degrading enzymes, and hence on the fermentability of the wort. 'Model' experiments

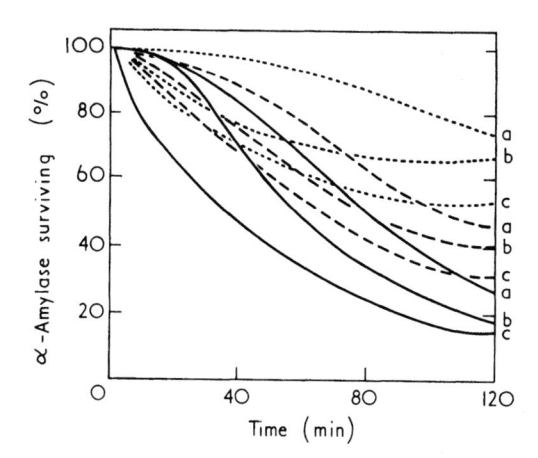

Fig. 9.12 The survival of α-amylase in model brewery mashes made at three different temperatures, and three grist concentrations, using ethanol-extracted malt grist {Young and Briggs (1969) [90]}. (a) 20% mash; (b) 16·7% mash; (c) 14·3% mash; ·····, 64°C (147°F); – – – –, 65·5°C (150°F); ——, 68°C (154·5°F).

in which mixtures of α-amylase and β-amylase were allowed to degrade solutions of soluble starch emphasize the interrelationships between (*i*) the initial level and composition of the mixture of carbohydrases in mashing, (*ii*) the mash temperature, and (*iii*) the effects of these variables on the composition of the final worts (Fig. 9.13). Similar results have been obtained in experiments in which enzyme concentrates were allowed to act on the starch in a malt grist in which the enzymes had been inactivated with boiling ethanol. In British malts the ratio of β-amylase to α-amylase often approximates to 4:1, while in malts from continental Europe the figure is 5:1–6:1 [91].

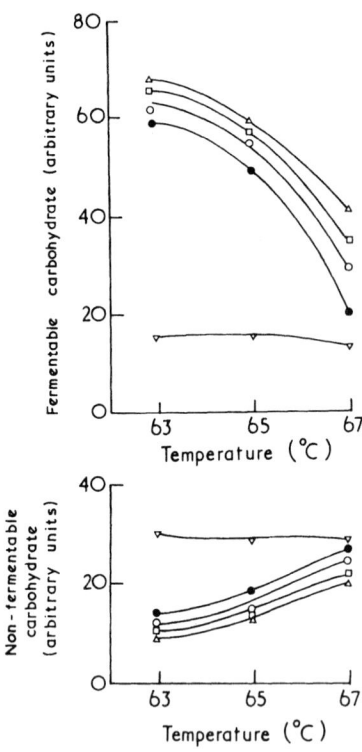

Fig. 9.13 The proportions of fermentable and non-fermentable dextrins formed from soluble starch, digested with different mixtures of α-amylase and β-amylase for 1 hr at the temperatures shown {Preece and Shadaksharaswamy (1949) [91]}. In these experiments β-amylase was inactivated in about 15 min at 65°C (149°F), as against the longer times of 40–60 min found in all-malt mashes. Ratios of β-amylase/α-amylase: 6·25 : 1 (△); 5 : 1 (□); 3·75 : 1(○); 2·5 : 1 (●); 0 : 1 (▽).

In mashing it is the liquefying action of α-amylase that is chiefly responsible for the dissolution of the starch granules, which only begin to gelatinize at about 65°C (149°F), but which slowly swell at lower temperatures [55–60°C,

(131–140°F)], and become progressively more susceptible to enzymic degradation. The subsequent formation of the characteristic spectrum of sugars is due to the concerted action of all the relevant enzymes on the products of liquefaction. Starch that is closely invested with protein is comparatively resistant to amylolysis, and the breakdown of this protein by proteases reduces its 'blocking' effect [92]. The survival of quantities of fine starch granules through mashing, due to their resistance to amylolysis, may cause run-off problems during lautering [93].

When malts having different properties are blended in various proportions and are mashed it is found that in many instances the wort analyses are intermediate between those found with the malts mashed alone and are directly dependent on the proportions of the malts in the mixture. However, in one case at least wort attenuation was higher than had been predicted from a knowledge of the calculated 'intermediate value', owing to unexpectedly high levels of the 'maltose *plus* maltotriose' carbohydrate fraction [94]. The same relationship held when the grist contained an adjunct mixed with various malt mixtures. Presumably the malt giving the more fermentable wort had an excess of at least one enzyme which was deficient in the other, but was adequate in the mixed grist.

9.18 Sparging

First worts may be collected from the mash at the final 'stand' temperature [e.g. about 65°C (149°F)] or after heating to about 75°C (167°F). Sparging, the process of washing the last extractives from the mash, is performed by using water that is hotter than the mash, for example, at 75–80°C (167–176°F) by spraying it on to the goods, that is the solids in the mash tun, and allowing it to percolate through. The higher temperature results in (*i*) a reduced wort viscosity and hence a faster run-off, (*ii*) a more rapid leaching of soluble extractives from the residual insoluble particles and (*iii*) almost complete inactivation of the enzymes. At higher temperatures carbonate waters become more alkaline and this, combined with the reduced buffering capacity of the mash, results in a rise in pH of the last runnings of 0·2–0·7 pH unit. As the pH rises, the composition of the wort solids alters and becomes less desirable as proportionately more silicates, phosphates, polyphenols, high-molecular-weight nitrogenous materials and possibly polysaccharides come into solution (Figs. 9.14 and 9.15). Most of the silicates found in wort come from the malt husk [95]. Because of the undue proportion of polyphenols, silicates and other unwanted materials in the last runnings (Table 9.12), it has been suggested that these should be treated with active charcoal, at 10–50 g/hl of wort to remove some of the polyphenols, tannins, coarsely flavoured substances and the nitrogenous materials having high-molecular-weights to improve the flavour of the beer and reduce the levels of haze precursors. Charcoal added

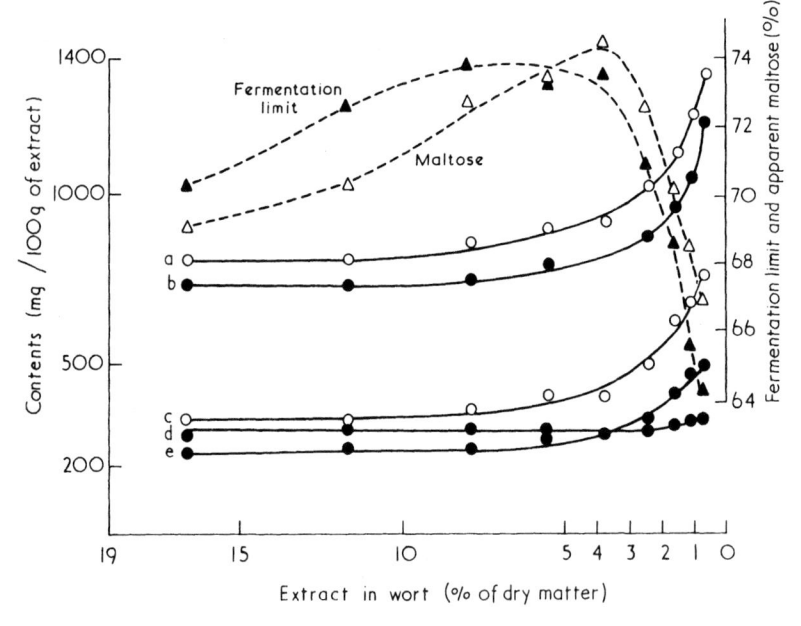

Fig. 9.14 Some analyses of sweet-wort solids at different stages of run-off {Schild (1936) [55]}. ▲, Fermentation limit (%). △, Apparent maltose (%). (a) ○, Total soluble nitrogen (mg/100 g). (b) ●, Permanently soluble nitrogen (mg/100 g). (c) ○, Soluble nitrogen not precipitated with phosphotungstic acid (mg/100 g). (d) ●, Formol nitrogen (mg/100 g). (e) ●, Nitrogen not precipitated with tannic acid (mg/100 g).

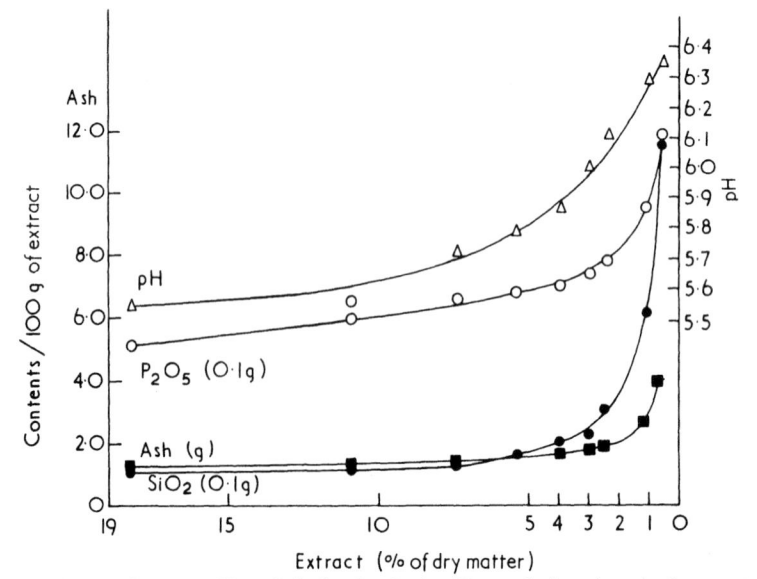

Fig. 9.15 Changes in wort pH, and the levels of ash, silica and phosphate in the sweet-wort solids (g/100 g) at different stages of run-off {Schild (1936) [55]}; (see also Fig. 9.14).

TABLE 9.12

Silica and tannin contents of laboratory brews made from whole malt, and malt 'dehusked' in a pearling machine
{STONE and GRAY (1948) [95]}

	Normal malt			Dehusked malt		
	Extract (%)	Silicates (SiO$_2$, mg/l)	Tannins (mg/l)	Extract (%)	Silicates (SiO$_2$, mg/l)	Tannins (mg/l)
First wort	20·5	37	78	25·4	nil	66
Spargings 1	16·9	39	49	13·8	nil	—
2	8·3	17	25	7·6	nil	24
3	6·0	18	24	6·4	nil	25
4	5·0	31	18	—	nil	—
Unhopped wort	12·0	30	—	12·1	nil	—
Hopped wort	12·0	35	59	12·1	3	46
Beer at end of fermentation	—	45	46	—	15	—
Beer after storage	—	51	47	—	16	43

to the mash or to the copper before the boil usefully reduces the tendency of the derived beer to form a haze [76, 96–98]. Thus the solids in the last runnings are rich in potential haze-forming materials, and beer prepared from them has a higher tendency to form haze and has a harsher flavour than that prepared from the first runnings. In the first spargings the percentage fermentability of the extract may increase, probably because, as the mash is diluted, the simpler carbohydrates are less-effective competitive inhibitors of carbohydrase action, and further hydrolysis takes place. However, in the later stages of sparging the fermentability of the wort falls sharply. High-molecular-weight viscous β-glucans can be extracted from the grist during sparging and may slow the filtration of strong worts to a serious extent, and later may even separate as gelatinous precipitates in strong beers [99].

9.19 Spent grains

At the end of sparging, spent grain (draff) remains, equivalent to about 20–22% by dry weight of the original grist. The steady production of large quantities of this potentially valuable by-product can be an embarrassment to brewers. This material retains a little of the potential extract, phosphates, lipids and some proteins. Approximate analyses are shown in Table 9.13. Spent grains may be disposed of while wet (70–80% moisture), but in this state they deteriorate and become mouldy within hours. If they are intended for use in cattlefeed, they could usefully be treated with a preservative or ensiled [31]. Often these materials are pressed to express surplus moisture and are then dried in a stream of hot air. They may leave the brewery for the merchants either direct from the mash tun or ready dried. The dried materials are used for cattlefeed, either after being compounded or being made into

TABLE 9.13

Analyses (approximate) of dried spent brewer's grains
{PREECE (1931) [100]; HOPKINS and KRAUSE (1947) [23]}

		(i)	(ii)
Water content	(%)	10·9	9
Fibre	(%)	16·9	18
'Protein' (nitrogen × 6·25)	(%)	14·4	21
Digestible 'protein'		—	15
Nitrogen-free extract	(%)	37·9	40
Starch	(%)	11·6	—
Fat, oil	(%)	4·5	9
Ash	(%)	3·9	4

silage. In this connection, they have about three-quarters of the nutritive value of barley, on a dry-weight basis. Nevertheless, they must not constitute more than one-third of the total ration of cattlefood. Means of increasing the value of spent grains might include treating with caustic alkali or ammonia to improve their feed value for ruminants, or increasing their complement of essential amino acids by using them as a basis for culturing filamentous fungi to feed to pigs [31]. They have also been used in preparing compost for growing mushrooms on a commercial scale. Experimentally, spent grains have been fractionated, and dried fractions have been incorporated into doughs used in bakeries. It is possible to use a wet separation process to divide spent grains into a protein-rich fraction suitable for feeding to monogastric animals, and a fibrous fraction that may be fed to ruminants.

The water pressed from spent grains has a high biological oxygen demand (BOD) and constitutes an undesirable effluent. Efforts to utilize this liquor include: (i) recovering its extract by incorporating it in a subsequent mash (with or without prior treatments to remove lipids and polyphenols); (ii) using it as an anti-foaming agent in deep fermentors; (iii) a proposal to use it as a nutrient medium for supporting useful micro-organisms, e.g. fungi that produce citric acid; (iv) solids collected from it by centrifugation have been added to animal feeds and (experimentally) to bakehouse products.

9.20 Other grist materials

A variety of materials, other than conventional malt, may be added to the grist. Special barley malts are described in Chapter 5 and unmalted adjuncts and industrial enzymes are described in some detail in Chapter 8. Malted wheat, oats or rye may be added to confer special qualities to the beer. The use of wheat malt at up to 20% of the total grist has been claimed to 'strengthen' the yeast, and improve the clarity and head retention of the beer. In the case of Munich Weissbier, wheat malt constitutes the greater part of the grist.

A typical wheat malt analysis might be: moisture, 1·7%; HWE 304·5 litre°/kg, 20°C; (103 lb/336 lb); colour, 10 E.B.C. units; DP, 35–55 (°Lintner); total nitrogen 1·49–1·60 (% dry); permanently soluble nitrogen 0·52–0·61 (% dry) [101].

In general, unmalted adjuncts, highly coloured malts and roasted barley are partly or wholly deficient in the enzymes needed in mashing, so these must be supplied by enzymes from malt or from micro-organisms. Common mash-tun adjuncts include raw whole barley, roasted barley, torrified or micronized barley or wheat, barley and wheat flours, pelleted wheat flours, maize or potato starches (which may be pelleted or granulated), grits prepared from maize, sorghum or rice, flaked maize, flaked rice, flaked barley and (rarely) flaked oats or flaked pearled barley.

The changes that occur during malting make barley starch more susceptible to enzymic degradation during the mashing process. The chemical and physical nature of the starch is altered and the investing endosperm structures are partly degraded (Chapter 4). Many physically isolated granular starches are comparatively resistant to attack by enzymes until they have been either swollen or completely dispersed in hot water, that is gelatinized (Chapter 8), or they have been abraded, chipped or broken by milling. Milled flours, for example barley flour, yield their extract more readily if they are first presoaked or gelatinized but these processes are not essential [102–107]. Indeed, cooking barley adjuncts is generally avoided because of the increased quantities of viscous β-glucans that cooking brings into solution. Flours have been successfully used in infusion breweries, in spite of early discouraging results. To obtain a good carbohydrate recovery it is essential to mix the flour and malt very thoroughly. Worts from grists containing unmalted adjuncts do not have the same properties as those from all-malt grists. For example, in many instances the nitrogen content of the wort is reduced, as are the polyphenol levels and the spectrum of carbohydrates that are present are at least marginally different. Worts prepared from mashes containing some adjuncts, such as cooked maize or rice grits, will be comparatively poor in β-glucans, while worts prepared from grists containing raw barley will contain the non-fermentable trisaccharide raffinose and will be more viscous and slower to filter because of the presence of undegraded β-glucan gums (Table 9.14). This problem may be overcome by using a malt rich in β-glucanase, or by adding bacterial β-glucanase to the mash, and using a low-temperature rest in the mashing sequence. In contrast, wheat flours are less prone to give β-glucan-dependent run-off problems. Unmalted adjuncts may give rise to run-off problems, owing to the formation of an unduly large proportion of fine grist particles composed of varying proportions of proteins, hemicelluloses, small and undegraded starch granules and lipids. Additions of β-glucanase to such mashes improve run-off, but do not reduce the volume of the particles [109].

TABLE 9.14

Analyses of worts and beers prepared from malt and various adjuncts
{HUDSON (1963) [108], HUDSON et al. (1963) [107],
BIRTWISTLE et al. (1962) [105]}

Grist material ...	Malt (100%)	Malt (75%) Wheat flour (25%)	Malt (75%) Barley flour (25%)	Malt (50%) Wheat flour (50%)	Malt (50%) Maize flakes (50%)	Malt (50%) Rice flakes (50%)
Wort						
Carbohydrate recovery (%)	>98	>98	—	90–95	98	97
Fermentability (%)	75–77	75–77	72–74	65–70	65–70	65–70
Nitrogen (mg/100 ml)	90–95	72–82	70–73	50–70	50	50
Beer						
Bitterness (mg/l)	23–24	26–27	26–29	28–30	30	28
Total soluble nitrogen (mg/100 ml)	44–48	34–40	33–36	25–40	25–30	25–30
Anthocyanogens*	0·7	0·5–0·6	0·6–0·7	0·4–0·6	0·4	0·5
Shelf life (weeks)	approx. 12	>20	16	20	20	15
Head retention (half-life, sec)	approx. 90	100–110	95–100	92–96	98	96

* Arbitrary units.

Traditional British mashing methods may be extended by using highly enzymic malts to convert adjuncts. In North American brewing, and in the preparation of grain whiskeys, enzymic malts are prepared from six-rowed barleys of high nitrogen content (e.g. 2·0–2·2% N) that, after a long and cool germination period, are dried and kilned slowly at relatively low temperatures, to maintain a high level of enzymes in the product. In Britain interest has revived in the use of malts prepared from low-nitrogen barleys in which high levels of enzymes are maintained either (*i*) by kilning very lightly or (*ii*) by using the green unkilned malt [106, 112]. Green malt is sticky when broken up, and needs to be ground in special mills. Also it is not stable on storage, so must be used as soon as it is ready. Therefore malt carefully kilned to the hand-dry stage that (*i*) may be ground in conventional mills and (*ii*) can be stored would seem to be more convenient. Such malt will readily convert a mixed grist containing 25% wheat flour. The use of green malt in mixed grists containing various unmalted adjuncts has also been investigated (Table 9.15). Carbohydrate recovery and wort analyses have been satisfactory, but the flavours of the products made in this manner have sometimes been unacceptable. Steam-stripping the wort or allowing sufficient evaporation in the hop-boil may remove the components with the objectionable flavours, which are also removed when 'green malt worts' are concentrated to syrups (Chapter 8). Presumably these volatile flavour components are normally removed on the kiln.

TABLE 9.15
Composition of worts prepared using green malt
{MACWILLIAM et al. (1963) [106]}

Grist material ...	Green malt	Freeze-dried green malt*	Green malt (50%) Wheat flour (50%)	Green malt (50%) Barley flour (50%)	Average kilned pale ale malt
Carbohydrate recovery (%)	99	98·8	97·0	96·5	98
Extract ⎰(litre°/kg, 20°C)		304·5	305·3	300·5	296·0–301·7
⎱(lb/Q.)		103·0	103·3	101·6	100–102
Specific gravity (SG) of wort	1028·0	1029·7	1029·9	1028·7	1028·0
pH	5·75–6·0	6·1	5·7	5·7	5·6
Total soluble nitrogen (mg/100 ml)	60–75	66·1	42·2	40·2	50
Amino nitrogen (% of total N)	40	—	36·6	35·3	33
Fermentability (%)†	86	82	80	78·2	75
Anthocyanogens‡	0·05–0·15	0·06	0·06–0·08	0·11	0·4
Colour, EBC units	8–10	6	6	6	4–7

* Moisture 7%.
† Green malt gave wort higher in maltose and maltotriose, and lower in dextrins.
‡ Arbitrary units.

The use of adjuncts may, in addition to providing cheaper extract, give a bonus in terms of increased shelf life and head retention in the final beer, an increased hop utilization associated with a smaller copper break, and a 'cleaner' more bland flavour. The reduced soluble nitrogen level in worts prepared from mixed barley–malt grists is not only due to a dilution effect of the barley, but also to the action of high-molecular-weight proteolytic inhibitors found in the barley, but not malt, that check the action of the malt, and microbial, proteases during mashing [110]. Both barley and malt contain inhibitors of bacterial proteases that are progressively inhibited during mashing [111].

The caution traditionally shown by British brewers in the use of starchy adjuncts has been due to a number of causes. Among these may be mentioned: (*i*) difficulties in handling adjuncts, particularly adequately mixing them with other components of the grist to ensure complete starch conversion and easy wort separation; (*ii*) difficulties in achieving complete saccharification; (*iii*) slower wort run-off; (*iv*) the development of unusual (not necessarily unpleasant) flavours in the beer; (*v*) a disinclination to alter a traditional process. It is notorious that extract recovery from adjuncts in the brewhouse can fall far below that obtained in the laboratory. The reasons for this include milling and mixing problems, and variations in the mash thickness and mashing temperature sequences. In any case such alterations cannot be introduced unless the products are readily acceptable to the consumers.

The variability of the raw materials and the large number of simultaneous processes that occur in the mash makes it impossible to generalize about the detailed biochemistry that is involved. When the numerous variations in brewing practice are also considered it becomes apparent that each brewer must test mashing plant and grist materials for their suitability for his own needs. Small-scale brewing plant is cheaper to use in testing new raw materials and techniques, but obviously it must match the performance of the production plant exactly, so that all the characteristics of the final beer may be evaluated and compared with the 'test normal' control product. Trials on the production scale are always expensive.

REFERENCES

[1] HUDSON, J. R. (1960). *Development of Brewing Analysis*, The Institute of Brewing, London.
[2] WAINWRIGHT, T. and BUCKEE, G. K. (1977). *J. Inst. Brewing*, **83**, 325.
[3] Institute of Brewing, Analysis Committee (1977). *Recommended Methods of Analysis of Barley, Malt and Adjuncts*, The Institute of Brewing, London. (Previous – (1971). *J. Inst. Brewing*, **77**, 181).
[4] European Brewery Convention, Analysis Committee (1975). *Analytica-EBC* (3rd edn.), Elsevier, London.
[5] Technical Committee and Editorial Board of the American Society of Brewing Chemists (1976). *Methods of Analysis of the American Society of Brewing Chemists* (7th edn.), Madison, New York.
[6] ESSERY, R. E. (1954). *J. Inst. Brewing*, **60**, 21; 193; 303.
[7] BISHOP, L. R. and HICKSON, W. (1948). *J. Inst. Brewing*, **54**, 189.
[8] BISHOP, L. R., CUFF, C. M. and HICKSON, W. (1948). *J. Inst. Brewing*, **54**, 194.
[9] Analysis Committee Newsletter (1978). *J. Inst. Brewing*, **84**, 215.
[10] HYDE, W. R. and BROOKES, P. A. (1978). *J. Inst. Brewing*, **84**, 167.
[11] SCHNELLER, M. A. (1949). *Proc. Annu. Meet. Am. Soc. Brewing Chem.*, p. 74.
[12] BISHOP, L. R. and WHITLEY, W. A. (1943). *J. Inst. Brewing*, **49**, 223.
[13] HARGITT, R. and BUCKEE, G. K. (1978). *J. Inst. Brewing*, **84**, 224.
[14] BATHGATE, G. N. (1979). *J. Inst. Brewing*, **85**, 92.
[15] WINDISCH, W., KOLBACH, P. and VOGEL, C. (1929). *Woch. Brau.*, **46**, 403; 417.
[16] BROWN, H. T. (1909). *J. Inst. Brewing*, **15**, 169.
[17] BISHOP, L. R. (1947). *J. Inst. Brewing*, **53**, 129.
[18] BUCKEE, G. K., HICKMAN, E. and BROWN, D. G. W. (1976). *J. Inst. Brewing*, **82**, 299.
[19] Analysis Committee Newsletter (1979). *J. Inst. Brewing*, **85**, 78.

[20] WILLIAMS, D. O., *Irish Maltsters' Conference, 1952-1961, Selected Papers,* Arthur Guinness, Dublin, p. 301.
[21] HUDSON, L. E., *Irish Maltsters' Conference, 1952-1961, Selected Papers,* Arthur Guinness, Dublin, p. 286.
[22] HARRIS, G. (1962). *Barley and Malt* (ed. A. H. COOK), Academic Press, London, p. 583.
[23] HOPKINS, R. H. and KRAUSE, B. (1947). *Biochemistry Applied to Malting and Brewing,* George Allen & Unwin, London.
[24] MACWILLIAM, I. C. (1968). *J. Inst. Brewing,* **74,** 38.
[25] Various (1974). *Eur. Brewery Conv. Monograph-I, EBC Wort Symposium, Zeist,* 383 pp.
[26] NARZISS, L. (1976). *MBAA Tech. Q.,* **13**(1), 11.
[27] BUTTON, A. H. and PALMER, J. R. (1974). *J. Inst. Brewing,* **80,** 206.
[28] HIND, H. L. (1950). *Brewing Science and Practice,* Vols. 1 and 2, Chapman & Hall, London.
[29] DE CLERCK, J. (1957). *A Textbook of Brewing,* Vol. 1 (translated by K. BARTON-WRIGHT), Chapman & Hall, London.
[30] HALL, R. D., HARRIS, G. and MACWILLIAM, I. C. (1956). *J. Inst. Brewing,* **62,** 232.
[31] BRIGGS, D. E. (1978). *Barley,* Chapman & Hall, London, 612 pp.
[32] ENEVOLDSEN, B. S. and SCHMIDT, F. (1974). *J. Inst. Brewing,* **80,** 520.
[33] ENEVOLDSEN, B. S. (1974). *Eur. Brewery Conv. Monograph-I, EBC Wort Symposium, Zeist,* p. 158.
[34] ANDERSON, F. B. (1966). *J. Inst. Brewing,* **72,** 384.
[35] BARRETT, J. and KIRSOP, B. H. (1971). *J. Inst. Brewing,* **77,** 39.
[36] MIKOLA, J., PIETILÄ, K. and ENARI, T.-M. (1972). *J. Inst. Brewing,* **78,** 384.
[37] BURGER, W. C. and SCHROEDER, R. L. (1976). *J. Am. Soc. Brewing Chem.,* **34,** 133; 138.
[38] JONES, M. and PIERCE, J. S. (1963). *Proc. Eur. Brewery. Conv. Congr., Brussels,* p. 101.
[39] HARRIS, G. and PARSONS, R. (1955). *J. Inst. Brewing,* **61,** 29.
[40] HARRIS, G. and PARSONS, R. (1957). *J. Inst. Brewing,* **63,** 227.
[41] HARRIS, G. and PARSONS, R. (1958). *J. Inst. Brewing,* **64,** 308.
[42] BOLLINDER, A., KURZ, W. and LUNDIN, H. (1956). *J. Inst. Brewing,* **62,** 497.
[43] ZIEGLER, L. and PIENDL, A. (1976). *MBAA Tech. Q.,* **13,** 177.
[44] DREWS, B., JUST, F. and DREWS, H. (1957). *Proc. Eur. Brewery Conv. Congr., Copenhagen,* p. 167.
[45] NORRIS, F. W. and DARBRE, A. (1956). *Analyst,* **81,** 394.
[46] SANDEGREN, E. (1948). *J. Inst. Brewing,* **54,** 200.
[47] WEINFURTNER, F., ESCHENBECHER, F. and HETTMAN, K. H. (1966). *J. Inst. Brewing,* **72,** 593.
[48] HOPKINS, R. H. and WEINER, S. (1944). *J. Inst. Brewing,* **50,** 124.
[49] TULLO, J. W. and STRINGER, W. J. (1945). *J. Inst. Brewing,* **51,** 86.
[50] HOPKINS, R. H. and WEINER, S. (1945). *J. Inst. Brewing,* **51,** 34.
[51] NORRIS, F. W. (1945). *J. Inst. Brewing,* **51,** 177.
[52] HOPKINS, R. H., WEINER, S. and RAINBOW, C. (1948). *J. Inst. Brewing,* **54,** 264.
[53] HOPKINS, R. H. and PENNINGTON, R. J. (1947). *J. Inst. Brewing,* **53,** 251.
[54] LYNES, K. J. and NORRIS, F. W. (1948). *J. Inst. Brewing,* **54,** 150; 207.
[55] SCHILD, E. (1936). *Woch. Brau.,* **53,** 345; 353.

[56] WOOF, J. B. and PIERCE, J. S. (1966). *J. Inst. Brewing*, **72**, 40.
[57] WOOF, J. B. and PIERCE, J. S. (1968). *J. Inst. Brewing*, **74**, 262.
[58] KIRBY, W., WILLIAMS, P. M., WHEELER, R. E. and JONES, M. (1977). *Proc. Eur. Brewery Conv. Congr. 16th, Amsterdam*, p. 415.
[59] ENEBO, L., BLOMGREN, G. and JOHNSSON, E. (1955). *J. Inst. Brewing*, **61**, 408.
[60] ÄYRÄPÄÄ, T., HOLMBERG, J. C. and SELLMANN-PERSSON, G. (1961). *Proc. Eur. Brewery Conv. Congr., Vienna*, p. 286.
[61] EDWARDS, R. and THOMPSON, C. C. (1968). *J. Inst. Brewing*, **74**, 257.
[62] RINKE, W. (1965). *J. Inst. Brewing*, **71**, 62.
[63] THOMPSON, C. C. and RALPH, D. J. (1967). *Proc. Eur. Brewery Conv. Congr. 11th, Madrid*, p. 177.
[64] TRESSL, R., BAHRI, D. and SILWAR, R. (1979). *Proc. Eur. Brewery Conv. Congr. 17th, Berlin (West)*, p. 27.
[65] ROBERTS, R. J. (1971). *J. Inst. Brewing*, **82**, 96.
[66] POMERANZ, Y. and DIKEMAN, E. (1976–July). *Brewer's Digest*, p. 30.
[67] HOLZMANN, A. and PIENDL, L. (1977). *J. Am. Soc. Brewing Chem.*, **35**(1), 1.
[68] HELIN, T. R. M. and SLAUGHTER, J. C. (1977). *J. Inst. Brewing*, **83**, 17.
[69] HOPKINS, R. H. and AMPHLETT, P. H. (1939). *J. Inst. Brewing*, **45**, 365.
[70] HOPKINS, R. H. and BERRIDGE, N. J. (1949). *J. Inst. Brewing*, **55**, 306.
[71] SANDEGREN, E. (1948). *J. Inst. Brewing*, **54**, 200.
[72] ESSERY, R. E. (1950). *J. Inst. Brewing*, **56**, 319.
[73] ESSERY, R. E. (1951). *J. Inst. Brewing*, **57**, 170.
[74] BERRY, A. E. and BARTRIPP, G. F. (1905). *J. Fed. Inst. Brewing*, **11**, 451.
[75] MACEY, A., STOWELL, K. C. and WHITE, H. B. (1966). *J. Inst. Brewing*, **72**, 29.
[76] WHATLING, A. J., PASFIELD, J. and BRIGGS, D. E. (1968). *J. Inst. Brewing*, **74**, 525.
[77] DEMPSEY, E. C. and BRIGGS, D. E., Unpublished results.
[78] KOLBACH, P. and HAASE, G. W. (1939). *Woch. Brau.*, **56**, 143.
[79] PREECE, I. A. (1954). *The Biochemistry of Brewing*, Oliver & Boyd, Edinburgh, p. 175.
[80] MATTHEWS, C. G. and AUTY, R. A. (1907). *J. Inst. Brewing*, **13**, 685.
[81] SCHUR, F., PFENNINGER, H. B. and NARZISS, L. (1973). *Proc. Eur. Brewery Conv. Congr. 14th, Salzburg*, p. 149.
[82] SCHUR, F., PFENNINGER, H. B. and NARZISS, L. (1975). *Proc. Eur. Brewery Conv. Congr. 15th, Nice*, p. 191.
[83] NARZISS, L. (1977). *Brauwelt*, **117**(37), 1420.
[84] LÜERS, H., KRAUSS, G., HARTMANN, O. and VOGT, H. (1934). *Woch. Brau.*, **51**, 361.
[85] HALL, R. D. (1957). *Proc. Eur. Brewery Conv. Congr., Copenhagen*, p. 319.
[86] HARRIS, G. and MACWILLIAM, I. C. (1961). *J. Inst. Brewing*, **67**, 144.
[87] HALL, R. D., HARRIS, G. and MACWILLIAM, I. C. (1961). *J. Inst. Brewing*, **67**, 151.
[88] HARRIS, G. and MACWILLIAM, I. C. (1961). *J. Inst. Brewing*, **67**, 154.
[89] BROWN, B. M. (1943). *J. Inst. Brewing*, **49**, 140.
[90] YOUNG, R. A. and BRIGGS, D. E. (1969). *J. Sci. Food Agric.*, **20**, 272.
[91] PREECE, I. A. and SHADAKSHARASWAMY, M. (1949). *J. Inst. Brewing*, **55**, 298; 373.
[92] SLACK, P. T., BAXTER, E. D. and WAINWRIGHT, T. (1979). *J. Inst. Brewing*, **85**, 112.

[93] KANO, Y. (1971). *Proc. Annu. Meet. Am. Soc. Brewing Chem.*, p. 27.
[94] YAMADA, K. and YOSHIDA, T. (1976). *Rep. Res. Lab. Kirin Brewery Co. Ltd.*, **19**, 25.
[95] STONE, I. and GRAY, P. P. (1948). *Wallerstein Labs. Commun.*, **11**, 301.
[96] MORRAYE, C. (1938). *J. Inst. Brewing*, **44**, 341.
[97] ENDERS, C. and NOWAK, G. (1939). *Woch. Brau.*, **56**, 243.
[98] VERMEYLEN, J. (1962). *Traité de la fabrication du malt et de la bière*, Vol. 2, Assoc. Roy. des Anciens Élèves de l'Institut Sup. des Fermentations, Gand, Belgium.
[99] ERDAL, K. and GJERTSEN, P. (1967). *Proc. Eur. Brewery Conv. Congr. 11th*, Madrid, p. 295.
[100] PREECE, I. A. (1931). *J. Inst. Brewing*, **37**, 410.
[101] COLLETT, J. H. and GREEN, J. W. (1939). *J. Inst. Brewing*, **45**, 48.
[102] BAKER, J. L. (1942). *J. Inst. Brewing*, **48**, 109.
[103] SCULLY, P. A. S. and LLOYD, M. J. (1965). *J. Inst. Brewing*, **71**, 156.
[104] SCULLY, P. A. S. and LLOYD, M. J. (1967). *Brewer's Guild J.*, **53**, 29.
[105] BIRTWISTLE, S. E., HUDSON, J. R. and MACWILLIAM, I. C. (1962). *J. Inst. Brewing*, **68**, 467.
[106] MACWILLIAM, I. C., HUDSON, J. R. and WHITEAR, A. L. (1963). *J. Inst. Brewing*, **69**, 467.
[107] HUDSON, J. R., MACWILLIAM, I. C. and BIRTWISTLE, S. E. (1963). *J. Inst. Brewing*, **69**, 308.
[108] HUDSON, J. R. (1963). *Proc. Eur. Brewery Conv. Congr.*, Brussels, p. 422.
[109] BARRETT, J., BATHGATE, G. N. and CLAPPERTON, J. F. (1975). *J. Inst. Brewing*, **81**, 31.
[110] ENARI, T.-M., MIKOLA, J. and LINKO, M. (1964). *J. Inst. Brewing*, **70**, 405.
[111] MIKOLA, J. and ENARI, T.-M. (1970). *J. Inst. Brewing*, **76**, 182.
[112] DUFF, S. R. (1963). *J. Inst. Brewing*, **69**, 249.

Chapter 10

PREPARATION OF THE GRIST

10.1 *Storage, cleaning and weighing*
Batches of malt and unmalted cereal arriving in bulk transporters (or more rarely, in modern times, in sacks) are conveyed into separate stores by elevator or conveyor. Except for special materials such as amber, crystal and chocolate malts, the grain is usually held in silos or storage bins, of steel and concrete construction, with smooth walls and hopper bottoms. The malt and unmalted cereal is maintained at the moisture level at which it arrived at the brewery in order (*i*) to discourage the breeding of insects and (*ii*) to prevent the malt or adjunct from changing biochemically because of the presence of water, that is from becoming slack. By the use of suitable electrodes, the levels of malt in the silos, the temperature and the humidity can be indicated or recorded automatically, preferably at the central panel where the various conveyor/silo-filling and emptying systems are also controlled. Adequate facilities for receiving and storing malt and unmalted cereal are required to obviate unforeseen delays in delivery but not exceeding a few days' requirements. Otherwise capital is held unnecessarily in stocks of malt and in storage equipment.

Syrups are delivered either in bulk or in drums. Bulk syrup is usually held at 50°C (122°F) from the sugar factory to the brewery using specially designed tanker trucks, so that it can readily be pumped into the bulk syrup tanks of the brewery. Again in order to keep the viscosity lower the syrup is held at this elevated temperature in the storage tanks. Therefore the tanks are insulated and have facilities for heating the syrup with steam or high-pressure hot water. Syrup in drums is heated up before being added to a special sugar-dissolving vessel or directly to the wort copper.

To obtain uniformity of milling, it is necessary to have reasonable consistency in size of corns. In some breweries, all samples of malt are graded or screened to obtain such consistency; while in others only those batches which fail to meet the specifications laid down for a range of corn size are so treated. The malt is conveyed by pneumatic or mechanical means past magnetic separators to rotating, cylindrical, oscillating or flat-bed screens. Not only are corns of abnormal size rejected, but foreign matter such as

PREPARATION OF THE GRIST

straw, stones, string, sacking and metal particles are removed. After screening the malt is weighed – often on a continuous basis until a predetermined total is reached. Wherever possible, the rejected corns are sold for animal feed.

Dust arising from malt-handling and -processing is an important hazard in breweries because of its long-term effects upon the mucous membranes of those that inhale it and because of explosion risk. Where possible, dust is aspirated by air cyclones and trapped in cloth-sleeve filters. It is often sold for animal feed. Sparks may be avoided by removing stones and metal from the malt before it enters the mill; careful attention should also be paid to mechanical and electrical installations wherever there is malt dust. Easily collapsible sections of the millroom structure, of chutes, elevators and conveyors, are provided so that should there be an explosion, blast is vented at convenient locations. Smoking and the wearing of nailed boots must be forbidden and metallic equipment is electrically earthed.

The maltroom where sacks of special malts are held is often situated above the millroom and is kept warm and free from steam or humidity. The millroom itself is frequently located fairly high in the brewery, so that in the case of explosion, damage to the rest of the brewery is minimized. The mill commands the grist cases which in turn feed the mash tuns by gravity (Fig. 10.1). With the development of 'bungalow' or single-storey breweries, mills may be at a lower level than the top of grist cases; the grist is then transported either pneumatically or by chain-conveyor to the grist cases.

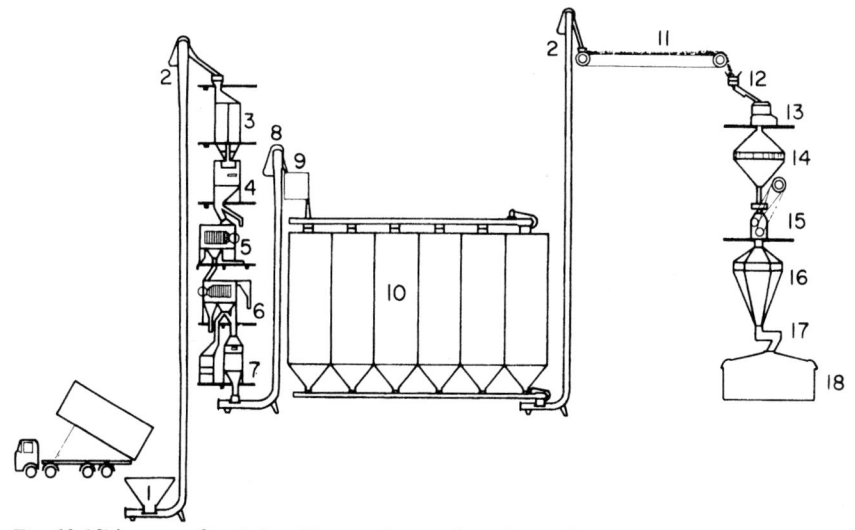

Fig. 10.1 Diagram of malt-handling equipment from intake hopper to mash tun.
1. Intake hopper. 2. Elevator. 3. Grain receiver. 4. Weigher. 5. Magnetic separator. 6. Screens. 7. Weigher. 8. Elevator. 9. Conveyors. 10. Grain silos. 11. Conveyors to brewhouse. 12. Magnetic separator. 13. Weighers. 14. Malt hoppers. 15. Malt mills. 16. Grist hoppers. 17. Steel's mashers. 18. Mash tuns.

10.2 Dry milling of malt

In milling, there are two basic (unit) operations, namely (*i*) size reduction and (*ii*) particle size control by screening. The object of milling is to obtain that mixture of particle sizes in mashing which will secure the highest yield of extract as quickly and efficiently as possible. Milling procedures are determined by the size distribution of malt corns, their modification, moisture content, the mashing methods and wort-separation method. In more detail, the milling splits the husk, preferentially longitudinally, in order to expose the endosperm. The latter should be so disintegrated that it is readily penetrated by water and enzymes. Furthermore, fine (floury) particles should be kept to a minimum. There should be no uncrushed kernels, the endosperm particles should be reasonably uniform in size, and the majority of husks should be entire, except for the longitudinal split, with no endosperm adhering (especially hard end material).

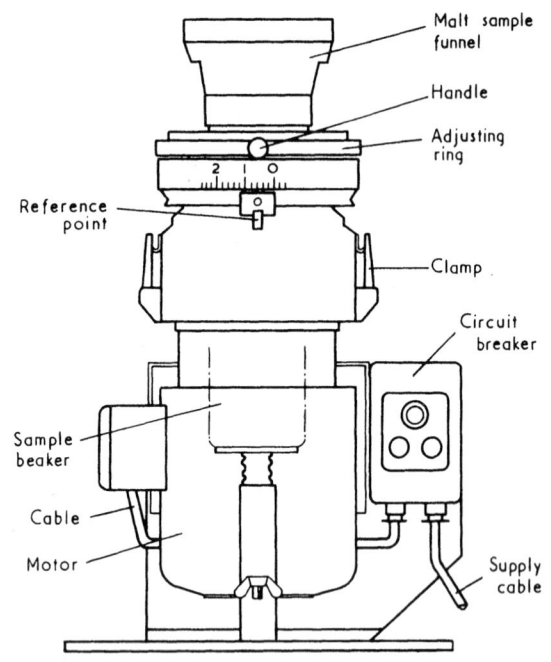

Fig. 10.2 Bühler–Miag disc mill.

Size reduction in the food industries is normally achieved by one of four types of mill, namely (*i*) ball mills, (*ii*) pin-disc mills, (*iii*) hammer mills, and (*iv*) roller mills. A ball mill basically comprises a cylindrical hollow vessel rotating on a horizontal axis. The vessel receives not only a charge of the material for crushing but also a collection of balls (usually of hardened metal) which crush the charge by their movement in the rotating vessel. Pin-disc mills have two discs, a lower rotating disc and, placed immediately above it, a stationary one. Alternatively the discs can be contra-rotating. Each disc has sets of vertical pins or teeth of hardened metal set a regular distance apart, disposed in a series of circular rows, in such a way that the circles of pins of one disc fit into the spaces left by the pins of the other disc (Fig. 10.2). The Bühler-Miag disc mill has been accepted as the standard mill for producing extracts by the Institute of Brewing [1]. The gap between the revolving disc and the stationary disc is 0·2 mm for fine grinding and 0·7 mm for coarse grinding (Table 10.1). Hammer mills have a horizontal rotating spindle bearing one or more discs which have either fixed or pivoted on their periphery a series of vanes or hammers. Surrounding the discs is a housing, the lower portion of which has either hardened screen bars or a sieve (Fig. 10.3).

TABLE 10.1

Extract values using twelve malts with different settings in a Bühler-Miag disc mill
{MARTIN (1979) [1]}

Gap between mill surfaces (mm)	0·2	0·5	0·7
Extract (litre °/kg)	296·22	293·48	291·44
95% confidence limits (±)	±2·02	±1·92	±1·92

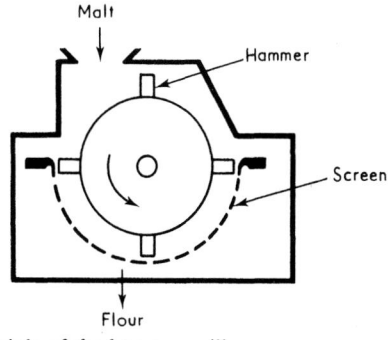

Fig. 10.3 Working principle of the hammer mill.

In breweries, dry milling is almost exclusively carried out in roller mills and these will be described in greater detail. Disc mills are used in malt analysis as mentioned above, in milling green malt and in the manufacture of maize products such as syrups. Hammer mills are used for providing grist in the Centribrew process (Alfa-Laval) [2], and for use in high-pressure mash filters (Nordon–van Waesberghe) [3] and ball mills are used in malting and brewing laboratories for certain analyses. The EBC or Casella mill for obtaining standardized coarse and fine grinds resembles a hammer mill, but the 'hammers' are not pivoted and are replaced by fixed blades with sharp cutting faces or edges [4]. In the Comparamill a flywheel drives hammers, and the energy required to mill can be measured. Except for exceptional varieties of barley, e.g. Mazurka, the milling energy for a barley sample is reported to correlate well with the hot-water extract after micromalting [5].

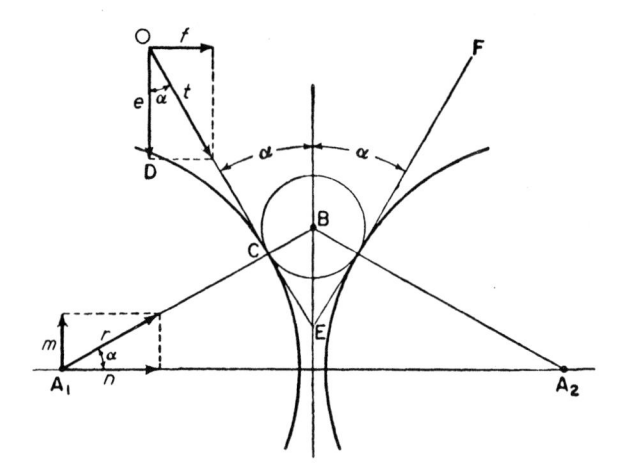

Fig. 10.4 Action of crushing rolls (for explanation please see the text).

Roller mills operate by using both direct pressure and shear [6]. Fig. 10.4 shows the forces operating when a particle whose centre is B has just been caught between two rolls, with centres A_1 and A_2. Force t tends to draw the particle between the rolls and depends on (*i*) force r and (*ii*) the coefficient of friction between the particle and the roll surface. If μ is this coefficient, $t = \mu r$. Forces e and m are opposed, the former tending to draw the particle between the rolls and m to eject it. For the particle to be crushed, e must be the greater of the two, or expressing it in another way $\mu r \cos \alpha > r \sin \alpha$ from which it follows that $\mu > \tan \alpha$. Thus $\tan \alpha$ must be less than the coefficient of friction. In many applications a typical value for α is 16°. The angle OEF, which is twice the angle α, is called the angle of nip. A definite relationship between roll diameters, feed and product can be demonstrated:

$$\cos \alpha = \frac{\text{radius of the roll} + \text{half the gap between rolls}}{\text{radius of the roll} + \text{radius of the particle}}$$

(If α is 16°, $\cos \alpha$ is 0·961.)

The theoretical capacity of a mill can be calculated as peripheral speed of rolls × width of rolls × roll setting (in appropriate units). In practice only 10–30% of this capacity is achieved and peripheral speed and angle of nip are both critical. If the speed is too high, the particles ride the rolls and are not drawn between them. In industries outside brewing, peripheral speeds of

1·8–3m/s (6–10 ft/s) are usual. This compares with 2·4–4·0 m/s (8–13 ft/s) commonly achieved in brewing. In batch brewing, it is usually required that the mill grinds sufficient for a mash in less than 2 hr. In a large brewery with many brews per day and several mash mixers, the mills operate on a more or less continuous basis and are therefore proportionately smaller. Malt endosperm is crushed so that it is in the most suitable state for efficient extraction and subsequent filtration of the wort. It is necessary to compromise between the requirements for extraction and filtration because, although a fine grind potentially yields more extract, it permits only very slow filtration and much of the extract may be left in the spent grains. A further requirement already mentioned is to preserve husks as intact as possible because filtration rates are enhanced. There is also less opportunity for leaching of materials from the husks which have undesirable effects on flavour and shelf life of the beer, although beer entirely produced from dehusked malt has a very unsatisfactory flavour. It is now possible to partly dehusk barley and use the derived malt as a replacement for some 10–30% of the grist and, in this way, reduce the total amount of material leached from the husks [7]. Subject to the husks remaining substantially intact, the choice of grind will depend on the malt and the filtration plant used. Well-modified malts yield over 95% of their potential extract to the wort during mashing even when coarsely ground, whereas poorly modified malts require fine grinding to achieve the same efficiency of extraction. Indeed, the difference in extract between fine and coarse grind is a most useful method of measuring modification (see Chapter 9 and Table 10.2). With regard to equipment, a mash filter permits the use of a much finer grist than the conventional mash tun or lauter tun (see Chapter 11).

TABLE 10.2
Stage of germination of Kenia barley in relation to the difference in extract between fine and coarse grind
{HOPKINS and KRAUSE (1947) [8]}

Days of germination	Percentage difference in extract between fine and coarse grind
2	8·4
4	2·0
6	1·1
8	1·0
10	0·6

Before 1880, British brewers were required by law to use only malt in the mash tuns, and the malt had to be crushed with smooth iron rollers. These outdated requirements have had important consequences in creating long-standing process preferences or even specifications, based on (*i*) well-modified malt and (*ii*) little use of adjuncts, except for flakes and flours requiring no milling. Thus, within the brewery, whole corns of unmalted barley ground to a fine flour have rarely been used as an adjunct despite the long-standing success of distillers in producing and using barley flour. Pulverized malt has been successfully used for small-scale trials. Its manufacture permits partial separation of 'endosperm flour' and 'aleurone flour'; the use of different proportions of these in the grist offers opportunities for modifying beer flavours [9].

The first reduction rolls of a modern dry mill receive the individual malt corns end-on from the feed rolls, so that crushing occurs over the whole length of the corn, thus preserving the husk almost intact. The least-modified part of the endosperm (that part distal to the embryo) may remain attached to the husk as a 'hard end' and it is necessary to detach it after the initial crushing without undue damage to the husk. The endosperm and embryo are crushed to produce fine grits rather than flour, although some flour production is inevitable. Care is taken to ensure that the proportions of flour, grits and husk conform to specification. This entails careful setting of the mills so that clearances between rolls are correct along their whole length. Samples may be removed in most mills below each pair of rolls, and analysed using sieves to assess the proportions of malt fractions. The sieving technique has received attention and several procedures have been advocated, probably the most important being those of the A.S.B.C. and the E.B.C. The former are based on six 'US Standard' sieve numbers, and the latter on the five sieves used in the 'Pfungstadt Plansifter' (Table 10.3).

TABLE 10.3

Comparisons of mesh widths (mm) between ASBC screens and those of the EBC Pfungstadt Plansifter
{DE CLERK (1958) [10]}

US standard numbers	Mesh width	Pfungstadt sieve	Mesh width	Characteristics
10	2·00	–	—	Husks held
14	1·410	1	1·270	Husks held
18	1·000	2	1·010	Husks held
30	0·590	3	0·547	Coarse grits held
60	0·250	4	0·253	Fine grits held
100	0·149	5	0·152	Ordinary flour held, fine falls through

Some laboratory and small pilot-scale mills are equipped with one pair of rolls (e.g. the Seck coarse-grind mill and the Boby laboratory mill). Unless the malt is extremely well modified, there is difficulty with either excessive

PREPARATION OF THE GRIST

fragmentation of the husk or, at the other extreme, getting a sufficiently fine grind to attain reasonable extract. There is, furthermore, difficulty with rolls less than 250 mm diameter, because (*i*) the malt falls at an obtuse angle and is therefore not snatched up by them and (*ii*) the crushing time is too short.

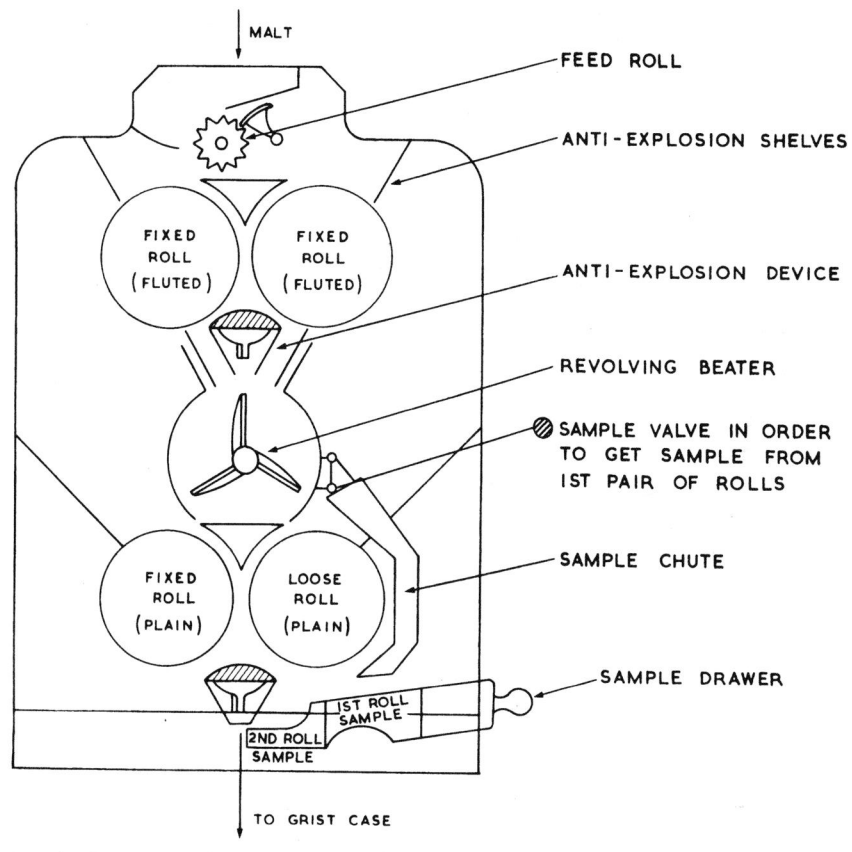

Fig. 10.5 Four-roll mill. 'Fixed' refers to driven rolls.

Two-high (or four-roll) mills are common in small breweries where well-modified malt is almost always used. A typical example (Fig. 10.5) will process 0·5–5·0 tonnes/hr, depending on the width of the rolls. Malt is fed through (*i*) a chute incorporating a magnetic separator to remove iron and steel objects, (*ii*) a deeply grooved feed roll, and (*iii*) the first pair of reduction rolls of 30 cm (12 in) diameter operating at 210–230 rev/min. These rolls have rounded flutes along their length which tear open the husks. Immediately beneath the rolls is a series of plates which keep the space full of grist and so reduce the risk of explosion by smothering any sparks generated if particles of flint or steel are struck. The ground malt is forced on to the second pair of rolls by a revolving beater which also serves to separate flour and grists from the husks. This second pair, equal in size to the first, but less fluted and operating at about 280 rev/min, breaks the hard unmodified ends. But the flour and grits which do not require further milling also have to pass between them, an inefficient arrangement. From the rolls, the grist is detained by further anti-explosion shelves before falling to a grist case. Thus this type of mill is simple, uses gravity feed and is relatively inflexible in performance. More sophisticated four-roll mills incorporate a pair of revolving beaters set in cylindrical screens (Fig. 10.6) which permit husks, fine grits, and flour to escape the second pair of rolls but coarse grits receive appropriate size reduction in these rolls.

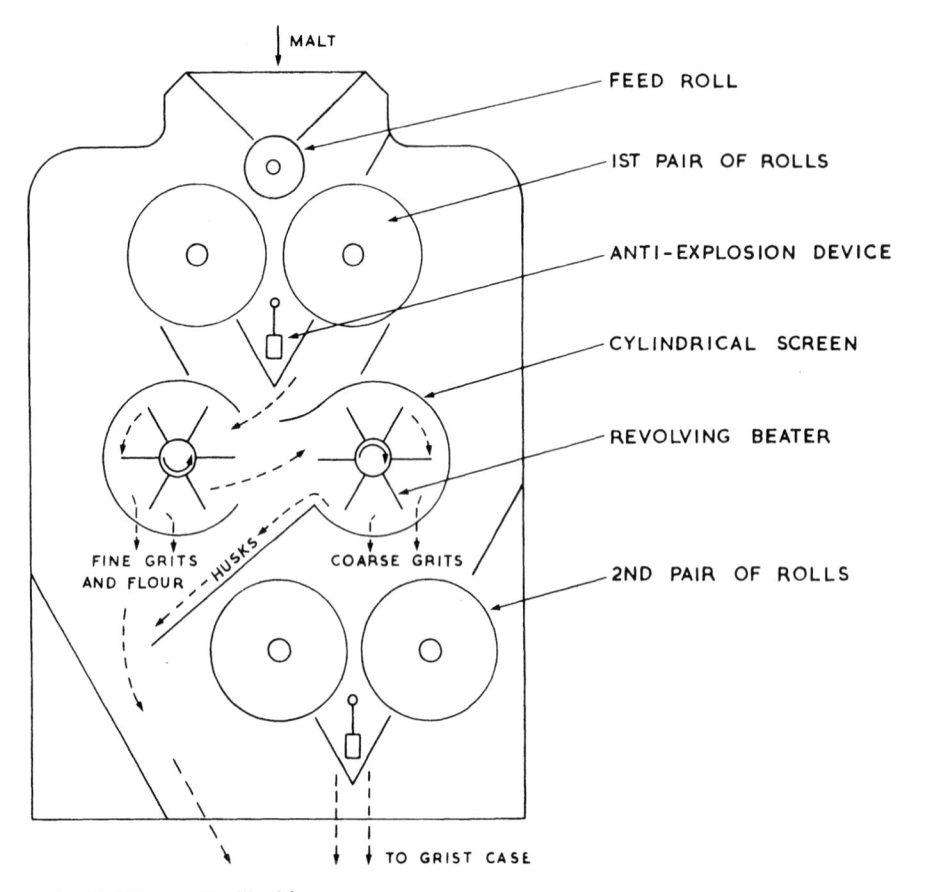

Fig. 10.6 Four-roll mill with screens.

PREPARATION OF THE GRIST   313

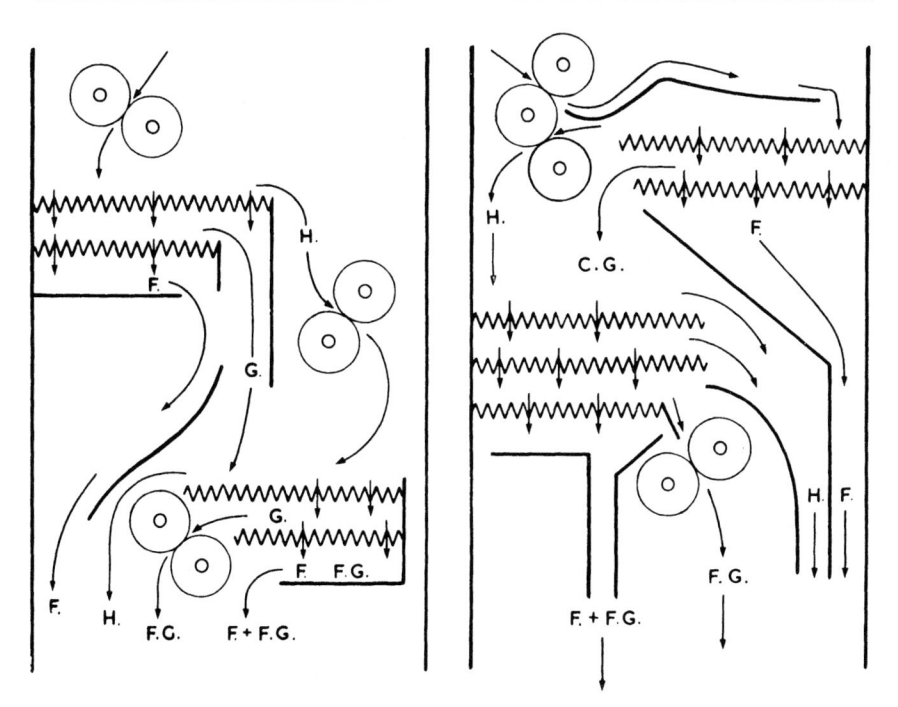

Fig. 10.7 Six-roll mill with screens. H., husks; G., grits; F.G., fine grits; F., flour.
Fig. 10.8 Five-roll mill with screens. For key see Fig. 10.7.

With six-roll mills, those with three pairs of rolls, more effective separation and differential size reduction can take place. This is particularly necessary with less-well-modified malts such as are commonly used in many European lager breweries. Flour produced by the first pair of rolls falls through to the grist case while fine grits are screened into the third pair of rolls (Fig. 10.7). In the second pair of rolls, the hard ends are crushed from the husks and coarse grits are reduced in size. Only grits from this second pair find their way to the third pair of rolls, the flour being delivered to the grist case. The screening is achieved either with inclined static screens or ones vibrating at about 400 oscillations per minute. Such mills allow both well-modified and poorly modified malts to be satisfactorily milled. A five-roll mill (Fig. 10.8) achieving the same result is available, as also is a four-roll mill providing three passages like the five- and six-roll types. More complicated mills involving four pairs of rolls are used by some breweries, particularly when they wish to mill relatively undermodified lager malts. Typical settings for the successive passages of a six- or five-roll mill are 1·3–1·5 mm (0·05–0·06 in), 0·7–0·9 mm (0·0275–0·0355 in) and 0·3–0·35 mm (0·012–0·014 in) [11]. The surfaces of the first pair of rolls are smooth for crushing and move slower than the second pair. The rolls of the third pair are channelled along their length and move fastest. Frequently there is a differential between peripheral speeds of the upper and lower rolls in a pair, and at its upper extreme this can be 2·3:1.

For a lauter tun, it has been suggested that the grist should contain 15% husks, 23% coarse grits, 30% fine grits and 32% flour, as shown by Plansifter analysis. For a mash filter (Chapter 11) on the other hand, there should be no coarse grits, while fine grits and flour should each be over 40%, and the proportion of husks remains the same. The need for care and extensive trials in milling so as to achieve the correct proportions is therefore apparent. From poorly modified lager malt the proportions by weight of the husks, coarse grits, fine grits and flour have been recorded as 15, 20, 30 and 35% respectively, but they yield 6, 11, 33 and 50% respectively of the total extract.

The ease with which the husks will fracture into pieces too small to be useful in mash-tun drainage has led to the development of processes in which the malt is steamed or briefly wetted. The husk takes up water to achieve 7% moisture and becomes more pliable and so is broken up less during milling. Care has to be taken to ensure that the endosperm remains dry, otherwise it would cling to the rolls. Nevertheless, many new breweries successfully spray or steam malt. A more extreme approach is to combine milling with mashing-in using a wet-milling process [12].

10.3 Wet milling [12–15]

The wet mill is fed from a malt hopper equipped at its base with water underlets and above with cold-water sprays (Fig. 10.9). Care is taken not to introduce air in the steep water and subsequent mash water. The steeped malt is received by a feed roller and then by a pair of 30·5 cm (12 in] diameter crushing rolls which are lightly patterned but not grooved and operate at 3000 rev/min. Below the rolls is a mash pump which has an extended shaft for rotating an homogenizer arm. The crushed malt falls into a mixing chamber and is sprayed with attemperated mashing-in water. The mash pump then conveys the mash to the receiving vessel. In some installations the mixing chamber, homogenizer and mash pump are omitted and the mash falls directly into a mash mixing vessel. The milling sequence consists of four stages (*i*) washing and steeping the malt in the hopper, (*ii*) draining steep water from the hopper and off to the mashing-in tanks, (*iii*) grinding and mashing and finally (*iv*) rinsing out the hopper and mill. Steeping takes about 15 min, during which time the malt moisture rises to 28–30%. The initial moisture content of the malt is therefore of little importance. During satisfactory grinding, husks are undamaged and even acrospires remain intact. The process is claimed to give uniform crushing to a slurry of fine

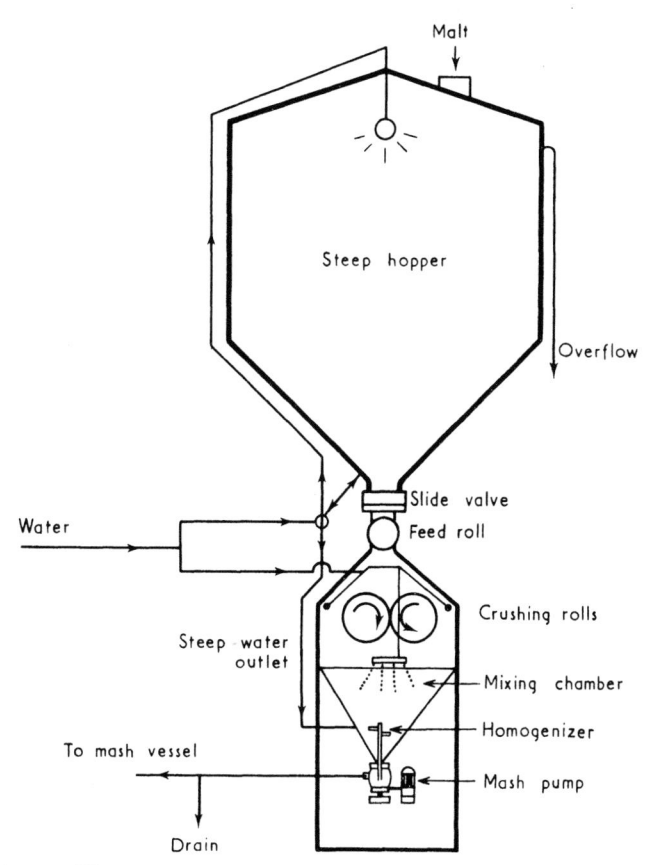

Fig. 10.9 A wet milling system.

grits and thereby give both yield of extract and speedy filtration. There are, of course, no hazards relating to dust or explosions. There are some disadvantages to wet milling. Thus steeping should not be longer than about 30 min and therefore milling has to take place for a limited time only. This leads to substantial demands from time to time for electrical power. In contrast, dry milling can proceed virtually continuously. Some breweries have also discovered that maintenance and particularly roll-wear are high, particularly when raw barley is being crushed. Again, where corn size is not uniform, it is easy for the smallest kernels to pass through the rolls of a wet mill without being fractured. Furthermore oversteeping of the malt is common, and some schedules therefore only provide for 5–10 min steep at 25°C (77°F).

It has been claimed that an attrition or pin mill can also be used satisfactorily for the milling of green malt or steeped barley. The grain is fed into the centre of the discs and is forced between the teeth by centrifugal force. If a stream of water is fed in with the grain, a mash is produced, and one derived from green malt and steeped barley will give rise to a wort similar to that from a conventional mash. Such a system has considerable attractions, because kilning of malt is not involved, the milling is simple and cheap, and a grist case and mashing-in device are not required. Differential roll mills have also been used successfully for the milling of green malt and steeped barley and indeed may well be more satisfactory than pin mills for this purpose. A number of different claims have been made that beer produced from green malt and steeped barley is of acceptable quality for commercial brewing and has special advantages in connection with haze stability [16–19].

10.4 The grist case

The density of ground malt and flaked adjuncts is extremely low and therefore grist cases are surprisingly large. Thus 100 kg of ground malt occupies up to 3 hl (about 20·8 lb/ft^3). The grist case normally holds the grist for a very short period, nevertheless the contents need protection from humid air or from steam. The grist case also serves as a station for the receiving and mixing of the various malts and dry adjuncts. Normally constructed of mild steel sheet, it commands a single mash tun, or provision may be made for it to serve several tuns, by (*i*) a movable chute, (*ii*) a movable mashing-in device, or (*iii*) by moving the grist case itself on wheels. It is highly desirable that the contents of the grist case are well mixed, and this is achieved by the use of a proportional feeder for the adjuncts before (or more usually during) milling. The grist case is normally equipped with a slider valve and the grist falls along an inclined chute, or occasionally is carried on a conveyor. Screw-conveyors are avoided because they tend to break up the husk. Movement from the grist case is in some cases accelerated by a vibrating device.

In some cases, the grist is heated in the grist case by steam pipes either to striking heat (retorrification) or to 130°C (266°F) for 15 min to increase coloration (aromatization).

TABLE 10.4
Details of various grist materials {LLOYD (1978) [20]}

	UK Cost (malt = 100)	Moisture (%)	Extract (malt = 100)
Flaked rice	131	10	109
Pale malt	100	3	100
Torrified barley	72	5	85
Flaked maize (corn)	88	9	106
Flaked barley	67	9	85
Torrified wheat	72	5	92
Flaked wheat	73	8	94
Maize (corn) grits	67	12	104

10.5 Grist composition

Table 10.4 gives details of various grist materials and Table 10.5 indicates typical compositions of grists used in brewing four different beers. If a brewer wishes to produce a given volume of wort at a prescribed gravity, it is necessary for him to multiply volume by the degrees of gravity (in °Plato, °Balling or excess gravity which is SG − 1000). Then, knowing the extract yield for the individual components of the grist and the ratios in which they are present by weight or extract, he must apportion them accordingly.

The calculations are useful but do not provide accurate results because the relations between specific gravity and concentrations of sugar and wort solutions are not linear. There is, furthermore, the possibility that the extract obtained from one component when mixed with a second component differs from the extract gained from the first component alone. This could be the case if a component inhibited the amylolytic activity of another component of the grist. In any case brewery extracts are normally different from those obtained in small-scale analytical mashes (see Chapters 8 and 9).

TABLE 10.5

Examples of grist compositions for various beers (OG, original gravity; specific gravity of wort before fermentation)

	% weight	% of total extract	Extract yield (%)
Lager, 11°Plato (SG 1044 approx.) with 24 EBU*			
Lager malt	74·5	70	80
Corn syrup	25·5 (31·9†)	30	100
Hops	0·10 kg/hl (at say 8% α-acid)		30‡
Lager 12°Plato (OG 1048 approx.) with 27 EBU*			
Lager malt	62	60	80
Corn grits	38	40	80
Hops	0·15 kg/hl (at say 6% α-acid)		30‡
Bitter ale 10·5°Plato (OG 1042 approx.) with 30 EBU*			
Ale malt	76·6	75	80
Crystal malt	3·5	3	70
Wheat flour	4·8	5	84
Flaked maize	6·8	7	84
Cane syrup	8·3	10	100
Hops	0·125 kg/hl (at say 8% α-acid)		30‡
Stout 12°Plato (OG 1048 approx.) with 45 EBU*			
Ale malt	80·1	80	80
Roasted barley	5·6	5	72
Flaked maize	14·3	15	84
Hops	0·22 kg/hl (at say 8% α-acid)		25·5‡

* European bitterness units or mg of isohumulones/litre.
† As is weight.
‡ These are percentage utilization figures for the hops.

Examples

EXAMPLE 1

2500 hl required at 11·5°Plato
Malt of extract yield 78% to give 2/3 of final extract
Corn grits of extract yield 80% to give 1/3 of final extract.

Calculations

Total extract = $2·5 \times 11·5 \times 10^3$ hl degrees = $2·875 \times 10^4$

Malt $\dfrac{66·66}{100} \times 2·875 \times 10^4 = 1·917 \times 10^4$ hl degrees

which at 78% extract is $2·46 \times 10^4$ kg.

Corn grits $\dfrac{33·33}{100} \times 2·875 \times 10^4 = 9·58 \times 10^4$ hl degrees

which at 80% extract is $1·20 \times 10^4$ kg.

EXAMPLE 2

1500 UK brl at SG 1035 required.
Prescribed proportions of grist constituents by weight:
Malt   75% grist with extract 294 litre degrees/kg
Flaked maize 10% grist with extract 310 litre degrees/kg
Sugar   15% grist with extract 355 litre degrees/kg

Calculations

Total extract = $1500 \times 35 \times 163·65 = 85·92 \times 10^5$ litre degrees

Malt = $\dfrac{75}{100} \times 85·92 \times 10^5 = 64·44 \times 10^5$ litre degrees

which at 294 litre degrees/kg is $21·9 \times 10^3$ kg

Flaked maize = $\dfrac{10}{100} \times 85·92 \times 10^5 = 8·592 \times 10^5$ litre degrees

which at 310 litre degrees/kg is $2·77 \times 10^5$ kg

Sugar = $\dfrac{15}{100} \times 85·92 \times 10^5 = 12·88 \times 10^5$ litre degrees

which at 355 litre degrees/kg is $3·63 \times 10^3$ kg

EXAMPLE 3

2000 brl required at 12·2 brewers' lb extract (SG 1037)
Prescribed proportions of grist by weight:
Malt 75% of grist, 100 brewers' lb of extract/Qr.
Syrup 25% of grist, 80 brewers' lb of extract/2 cwt.

Calculations

Total extract = 2000 × 13·3 brewers' lb = 26 600 brewers' lb.

Malt = $\dfrac{75}{100}$ × 2660 = 19 950 brewers' lb required

which at 100 lb of extract/Qr. = 199·5 Qr.

Syrup = $\dfrac{25}{100}$ × 2660 = 6650 brewers' lb required

which at 80 lb of extract/2 cwt = $\dfrac{6650 \times 2}{80}$ = 166·25 cwt.

REFERENCES

[1] MARTIN, P. A. (1979). *J. Inst. Brewing*, **85**, 290.
[2] ALFA LAVAL (1978). *Brewery Handbook*, Alfa Laval A. B., Tumba, Sweden.
[3] VAN WAESBERGHE, J. W. M. (1979). *Brewing Distilling Int.* **9**(9), 54.
[4] BISHOP, L. R. (1963). *J. Inst. Brewing*, **69**, 228.
[5] ALLISON, M. J., COWE, I. A., BORZUCKI, R., BRUCE, F. and MCHALE, R. (1979). *J. Inst. Brewing*, **85**, 262.
[6] BADGER, W. L. and BANCHERO, J. T. (1955). *Introduction to Chemical Engineering*, McGraw-Hill, New York.
[7] MACEY, A., SOLE, S. M. and STOWELL, K. C. (1969). *Proc. Eur. Brewery Conv. Congr., Interlaken*, p. 121.
[8] HOPKINS, R. H. and KRAUSE, B. (1947). *Biochemistry Applied to Malting and Brewing*, Allen & Unwin, London.
[9] KIENINGER, H. (1969). *Proc. Eur. Brewery. Conv. Congr., Interlaken*, p. 139.
[10] DE CLERK, J. (1958). *A Textbook of Brewing*, vol. 2 (translated by K. BARTON WRIGHT) Chapman & Hall, London, p. 321.
[11] VERMEYLEN, J. (1962). *Traité de la fabrication du malt et de la bière*, Vol. 1, Assoc. Roy. des Anciens Élèves de l'Institut Sup. des Fermentations, Gand, Belgium, p. 472.
[12] LAURIDSEN, P. B. (1966). *Tech. Q. Master Brewers Assoc. Am.*, **3**, 34.
[13] BUTTON, A. H. and PALMER, J. R. (1974). *J. Inst. Brewing*, **80**, 206.
[14] NARZISS, L. and LUTHER, G. (1974). *Brauwelt*, **114**, 40.
[15] KIENINGER, H. (1977). *J. Inst. Brewing*, **83**, 72.
[16] COOK, A. H. and HUDSON, J. R. (1964). *Proc. Annu. Meet. Am. Soc. Brewing Chem.*, 29.
[17] DUFF, S. R. (1963). *Proc. Eur. Brewery Conv. Brussels*, 422.
[18] SCULLY, P. A. S. and LLOYD, M. J. (1966). *J. Inst. Brewing*, **72**, 42.
[19] CURTIS, N. S. (1966). *J. Inst. Brewing*, **72**, 42.
[20] LLOYD, W. J. W. (1978) *The Brewer*, **64**(3), 84.

Chapter 11

MASHING

11.1 Traditional brewing

In traditional brewing in homes and inns, mashing was simply the mixing of warm water with ground malt and, after a suitable standing period, recovering as much liquid as was possible with the primitive equipment to hand, such as ladles and strainers. Sparging or washing the grains was probably not universal. Brewers must soon have realized that hotter water gave a different result from cooler water, but were unable to control the temperature accurately because thermometers were not invented. In traditional British brewing, the brewer is said to have overcome this difficulty by using brewing water (liquor) whose temperature was such that his face was best reflected by the water surface. Below this temperature 65–71°C (150–160°F), the ability of the water to reflect is less, while above this temperature water vapour fogs the air. The single-temperature mashing system, called infusion mashing, is best for the highly modified malt produced in Britain. A small percentage (5–20%) of adjuncts can be used in the mash, but are either finely ground or precooked.

In the absence of a thermometer, consistent temperatures can be achieved by mashing with water from a well [normally about 12°C (54°F)], boiling a predetermined fraction of this mash, and then mixing the hot and cold mashes. If necessary, this procedure of boiling a portion and adding it back to the main mash may be repeated. It was probably the need for consistent temperatures which first inspired this now traditional *decoction* mashing on the continent of Europe. The effects of the rising steps of temperature in decoction mashing are increased enzymic breakdown of protein and, later, carbohydrate. The malts produced in central Europe tend to be less well modified and less strongly kilned than British malts and therefore give better results with decoction rather than infusion mashing (Chapter 9).

From the point of view of unit operations, infusion mashing entails mixing, grist hydration, enzyme reactions, liquid-solid separation, elution and liquid-solid separation [1]. Nevertheless, only one vessel, the mash tun, is used, apart from a relatively small mixing device. In decoction mashing, the unit operations are extended by the additional heating, pumping and mixing steps; thus there is some variation from brewery to brewery in the number of

MASHING 321

Fig. 11.1 Two-vessel brewhouse with mash-mixing vessel on the left, copper for mash or wort in the centre and to the right elevated lauter tun. (Courtesy of A. Ziemann GmbH.)

vessels used. In the simplest arrangement for decoction mashing (Fig. 11.1), a vessel for mixing and one for boiling are used in conjunction with a vessel for liquid-solid separation (the lauter tun). It is, however, more usual to find the double brewhouse as shown in Fig. 11.2 comprising a mash mixing vessel (maischbottich), a mash copper (maischpfanne), wort copper (wurzepfanne) and a lauter tun (lauterbottich). Alternatively, there may be one lauter tun shared by two mash mixing vessels and two coppers, the idea being to utilize equipment to the greatest extent consistent with the flow pattern of the process.

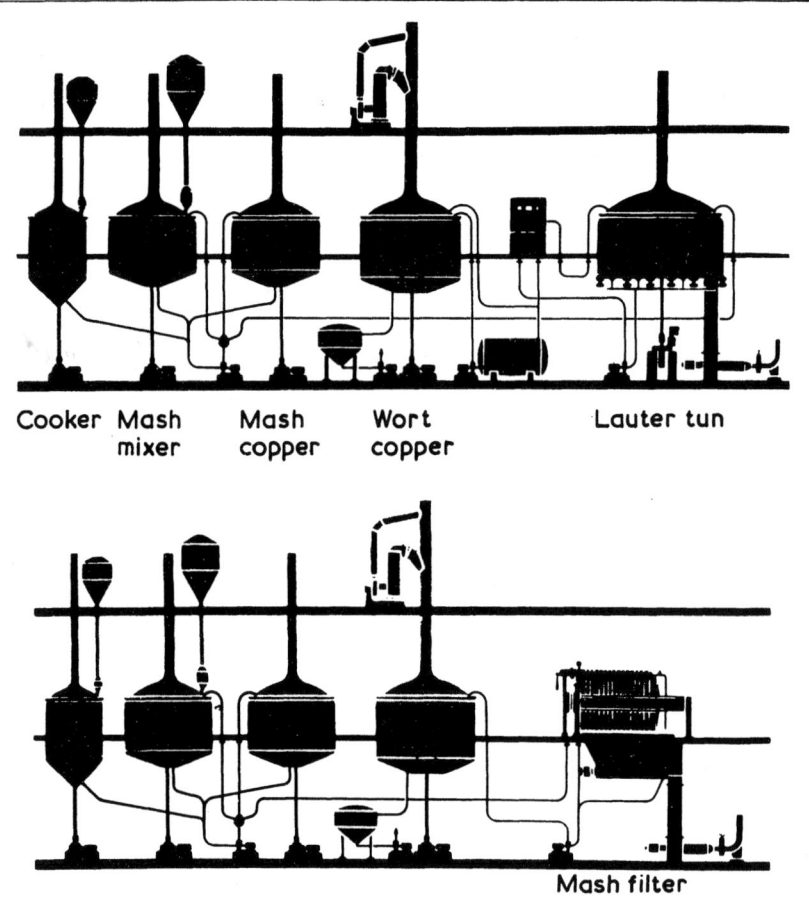

Fig. 11.2 Conventional four-vessel brewhouse with cereal cooker, mash-mixing vessel, mash copper, wort copper and, on far right, lauter tun in upper picture, mash filter in lower picture. (Courtesy of A. Ziemann GmbH.)

11.2 *Traditional infusion mashing* [1, 2]

Mixing of grist with hot liquor in infusion mashing is usually carried out with a Steel's masher. This piece of equipment was devised in 1853 in order to supplant the laborious mixing with mashing oars. It comprises a large bore tube usually of about 46 cm (18 in) diameter with a right-angle bend (Fig. 11.3). The flow of grist from the grist case is controlled with a simple slide valve and meets hot liquor before falling at the right angle on to a short screw-conveyor whose spindle bears mixing rods at the end of the masher. Alternative mashing devices, like the Maitland masher, although much less common than the Steel's masher, include the grist hydrator shown in Fig. 11.3 and the wet milling system described in Chapter 10.

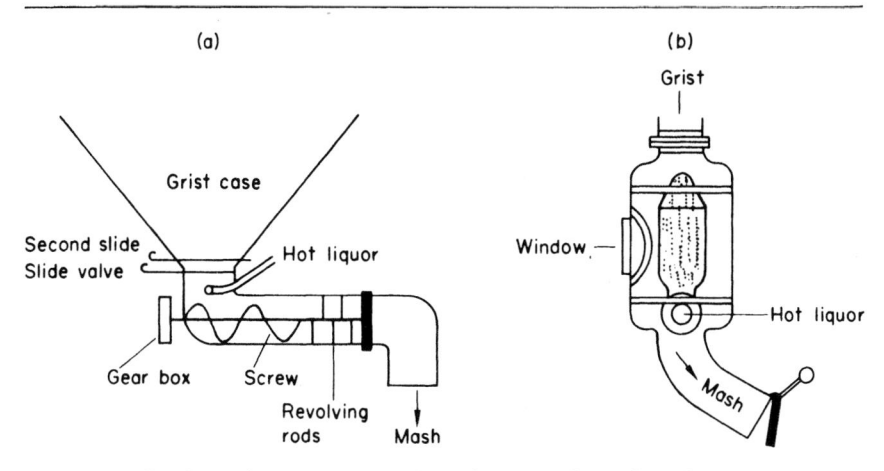

Fig. 11.3 (a) Steel's mashing machine used in infusion mashing. (b) Grist hydrator of premasher used in decoction mashing. Note the perforated cone.

Because the mash temperature is of immense importance in the enzymic degradation of grist constituents, considerable care is taken to achieve the temperature which is considered to give wort of the best composition, usually about 65·5°C (150°F). The most easily controlled variable is the liquor temperature, traditionally known as the 'striking heat'. Less easily controlled variables are the temperature of the grist and its moisture content. Losses of heat to and from the mashing equipment and the liquor mains can only be partly controlled by insulation and by preheating with hot water or steam. It is usual to estimate these losses on the basis of trials.

Ideally, the striking heat should be very close to the desired temperature of the mash, that is the 'initial heat', because there is inactivation of at least the more thermolabile enzymes when very hot liquor strikes the cold grist, and local overheating occurs. In order to reduce this inactivation, the grist can be warmed so that the 'striking heat' is less for a given 'initial heat'. At one time the grist case in some breweries was heated by steam pipes – a process called 'retorrification'. In its modern counterpart, the malt is attemperated with warm air before milling. Both procedures maintain a low moisture content in the grist or even succeed in reducing it. Nevertheless, it is becoming common in Britain to treat malt with steam before milling which increases the moisture content of the husk, making it less brittle. These various procedures have been claimed to improve the flavour and aroma of the beer, permit easier temperature control at mashing as well as giving protection to the enzymes of the mash against thermal shock.

The volume of liquor to weight of mash is usually in the order of 1·6–3·2 hl/100 kg (10–20 brl/ton) in infusion mashing, and this has to be taken into consideration in estimating the striking temperature. Ignoring the heat loss to and from the equipment, the striking heat can be estimated from the formula:

$$\text{Initial heat} = \frac{St + RT}{S + R} + \frac{\frac{1}{2}H}{S + R} \qquad (11.1)$$

where S is the specific heat of the malt and t the temperature of the malt; R is the ratio of weight of liquor to that of grist and T the temperature of the liquor (striking heat); H is the slaking heat of the malt in g-cal/degree temperature. The expression is applicable to both °C and °F provided that terms are expressed in the appropriate units.

TABLE 11.1

Specific heat (water = 1) and slaking heat of a malt at various moisture contents
{BROWN (1910) [31]; HOPKINS and CARTER (1933) [32]}

Moisture (%)	Specific heat	Slaking heat (g-cal) at mashing temp. of 65·5°C (150°F)
0	0·38	33·5
1	0·38	29·0
2	0·39	25·0
4	0·40	18·8
6	0·41	14·5
8	0·42	12·4

A slack malt has a slightly greater specific heat (Table 11.1) but a much reduced slaking heat and requires a striking heat some 1·1–2·2°C (2–4°F) higher than a malt of normal moisture content. Thus for an initial heat of 65·5°C (150°F), a grist moisture content of 2%, a grist temperature of 27°C (80°F) and a mashing rate of 2·7 hl/100 kg (16·7 brl/ton), the striking heat would be 69°C (156°F) assuming the volume of liquor to be measured at 15·5°C (60°F). Using a grist moisture content of 6%, the corresponding striking heat would be 70·5°C (159°F).

With the thick mashes of about 2·7 hl/100 kg (16·7 brl/ton) used in infusion mashing, it is difficult to raise the temperature once the mixing is completed and is usually not attempted. Heat transfer is particularly slow and the use of steam-jacketed vessels with the thick mash can lead to localized baking of the materials. It is, however, possible in some circumstances to pass free steam into the mash and this causes very little dilution of the mash, an important factor in the protection of mash enzymes. On the other hand, the temperature of the mash may be raised by pumping in hot liquor from underneath (underletting); thus a stand of 63°C (145°F) may be followed by underletting to raise the temperature to 68°C (155°F). A further formula (11.2) has been devised to assist in the calculation of the temperature of liquor underlet.

Again, it is approximate because it does not take into account heat losses from the equipment.

$$\text{Final mash temperature} = \frac{M(S + R) + QT}{S + R + Q} \qquad (11.2)$$

where S and R are the same as in equation (11.1), M is the mash temperature before underletting and T the temperature of the introduced underletting liquor; Q is the weight of water in the mash in appropriate units. In order to convert Q into volume of liquor, it must be multiplied by the weight of malt in the mash and divided by the weight of water in one unit volume, expressed in the same units of weight.

The infusion-mash tun (Fig. 11.4) is traditionally circular, varies very much in diameter, but is usually 2·0–2·5 m (6–8 ft) in depth and in some cases even deeper. Although the traditional wooden construction has the advantage of heat insulation, wood soon becomes spongy. The tuns are therefore constructed of copper or stainless steel and are well insulated on the vertical sides and the base. The old traditional open mash tuns have been replaced by enclosed ones in order to conserve heat and avoid steamy conditions in the brewing room. The dome-shaped lid to a mash tun is fitted with sliding panels for inspection purposes. Sometimes the steam is permitted to escape by a flue pipe. Inspection lamps and detergent-spraying equipment are often fitted to the lid. Standing some 5 cm (2 in) above the mash-tun base are a series of slotted plates which form a removable false bottom to the tun (Fig. 11.5). Before charging the mash tun, these filter plates are barely covered with hot liquor in order that the mash falling on to them does not block the slots which are typically 0·71–1·02 mm (0·028–0·040 in) wide at the top of the plates, widening to 4·6 mm (0·18 in) at the lower surface, 57 mm (2·25 in) in length and are cut with a 12·3 mm (0·5 in) staggered pitch. The total slot area represents about 11% of the filter-plate area. Alternatively they may be constructed from stainless-steel wedge wire which gives a greater percentage. Plates of the traditional type are usually constructed of gun-metal or stainless steel and interlock in a unique pattern so that, after they have been removed for cleaning, they can be reassembled correctly. One or more large diameter holes in the false bottom can be uncovered at the end of mashing. The spent grains are swept through the holes by a pair of large arms rotating horizontally. The discharged grains fall into a chute and are collected for animal feed. Other methods of grain discharge include manual digging or slurrying the grains in water and pumping the slurry out of the tun. Other rotating machinery within the mash tun comprises sparge arms (which are moved by water pressure) for spraying liquor evenly over the goods. Arms bearing rakes or knives for loosening the mash and driven by a central spindle are provided in some mash tuns. By adjusting the angle of the knives, this equip-

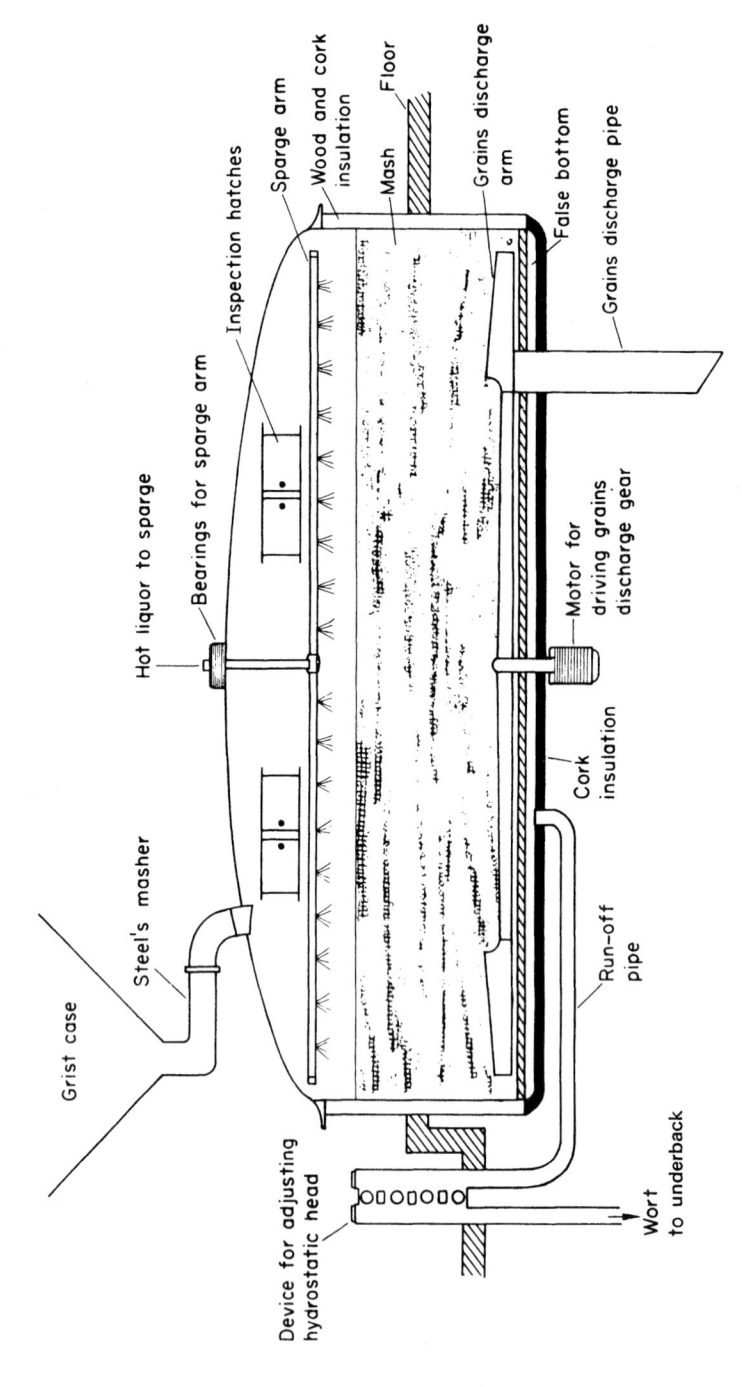

Fig. 11.4 Infusion-mash tun.

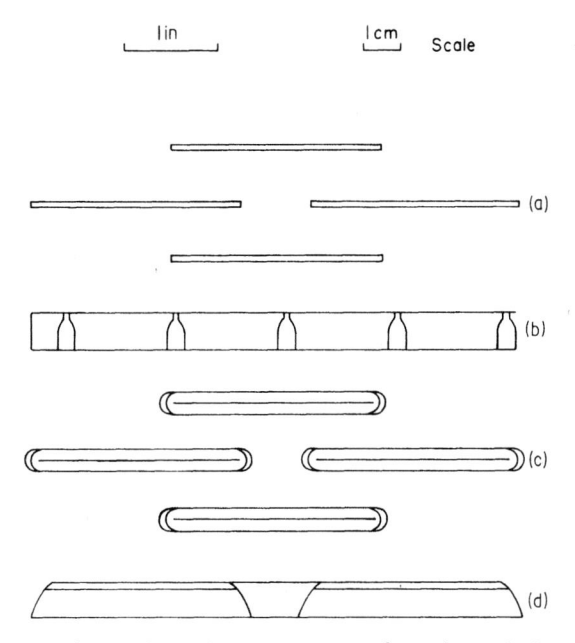

Fig. 11.5 Slots in mash-tun plates. (a) plan, upper surface; (b) vertical section; (c) plan, lower surface; (d) vertical section (at right angles to (b)).

ment can also serve to discharge grains. Mash is loaded into the tun to a depth of 0·9–1·2 m (3–4 ft) in most cases, but considerably deeper in some breweries. After a standing period ranging from 15 min to 2 hr, the wort is allowed to run from the tun through one or more discharge pipes set in the true base of the tun, each pipe serving an approximately equal area of the base, say 1·9–2·3 m² (20–25 ft²) and terminating in a tap. The taps are grouped above an open trough (the spend safe or grant) and are manually controlled with care so that there is very gentle even withdrawal of wort from the mash bed. Wort in the trough is in some cases recycled back to the bed until it runs clear (a process sometimes referred to as 'vorlaufvehrfahren'). This recycling has often been criticized on the grounds that it reintroduces 'fines' on top of the bed and thus impedes run-off. It is advocated that the turbid worts should be clarified away from the tun, if, in fact, starch-free worts need to be completely clarified.

A manometric device such as a Valentine tube may replace the battery or taps. In this case the gentle withdrawal is ensured by raising the swinging U-tube or adjustable swan neck, so that the hydrostatic head is relatively small. A variant of this equipment is shown in Fig. 11.4 where the device is equivalent to a set of inverted U-tubes, each tube having its own valve. If care is taken to run-off the worts gently, there is less chance of pulling down the goods hard upon the false bottom by suction, or of channelling or of cracking the goods. The advantage of individual taps is that the appearance and SG values for worts running from the different areas of the plates can be examined. The presence of suspended solids and low SG values from wort coming out of one or more taps suggests that there is channelling and cracking of the goods. An important difference between infusion and decoction mashing is that during the running-off of worts, the goods float in the infusion-mash tun because of entrained air in the mash. In contrast, the goods in a lauter tun rest on the slotted plates.

Once the wort is bright it is no longer recycled but run into an underback (a holding vessel between mash tun and copper) which maintains the wort at temperatures between 71 and 82°C (160 and 180°F) by means of heaters. The underback, and in some cases an additional vessel surmounting the copper called the upperback, is essential for temporary storage of worts when the copper has to be charged more than once with the worts produced from a single vessel. The first charge of the underback is referred to as 'first worts' and the next charge as 'second worts'. As the underback is filled, the level in the mash tun is maintained by spraying liquor over the goods by means of the revolving sparge arms. The sparge temperature is often as high as can be achieved without extracting any unconverted starch from the goods, usually around 77°C (170°F). This means that the temperature of goods, which falls slightly during the stand period, gradually rises to above 74°C (165°F) by the end of sparging. The efficiency of extraction is enhanced by the increase in temperature because both viscosity of the wort is reduced and the rate of diffusion of dissolved substances from within the grist particles is enhanced. About 2–4 hr of sparging are needed to extract the soluble materials in the mash economically. The specific gravity of early worts may rise to as high as 1100 if the mash is initially thick. It is not generally considered worth while to run-off worts below SG 1005 (1·2°Plato), because of the extra time required and extra steam needed to evaporate the additional water from the brewery copper. The calcium content and hardness of the final sparge liquor and the pH of the mash are very important in regulating the quality of the last runnings (Chapter 9). Furthermore, undesirable materials such as fats and tannins are extracted at low SG (see Chapter 9). About 5·35 hl of sparge liquor per 100 kg (33 brl/ton) of malt is used to reach this situation. A total of about 8·1 hl of liquor is therefore required for mashing-in and sparging for each 100 kg of malt (50 brl/ton). Of this, rather less than 2% is retained by the spent grains.

In the past there was considerable difficulty in producing satisfactory mashes, and several means were adopted to overcome the blocked or blinded filterbeds. One out-dated method was to ensure that there was adequate coarse filtering material by using malts from brewing barleys having large

MASHING

coarse husks or even adding chopped straw or oat husks to the mash. In practice, additional husk material gives rise to beer of increased colour, harshly bitter after-taste and reduced shelf life. Another, more recent, method is to use rakes or knives to disperse the goods and so make the filterbed lighter and less channelled. Probably better, however, is underletting, which tends to float the filterbed off the plates. The best practice is to maintain the floating bed consistently. In modern mashing it is unusual to encounter difficulties in running worts regularly. Slow running of worts may be caused by the use of too finely ground materials, undermodified malt or particular adjuncts or mash-tun plates with clogged slots. Too long a stand, rapid initial withdrawal of wort or excessive early sparging can have the

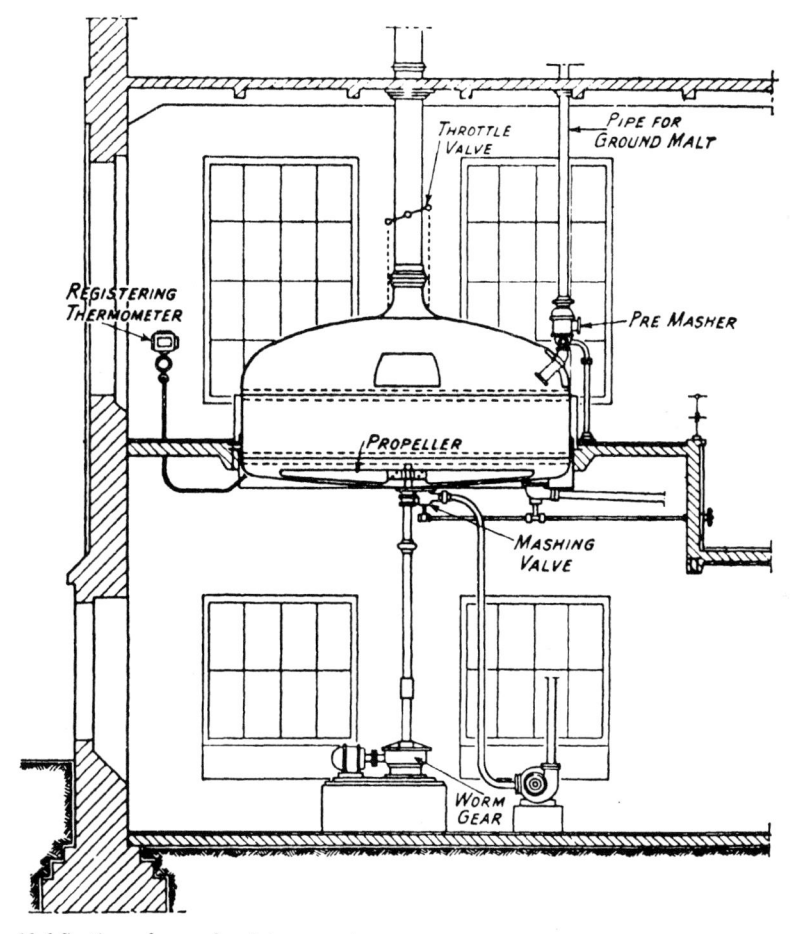

Fig. 11.6 Section of a mash-mixing vessel.

same effect. Alkaline liquor may also give rise to similar difficulties. If worts virtually cease running, then the mash has to be reslurried with rakes, or even oars if rakes are not fitted, and allowed to form a fresh bed, preferably after the addition of β-glucanase and protease to the mash.

11.3 *Clarity of worts* [3, 4]

Over the past eighty years it has been considered desirable to produce bright sweet worts, containing little or no suspended matter, direct from the mash tun. Modern methods, which substantially raise the rate of run-off often yield slightly cloudy worts (which was probably also the case before, say 1880). If the cloudiness is not due to gelatinized unconverted starch and if the turbidity is removed by boiling, then there is little danger to subsequent beer quality, although the wort will have an increased content of lipids. Incidentally, starch can readily be detected in cooled worts by means of the iodine test. This view is nevertheless contentious; some authorities strongly hold the view that poorer quality beers arise from cloudy sweet worts. Somewhat cloudy worts are usual in vinegar brewing and in the wash from distillers' mash tuns. In the latter, wort may be run-off not only from below but also from the surface of the mash where it escapes through vertical slits in the tun wall.

11.4 *Decoction and double-mash systems* [3, 5, 6]

Traditionally, decoction mashing employs malt which is less well modified than that used in infusion mashing and is only lightly kilned. It is milled with great care in five- or six-roll mills to give intact husks and grits which are considerably finer than those of an infusion grist. A slide valve controls the flow of the grist into a premasher or hydrator (Fig. 11.3), which is simply a swollen portion of the tube connecting grist case and the mash-mixing vessel. The liquor pipe enters the premasher and terminates in a perforated region. The mashing rate is 3·3–5·0 hl/100 kg (20–30·3 brl/ton), which gives thinner mashes than is usual in infusion mashing. Mixing is completed by the large impellor present in the base of the mash-mixing vessel operating at between 5 and 50 rev/min. The mash mixer (Fig. 11.6), mash copper (Fig. 11.7) and lauter tun (Fig. 11.8) are usually of copper but may be of cold-rolled stainless steel. In fact, the mash copper is very similar indeed to the mash mixer except that it is about half the size. Heating is by coils or steam jackets or by the direct tangential introduction of steam. In order to accommodate smaller volumes of mash than normal, the heaters may comprise independent lower and upper portions. Some mash coppers are operated at a small positive pressure.

In double mashing, the mash copper is replaced by a cereal cooker which bears many similarities to it. The cereal cooker receives a charge of malt

MASHING 331

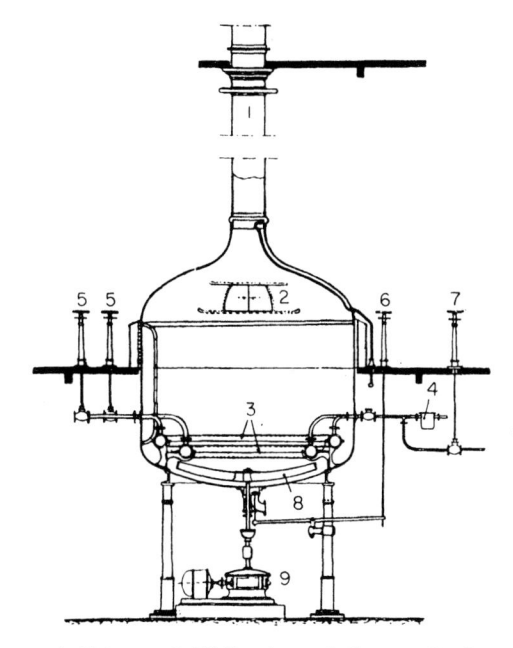

Fig. 11.7 Mash copper. 1. Chimney. 2. Sliding doors. 3. Steam coils. 4. Outlet for condensate 5. Steam valves. 6. Discharge valve. 7. Valve for discharge of condensate. 8. Propeller. 9. Motor for propeller.

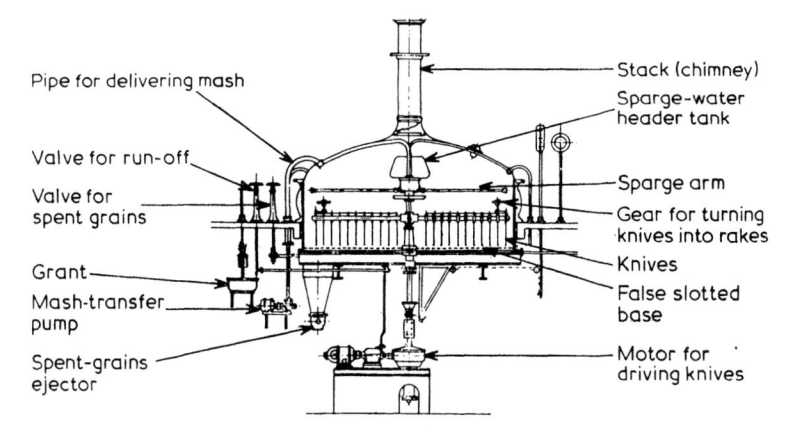

Fig. 11.8 Lauter tun (Grant: wort inspection sink).

(and maize or rice grits in the proportion of say 1:9). Usually the cooker is of stainless steel construction (or clad with stainless steel) and is a vertical cylinder with dished or conical base, equipped with either a bottom- or top-drive variable-speed mixer. In some cookers, there is an eccentric positioning of the shaft to aid mixing and a system of baffles may be provided. Heating is by either double-wall steam jacketting or direct steam injection. The latter leads to some dilution of the mash because of condensation of steam, but in most other respects it is desirable, achieving a precise control of temperature, rapid heating and additional mixing. Injection nozzles are placed so that the steam is directed at an angle to the dished bottom, thereby reducing the impact on the mash. The lauter tun is rather like an infusion-mash tun in general appearance (Fig. 11.8). It is placed at a higher level than the other brewing vessels or at least above the wort copper which receives wort from it. The depth of goods in a lauter tun is usually about 46 cm (18 in), so that rapid filtration of the wort can occur. The shallow bed means that the lauter tun has to be about 50% greater in diameter than an infusion-mash tun of the same capacity.

Because the mash is thinner, the 20-30 min needed to fill the lauter tun permits segregation of fine material from the husks. The mash fails to float as it does in infusion mashing, because the air has been driven from it by the mixing, pumping and boiling. It assumes a water-logged appearance with a considerable depth of free liquid over the bed. The slots in the false bottom of the lauter tun are narrower than those of the infusion-mash tun – about 0·51 mm (0·02 in) – and represent about 8% of the area of the plates. Wort is run-off as in infusion mashing.

In order to run-off the second and subsequent worts, sparge water may be introduced rapidly, the whole mash reslurried with rakes and then permitted to resettle. After running-off worts for a period of time, the sparging and the raking is repeated. This batch system of washing contrasts with an alternative continuous system (which is like the method used for the infusion mashing). Here, liquor is sprinkled by rotating sparge arms or fixed sprays, and running-off the worts is accelerated by the use of rotating knives which cut through the mash bed and loosen it. When mashing is completed, the knives are

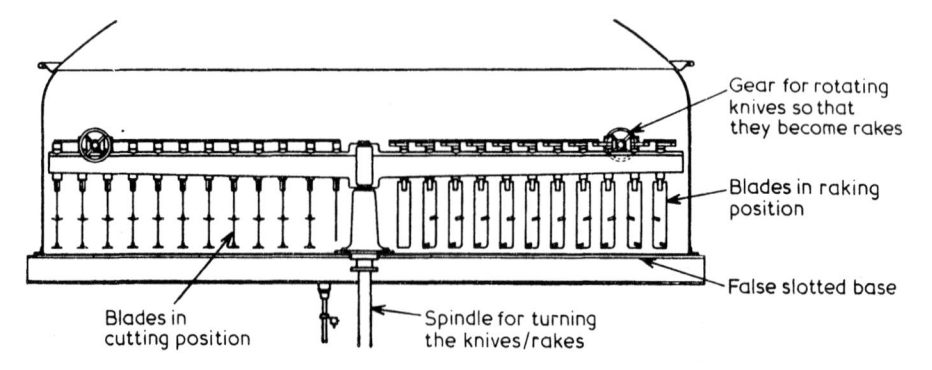

Fig. 11.9 Lauter-tun knives that can be rotated on their vertical axis. (Courtesy of A. Ziemann GmbH.)

twisted through 45–90° so that they present the maximum area to the mash (Fig. 11.9). They are then revolved slowly in order to drive the spent grains to the newly opened discharge ports in the false floor.

The lauter tun may be replaced with a mash filter press. The press is cheaper to construct, occupies less space and can operate with less sparge water,

TABLE 11.2

Grist compositions for lauter tuns and mash filters
{DOUGHERTY (1977) [6]}

Retention on screen number (American)	Mesh width (mm)	% when grinding for lauter tun separation	% when grinding for mash filter separation	Description of fractions
10	2·00	13	3	Husks held
14	1·41	20	4	Husks held
18	1·00	32	13	Husks and some grits
30	0·59	24	20	Coarse grits held
60	0·25	6	45	Fine grits held
100	0·15	2	10	Flour held
>100	—	3	5	
		100	100	

hence stronger worts are obtained [7]. Because the mash is retained by sheets of cloth rather than slotted plates, the grist can be ground more finely (Table 11.2). The filter press can also operate satisfactorily with grists containing high proportions of unmalted grain. To set against this, worts tend to be somewhat cloudy [3], and, because the mash filter can conveniently only receive a predetermined volume of mash, it is less flexible in throughput. (By blanking off portions of the filter, it is possible to accommodate smaller volumes, although usually this practice is inconvenient.) At the end of mashing, opening the filter and reassembling it is carried out hydraulically in modern equipment. Furthermore with the advent of plastic sheets (polypropylene) which can be quickly rinsed clean (unlike the traditional filter cloth), the down-time is only about 24 min. Filter cloths need renewing after about 200 brews at a cost of about £1500 for 60 chambers, but they yield cleaner worts than do plastic sheets (Table 11.3). The filter (Fig. 11.10) comprises (*i*) hollow frames alternating with (*ii*) metal-grooved grids covered with sheets of cloth. The frames and grids are fitted with gaskets and, when compressed, they provide a series of channels for mash, sparge liquor and worts. The filter is heated with steam before being charged. The mash is introduced into the frames which form compartments about 6·4 cm (2·5 in) thick and the first runnings find their way through the filter sheets and are drained from each grid. The sparge liquor is fed to alternate grids and travels

TABLE 11.3
Relative turbidity of worts from lauter tuns and mash filters
{KIENINGER (1977) [3]}

	First worts		Washing stages			Average
	Begin.	End.	1	2	3	turbidity
Lauter tun (190 kg of malt/m²)	13	4	1	5	0	4·6
Lauter tun (340 kg of malt/m²)	37	10	25	4	13	12·4
Mash filter (cotton cloth)		7	0	0	0	2·6
Mash filter (polypropylene cloth)		28	8	2	0	11·6

through the sheets and into the mash. Wort from the mash is then forced out through the sheets and into those grids not receiving sparge liquor; from these grids the wort escapes. When the mash filter is opened after use, spent grains fall from the frames directly into a trough containing a screw-conveyor. The mash filter may be regarded as a series of lauter tuns each with a 6·4 cm (2·5 in) bed.

11.5 *Temperature-programmed systems* [8, 9]

Some breweries have developed processes for mashing which are neither isothermal (like traditional infusion mashing) nor use decoctions to achieve increases in temperature of the mash. The grist may be either all malt or malt plus adjunct similar to that used in the infusion process. The mashing-in is achieved in a stirred mash-mixing vessel, and steam jackets, steam coils or direct injection of steam are used to raise the temperature of the mash through a predetermined programme. Such a programme may be carried out automatically and will be designed to achieve the required glucanolytic and protein rest and the amylolytic rest, based on the grist composition and the type of wort required (especially the enzymic complement of the malt). In one report, a lager mashing comprised (*i*) 40 min at 50°C (122°F), (*ii*) 35 min to reach 65°C (149°F), (*iii*) 45 min at 65°C (149°F), (*iv*) 20 min to reach 75°C (167°F) [8]. In some cases, the programme merely calls for a continuous rise in temperature. This type of mashing has been of particular value to those brewers wishing to use a malt and barley grist. In such cases, the temperature and duration of the various steps will reflect the use of malt enzymes alone or supplemented with bacterial enzymes for conversion (see Chapter 8). An example of three-programmed systems, two of them including decoctions, are shown in Fig. 11.11. Very similar worts and beers were produced but the energy required increased by 44% with one decoction and by 73% with two decoctions as compared with the system without decoctions [9].

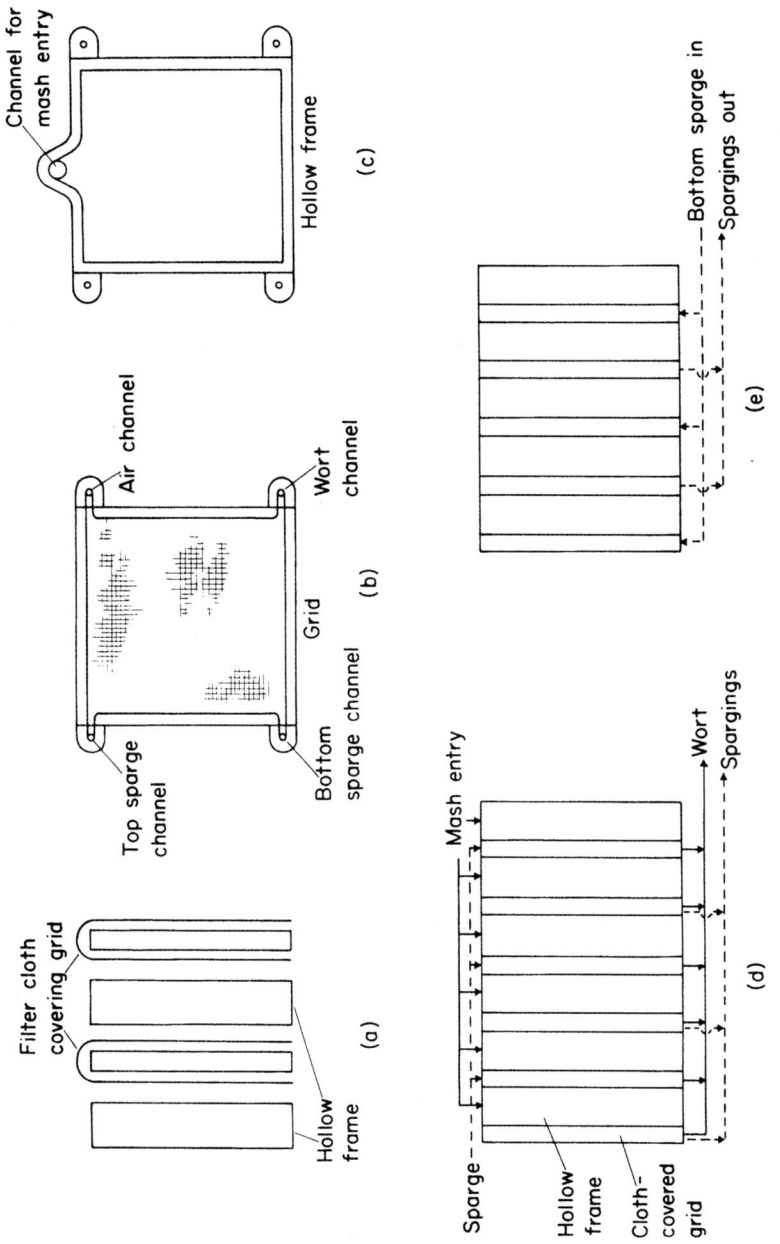

Fig. 11.10 Details of a mash filter. (a) diagram of vertical section; (b) elevation of grid; (c) elevation of frame; (d) early stages in mashing; (e) later stages.

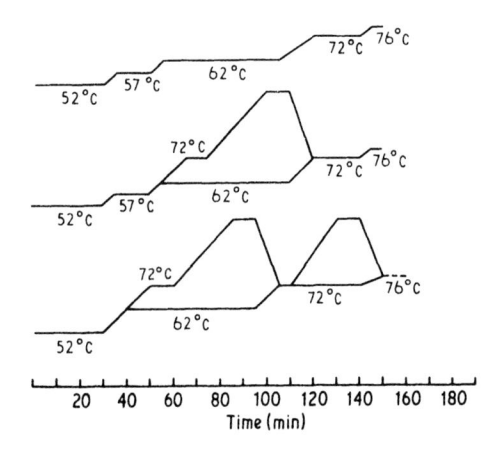

Fig. 11.11 Three-programmed mash procedures, two of which include decoctions, that yield similar worts and beers. {Hug and Pfenninger (1979) [9]}.

11.6 *Modern mashing equipment* [10, 11]

Considerable thought has been given to using brewing plant more fully during the working day, thus reducing capital expenditure on items of equipment and spreading the demand for steam, electricity, and other services. At the same time, high costs of labour encourage the development of plant and processes which can be readily automated. A great deal of research and development has been concerned with continuous methods of production, and complete continuous commercial breweries have been built. Few of them still exist because of their excessive complexity and cost plus the fact that the process is limited in speed by physical and biochemical processes just as the batch brewhouses are. There can be no doubt, however, that the attention paid to continuous wort production, particularly the chemical engineering aspects, has benefited the development of modern batch methods.

One of the simpler ideas aimed at spreading the demand for services is the confluent brewing technique in which the mash tuns are used in sequence so that the peak demands for services are reduced and a virtually continuous supply of sweet wort is made available [12].

Primarily with a view to reducing heat losses, 'block brewhouses' have been built in which the wort production vessels are grouped logically close together and are individually and collectively insulated against heat loss. Pipework may be reduced and the brewhouses are designed to allow several mashes per day and cut the requirement for manpower. Several arrangements of vessels are available but in the plant shown in Fig. 11.12 there are three main vessels comprising (*i*) the mash-mixing vessel, (*ii*) the lauter tun and (*iii*) the wort copper. The first vessel can be heated and therefore acts as a mash cooker if required. This assembly can be adapted for both decoction and infusion mashing. A programme for infusion mashing has been described which gives a fresh supply of wort every 4 hr [13]. Grist falls through a premashing device and into the slowly revolving cylindrical mash-mixing vessel. The mash is stirred by a spiral rake attached to the inside wall. After 90 min conversion, the mash is pumped to the rectangular lauter tun. The mash mixer is washed and receives the first runnings using sixteen draw-off taps from the 30–45 cm (12–18 in) mash bed in the lauter tun, a process completed in a second 90 min period. By this time, the wort copper is ready to receive the first wort runnings from the mash mixer and second runnings directly from the lauter vessel. When the 4 hr cycle is completed, the worts are boiled in the cylindrical wort copper, equipped with offset heater banks. Simultaneously, the mash mixer receives its next charge of mash. This complex but effective logistical system for maximizing use of vessels and giving greatest wort production has many features in common with many modern mashing methods. Such methods are often associated with automatic transfers, heating, mixing and cleaning cycles.

The vertical grouping of vessels in the block brewhouse contrasts strongly with the horizontal arrangement shown in Fig. 11.13. The wet milling system (Chapter 10) is well adapted to the positioning of the mill at the side of, rather than above, the brewing vessels. An inexpensive 'bungalow' building can therefore house the entire wort production plant [14, 15]. The equipment comprises the wet milling plant, a mash-mixing vessel, a mash copper or cereal cooker, a lauter tun, a wort copper and a hop separator. The mash-mixing vessel, the mash copper and the wort copper are rectangular in plan view and in one elevation. The base, however, of these vessels is made of two sloping sheets, giving a valley a little off-centre. Banks of steam or alternatively high-pressure hot-water tubes are provided on the larger of the sloping sheets, and the asymmetric heating so provided gives a rolling action to the contents. Specially shaped agitators, driven from above the vessels, ensure mixing both in the valley and in the body of the vessel during heating. The lauter tun uses a bed of 61–81 cm (24–32 in), but achieves fast run-off because, it is claimed, of the grinding technique and the novel cutting knives of the lauter. The first wort can be withdrawn above the bed or through the slotted false bottom. The false bottom plates are hinged for easy cleaning and there are jets beneath the plates for underletting.

All the vessels are fitted with large observation windows. The control of brewing operations is carried out at a central control console which affords a view of each of the windows. The wet milling takes 2 hr (including the conveyance and weighing of material) while the time-cycle for the lauter tun totals 4 hr. By drawing first runnings through a float from the surface during

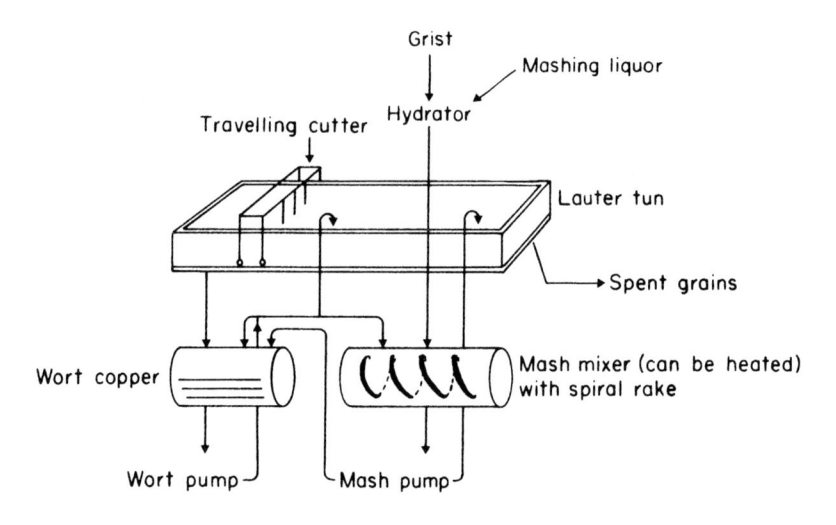

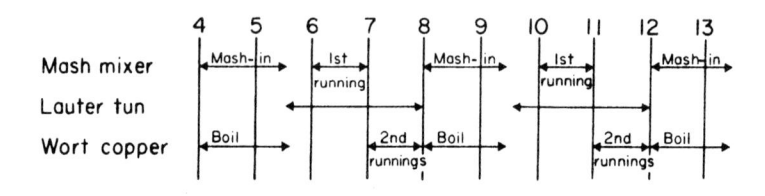

Fig. 11.12 Simplified diagram of a block brewhouse (Ziemann) and a schedule for 4 hr brewing cycles with infusion mashing. The numbers refer to hours.

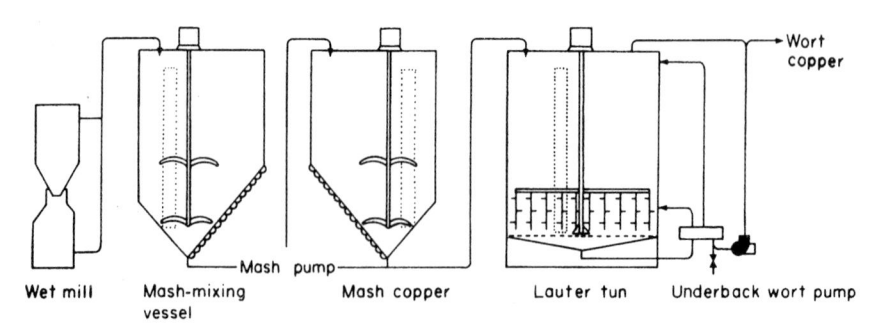

Fig. 11.13 Simplified diagram of a horizontal system of dispensing brewing vessels (Steinecker).

settling of the bed, about 45 min can be saved. Thus a 4 hr brewing programme can be carried out. For infusion mashing, it is possible to operate with only the wet milling plant, the lauter tun and wort copper.

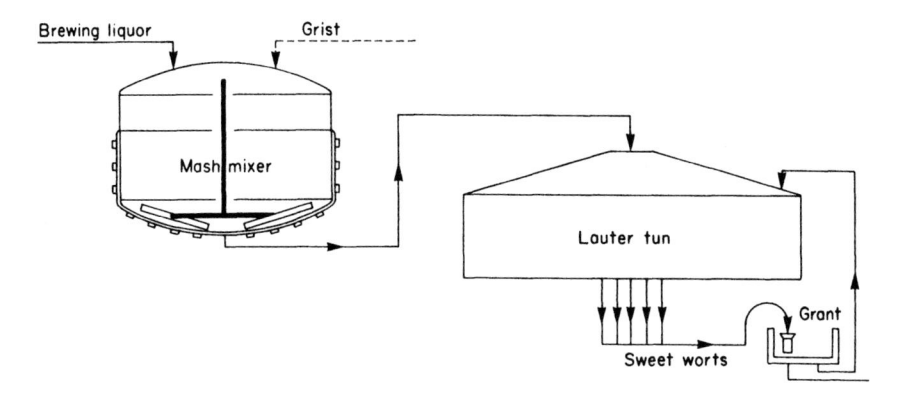

Fig. 11.14 Brewhouse for rapid wort production (Schock-Gusmer). A cereal cooker may precede the mash mixer. The lauter tun is characterized by having a slotted false base with 'valleys' arranged in concentric circles.

With infusion mashing or double-mashing, another variety of brewing plant is claimed to give 8–10 brews per day (Fig. 11.14). The grist and liquor are mixed in a mash-mixing vessel using a special sparge ring and an agitator which 'folds' the mash [15]. Heaters are provided for raising or maintaining temperature. After 10–30 min the mash is dropped into the lauter tun, and specially designed knives keep the bed of the lauter tun open to give particularly fast flow of wort. Alternatively with the double-mash systems, the contents of the cereal cooker are dropped into the main malt mash in the mash mixer, and, after mixing and a suitable conversion time, the entire mash is pumped over to the lauter tun. The false bottom has troughs which descend to the wort outlets and are known as 'valley bottoms'. Each valley receives wort from an equal volume of mash, and the arrangement is claimed to give more efficient withdrawal of worts. When the mash is dropped into the lauter, it is distributed by rotating the knives in the open position to give a 30 cm (12 in) bed. The mash is then allowed to settle and wort is withdrawn from below the false bottom to be recycled on top of the bed. Once the wort is clear, it is run to the copper and sparging of the bed begins along with slow rotation of the knives. Less than 2 hr dwell-time in the lauter is sufficient to withdraw the worts.

A further type of lauter tun has been designed with four times the straining area in half the floor space of a normal tun [16, 17]. The rapid run-off

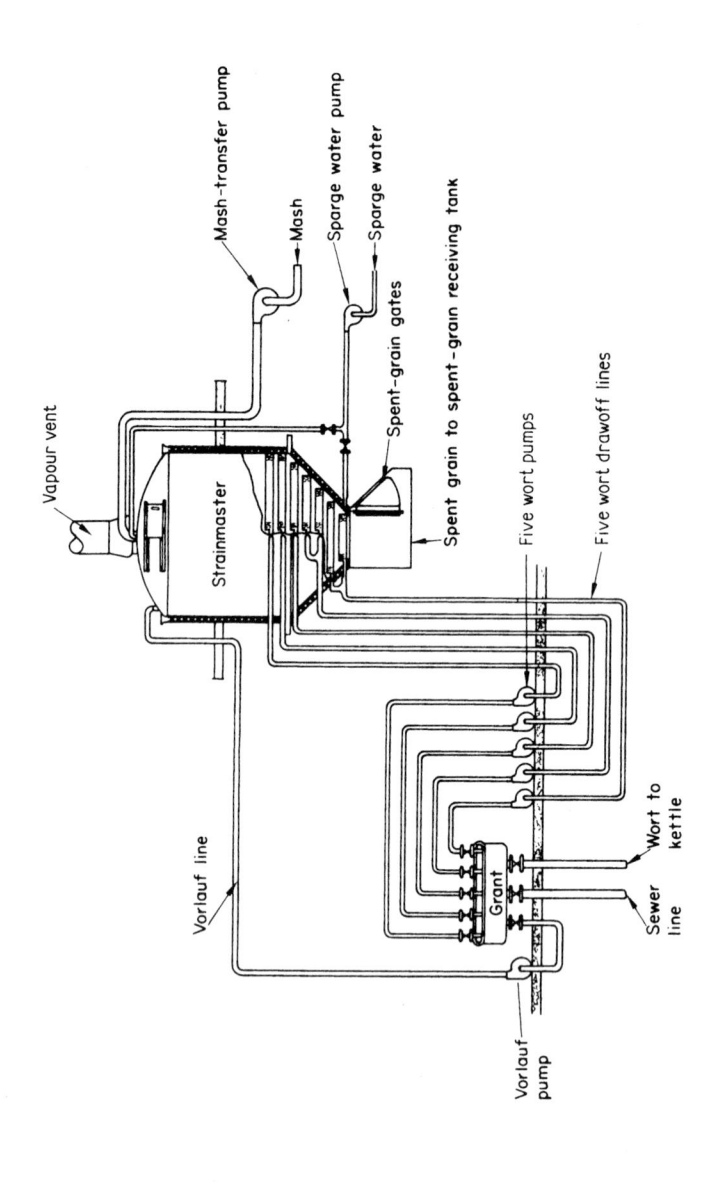

Fig. 11.15 Strainmaster mash tun. Patent owned by Anheuser Busch, Inc.

MASHING 341

achieved enables 11–13 brews per day to be put through. Stainless-steel straining tubes of teardrop cross-section are arranged in seven layers from the bottom of the vessel up to about half-volume mark (Fig. 11.15). The tubes are 13 cm (5.25 in) high and of 5 cm (2 in) maximum width, and their surface is perforated with vertical slots 0.63 mm (0·025 in) wide and 1·27 cm (0·5 in) long. The empty tank is preheated and mash discharged into it through two distribution heads in order to give even spread. As the tubes are covered, pumps connected with them are operated and this sucks mash on to the tubes to form a 'prefilter', analogous with the precoating of a kieselguhr beer filter. The wort is at first recycled, but quickly becomes clear. Sparging is automatic after much of the first runnings have been withdrawn. Finally, the 'bomb doors' at the bottom of the vessel are opened to release the spent grains. The grains tend to be wetter than with conventional lautering. They have therefore to be sieved or pressed, and the extract so recovered may be used for mashing in. Vibrating screens have proved of particular value in this respect [18]. The operations of preheating, mash receipt and circulation, withdrawal of first worts, of second worts and finally spent-grain removal take 3, 12, 15, 60 and 15 min respectively, totalling 105 min. However, it is claimed that this system is not as well suited as the traditional mash tun or lauter tun for the production of high-gravity worts [17].

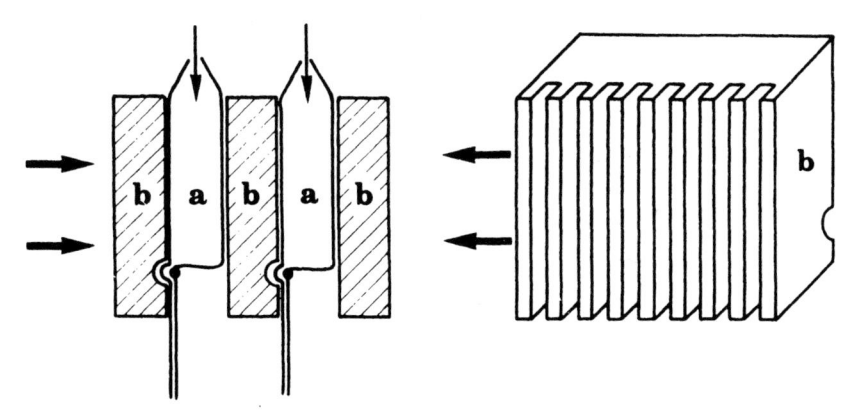

Fig. 11.16 Norden high-pressure mash filter. Left: diagram to show arrangement of filter pockets, (a) and vertically grooved plates (b) in section. Right: the grooves of the plate. Heavy arrows indicate compression of the filter components, lighter arrows show the loading of the mash {van Waesberghe (1979) [19]}.

The Norden high-pressure filter comprises a series of polyester or polypropylene filter pockets separated by vertically grooved plates, held together in a frame (Fig. 11.16). Each pocket is closed automatically at the base before receiving mash pumped in at 4·5–12·0 psig. (0·3–0·8 bar) from the mash mixer [19]. The first worts emerge bright and require no recirculation. Sparging water is introduced at 4·5–12·0 psig (0·3–0·8 bar), and towards the end of the sparging the distance between the plates is reduced from 7 to 2 cm. Table 11.4 gives results of a small-scale trial. Hammer-milled grist of 0·5–0·8 mm (0·02–0·03 in), sometimes with a proportion of the husk fractions removed, is used so that the bed within each pocket stratifies with the more permeable fraction surrounding the less permeable. A filter cake of 40–50% dry solids is produced which is readily discharged before the pockets are washed with hot water that can be used for the next mashing-in.

TABLE 11.4

Results with a Norden high-pressure filter [750 litres of mash with 135 kg of malt plus 45 kg of maize grits, sparge water 1 hl, final gravity 16°Plato (SG 1066)]
{VAN WAESBERGHE (1979) [19]}

Time from filling (min)	Wort volume (litre)	Wort °Plato running at time indicated
5	280	16·8*
10	450	16·8*
15	560	15·5
20	600	11·5
33	700	9·4
45	760	9·5

* First runnings, before sparging.

11.7 Continuous mashing [17–30]

A German patent of 1902 [20] described a circular tank divided by radial partitions into a number of individual cells. Each cell was fitted with a hinged perforated false bottom, and beneath it, an outflow pipe. The tank rotated on a vertical axis bringing each cell in turn to each of the stations for filling with mash, draining, sparging and finally releasing the spent grains by swinging out the false bottom. The lower ends of the outflow tubes were submerged in order to exert more suction through syphonic action. In 1938, an American patent [21] described a process in which mash passed continuously through conversion vessels and was then lautered on a vacuum drum filter with sparges. This method was adopted by the Brewing Institute of Waldkirch, Germany [22]. Continuous tube converters and gelatinizers were the subject of several patents about 35 years ago. Batteries of centrifuges for separating wort from mash have also been advocated from time to time, notably in American and Russian work [23, 24]. The use of a travelling filterbed for mashing has also been the subject of a number of patents [25] around 1960 (Fig. 11.17). Other devices for liquid–solid separation which have been advocated include cyclones and vibrating-screen extractors. Screw-conveyors working in inclined tubes have been used in which grains are propelled upwards and the wort separated and percolated downwards to the base of the tube where it is clarified by a spinning disc filter.

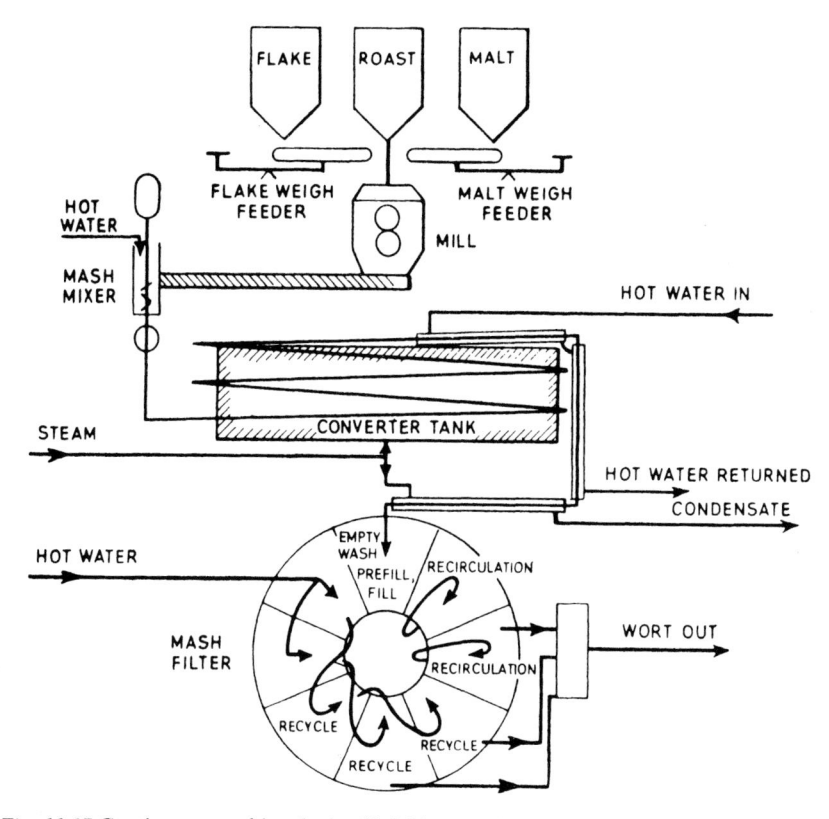

Fig. 11.17 Continuous mashing device (A.P.V. system).

An experimental small industrial continuous mashing plant was built comprising a mash mixer, mash converter, inclined sieves and 'contactor' sparges [27]. From the mixer, mash passes down in plug flow into a pump. Wort separation occurs on a series of inclined sieves over which sparge liquor is applied (Fig. 11.18).

In one system of continuous mashing, used in certain breweries, the mash is pumped through a series of spiral heat exchangers which permits temperature-programmed conversion (Fig. 11.19). The wort is separated in three stages, each separation being achieved using a rotating conical-sieve and a holding tank. After the wort is sieved off in the first stage, the separated spent grains are diluted and led to the holding tank before encountering the

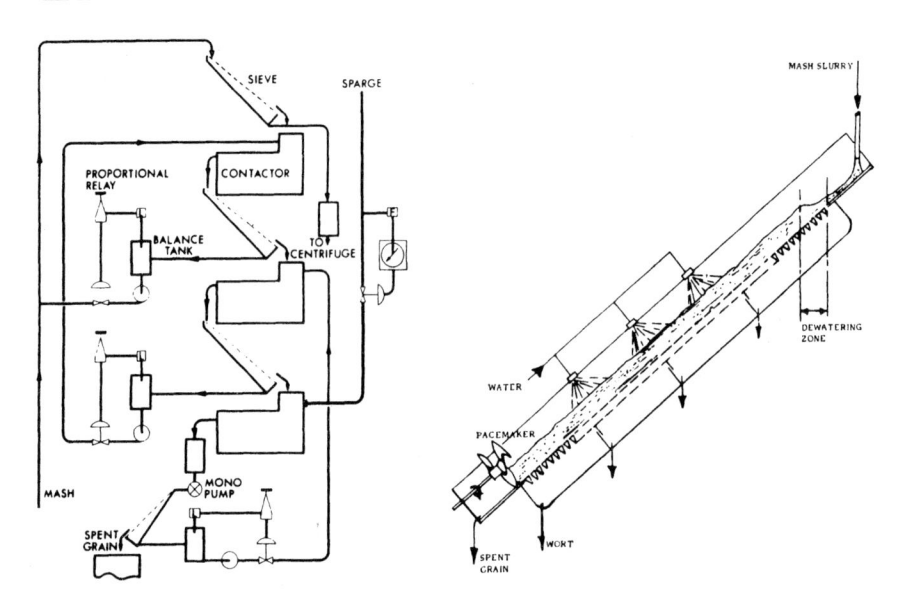

Fig. 11.18 Inclined sieves and contactors used for lautering in a pilot-scale continuous brewery.

second sieve. The process continues until the grains have been washed three times, and the wash liquid proceeds counter-current to the grains. Wort issuing from the extraction stages is centrifuged and led to a tank where hops or hop derivatives are added. It proceeds to a vessel where high-pressure steam raises the wort to 150°C (302°F) for 2 min before entering a vacuum vessel where the temperature is rapidly lowered by flash cooling to 90°C (194°F). Finally the wort is led through a centrifuge for separation of the trub and is then cooled [28].

11.8 *The theory of mash filtration* [10, 11, 29]

The lautering stage includes the two unit operations of (*i*) filtration and (*ii*) leaching or extraction.

The rate of filtration can be readily adduced from simple theory as being controlled by (*i*) pressure differential across the filterbed, (*ii*) permeability of the bed, (*iii*) the depth of the bed and (*iv*) the area of the bed. For a filterbed of specified diameter:

$$\frac{\text{Velocity of liquid through bed}} = \frac{\text{constant} \times \text{pressure differential} \times \text{bed permeability}}{\text{bed depth} \times \text{viscosity of liquid}}$$

The permeability of a bed of particles of uniform size is a function of (*i*) the square of the particle diameter and (*ii*) the particle shape. With grain beds,

MASHING 345

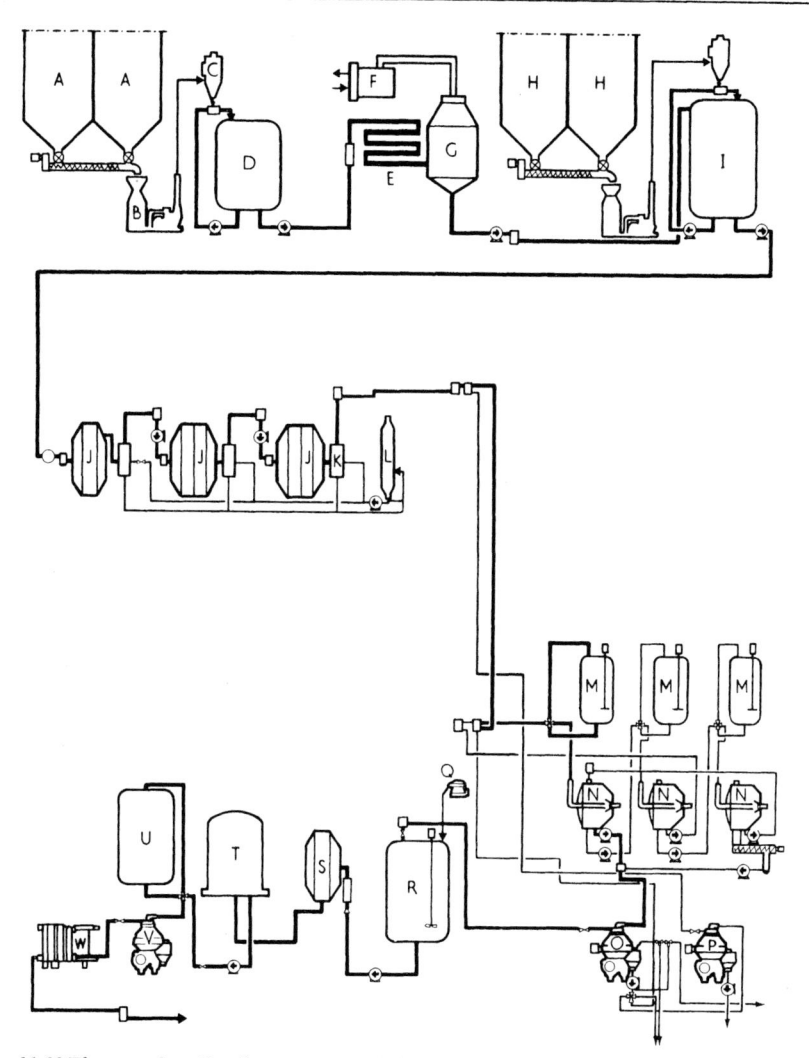

Fig. 11.19 The complete Centibrew process. {Ehnstrom (1977) [28]}.

Key: A Adjunct silos; B Tip-weigher, hammer mill and fan; C Cyclone for separating grist from transport air; D Adjunct mash-in vessel; E Temperature holding coil; F Vacuum (expansion cooling) vessel; G Box condensor; H Malt silos; I Malt mash-in vessel; J Spiral reactors; K Spiral heat exchangers; L Hot Water unit; M Holding tanks; N Rotating sieves; O Mash separator; P Wort recovery separator; Q Hops mill; R Extraction tank; S Temperature holding reactors; T Flashing-off of volatiles vessel; U Wort stabilizing tank; V Wort clarifier; W Wort chiller; X Cold water supply; Y CIP tanks.

these and other relevant parameters are dependent on the composition of the grist, the fineness of the grind and the method of mashing. It is assumed that the bed is incompressible.

Leaching efficiency is an inverse function of the particle size and directly proportional to the diffusion coefficient. It is also concerned with the number of extraction stages. The number of stages is simply the number of times the sparge liquor is applied discontinuously and the grains paddled with each charge of fresh liquor. For continuous sparging, the number of theoretical extraction stages can be estimated. It is a function of (*i*) the ratio of bed depth to bed diameter and (*ii*) the run-off rate. The number of stages is inversely proportional to (*i*) the permeability of the bed, (*ii*) the overall mass transfer coefficient, and (*iii*) the interfacial area. (The mass-transfer coefficient itself is a function of the diffusion coefficient and the liquid-film-transfer coefficient and an indirect function of particle diameter.)

The efficiency of leaching can be calculated from the following equation:

$$\text{Fraction of extract left in grains} = \frac{(M+W)}{U}\left[1 + \frac{W}{U} + \left(\frac{W}{U}\right)^2 + \left(\frac{W}{U}\right)^3 + \ldots \left(\frac{W}{U}\right)^n\right] \quad (11.3)$$

where M is mashing liquor rate, W is sparge liquor rate, U is liquor left in spent grains and n is number of extraction stages.

Results of carrying out batch washings of a mash are given in Table 11.5. It can be seen that for a wort of SG 1050, three stages of washing will give $> 99\%$ efficiency. If, however, the volume of mashing-in liquor is excessive or insufficient sparge (wash) water is used, the efficiency falls. For wort of SG 1100, more stages are required and wort must be recovered from the spent grains by dewatering.

TABLE 11.5

Loss of yield (%) in extraction of mash with liquor/grist ratio of 2:1 by weight
{ROYSTON (1968) [11]}

Spent grains Moisture content (%)	For wort of SG 1050 (12·4°Plato)							For wort of SG 1100 (23·7°Plato)						
	First worts	Washing stages						First worts	Washing stages				
		1	2	3	4	5		1	2	3	4	5
87	22	6	1·5	0·5	0·1	0·1	40	18	11	8	6	4
75	12	2	0·4	0·1	0·1	0·1	25	11	6	4	3	2
60	7	0·7	0·1	0·1	0·1	0·1	15	5	3	1·5	0·8	0·5

In assessing the performance of lauter tuns and infusion-mash tuns, reference must be made to filtration and leaching efficiencies. But as has been mentioned earlier in this chapter, another important difference between the two types of filtration vessel relates to the capacity of the goods to float in the infusion-mash tun but not in the decoction lauter vessel. The flotation arises from (*i*) the presence of entrained air or other gas, and (*ii*) the diffusion of strong worts as conversion proceeds, so that the specific gravity of the wort

MASHING

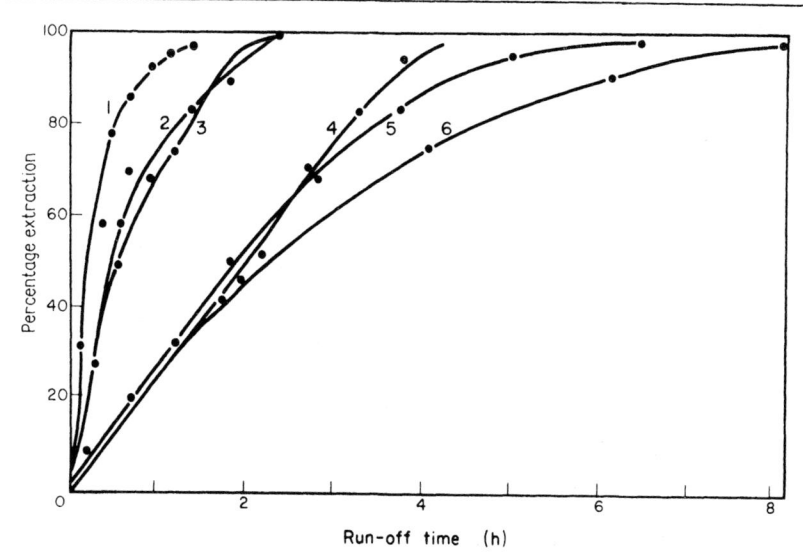

Fig. 11.20 Patterns of run-off for various systems used in mashing. 1. Strainmaster mash tun (Fig. 11.15). 2. Decoction mash in lauter tun. 3. Infusion mash in lauter tun. 4. Mash tun with medium depth of mash 1·5 m (5 ft). 5. Mash tun with deep bed 2·4 m (8 ft). 6. Mash tun with deep bed, using malt and maize grits. {Harris (1971) [10]}

between goods and the mash-tun plates tends to be high. The air becomes entrained because of the thick consistency of an infusion mash and the small degree of agitation it receives. Because of the flotation, the run-off of worts tends to be higher in the infusion-mash tun (Fig. 11.20) and this is aided by the even dispersal of fine particles throughout the goods. In the decoction-mash tun, the fines tend to accumulate on the top of the goods because of the thinner consistency of the mash. This accumulation leads to the development of a relatively impervious layer and is reflected in high differential pressures between the top of the mash and the bottom as measured by a manometer placed immediately below the lauter-tun plates (Fig. 11.21). The accumulation of the fines is favoured by recycling of worts to clarify them and is only temporarily relieved by the use of mechanical knives or rakes.

11.9 Choice of mashing systems [3, 10, 11, 13, 17, 29]

The selection of a mashing method involves a consideration of (*i*) the grist composition, (*ii*) the equipment available, (*iii*) the brewing liquor available, (*iv*) the type of beer to be brewed and (*v*) the number of brews required each day. With fully modified malts and a restricted level of starch adjuncts, say up to 30%, it is possible to achieve efficient saccharification by either infusion mashing or the short double-decoction system described. Little if any proteolytic action is required for malts which are well modified. For grists

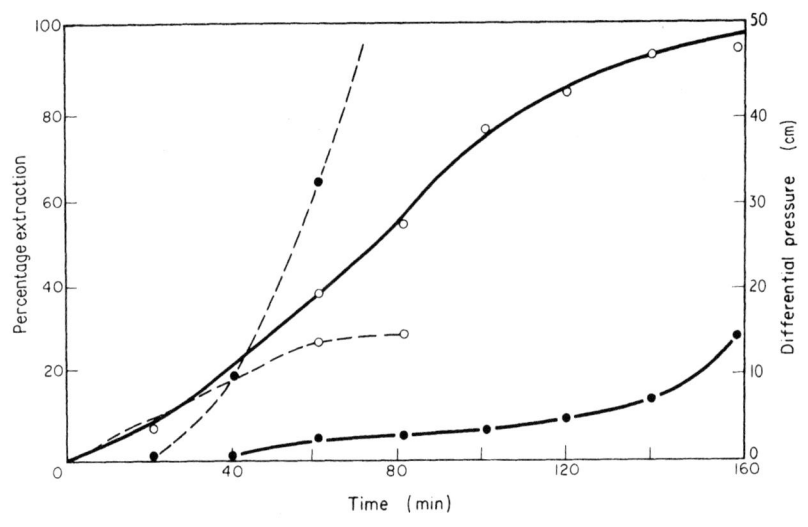

Fig. 11.21 Comparison of two infusion mashes: one at 2·18 hl/100 kg (13.3 brl/ton) mashed *in situ* (———) and the other at 4·36 hl/100 kg (26·6 brl/ton) transferred from a premasher (- - - -). ○, percentage extraction; ●, differential pressure (units: inches of water gauge pressure). {Harris (1971) [10]}

containing high proportions of adjuncts, either the double-mash system or the longer double-decoction system is employed. These secure complete gelatinization and saccharification of the high levels of starch present at mashing-in. Where malts are used which are less well modified, the protein rest assumes importance if the beers produced need a moderate or long shelf life. For dark Munich beers, the three-mash decoction system is used in conjunction with lightly modified malts. The influence of the brewing liquor is considered in Chapters 7 and 9.

One of the most important considerations in present-day brewing is increasing the utilization of vessels, often by getting a large number of brews through the plant each day. This leads to automated systems designed to give the greatest efficiency (Table 11.6). With the infusion-mash tun, at least 1 hr is occupied in converting the mash, while with decoction mashing, conversion takes place in a separate vessel, usually a mash-mixing vessel. There is therefore difficulty in reducing substantially the residence time in an infusion-mash tun despite its capacity to allow quick run-off of worts. The operation which can be hastened in both cases is the run-off, and this variable is denoted by T in Table 11.7. If the time available per brew is reduced, the demand on the mash tun for quick run-off is inordinately greater than it is on the lauter tun. Acceleration of run-off for a given size of mash could be achieved in the mash tun by (*i*) reducing the depth of goods, (*ii*) increasing the surface area and (*iii*) increasing the temperature at run-off so as to lower wort viscosity

TABLE 11.6

Duration (min) of various operations during conventional and modern lautering
{KOLLNBERGER (1979) [33]}

Operation	Conventional lauter tun	Modern lauter (6 brews/day)	Modern lauter (8 brews/day)	Modern lauter (10 brews/day)
Filling	10	5	5	5
Mash transfer	15	7	7	7
Mash rest	10	10	5	0
Wort recirculation	10	5	5	5
First wort running	70	70	50	40
Sparging	90	120	80	72
Wash and drain	5	8	13	0
Effective filtration time	165	198	143	112
Grains removal	20	10	10	10
Flush false bottom	10	5	5	5
Total ancillary time	75	42	37	32
Total residence time	240	240	180	144

TABLE 11.7

Comparison of times required for the various operations needed for infusion-mash tuns and decoction lauter tuns
{ROYSTON (1968) [11]}

Mash tun		Lauter tun	
	min		min
Underlet	5	Underlet	5
Mash-in	15	Pump-over	15
Stand	60	Recirculation	10
Recirculation	10	Run-off	T
Run-off	T	Drain	15
Drain	15	Grains out	15
Grains out	15		
Total	T + 120 min	Total	T + 60 min

and therefore achieve more rapid filtration. To achieve this would increase the cost of a mash tun and when schedules of about six brews per day are demanded, the overall cost of a lauter tun and mash-mixing vessel is similar to that of a mash tun which could maintain this rate of turnover. An alternative way of increasing the number of brews per day is to forego some of the extract by deliberately cutting off the last runnings which are of low gravity (Fig. 11.22). In North America, it is common to do this in recognition

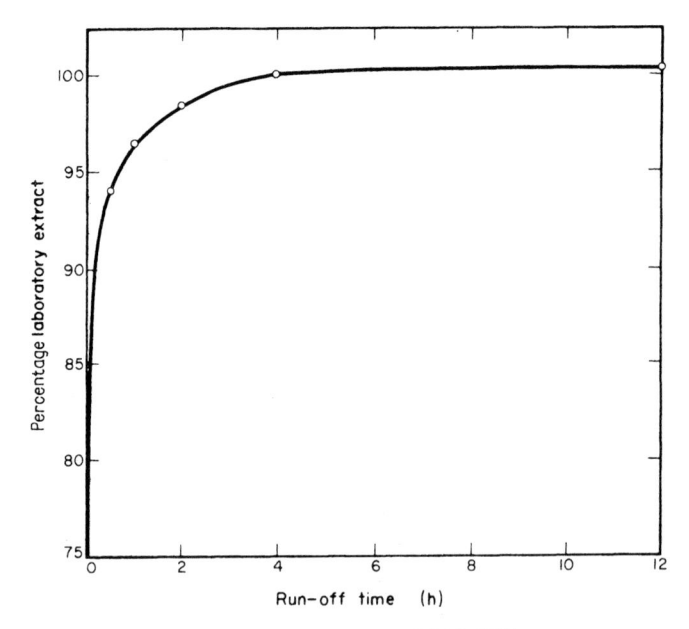

Fig. 11.22 Effect of run-off time on yield {Royston (1966) [29]}.

TABLE 11.8

Comparison of brewing practices
{ROYSTON (1968) [11]}

	Britain	Continental Europe	N. America
First runnings (SG)	1100 (23·7°Plato)	1090 (21·5°Plato)	1080 (19·3°Plato)
Last runnings (SG)	1004 (1·03°Plato)	1004 (1·03°Plato)	1008 (2·06°Plato)
Optimum bed depth	1·2 m (4 ft)	0·3 m (1 ft)	0·13 m (5 in)

of the (*i*) relatively low costs of raw materials and power, (*ii*) relatively high costs of plant and labour and (*iii*) recognition that the last runnings tend to affect flavour and head retention adversely. In Table 11.8, a comparison of brewing practices in Britain, continental Europe and North America is given.

In the light of these practices, a comparison of production costs has been made for (*i*) mash tuns securing maximum extract, (*ii*) lauter tuns yielding maximum extract and (*iii*) lauter tuns in which 5% of the potential extract yield is discarded in the last runnings. The results of the comparison (Fig. 11.23) reveal that the optimum for (*i*) is four brews/day while with (*ii*) and (*iii*), it is six and twelve brews/day respectively.

MASHING

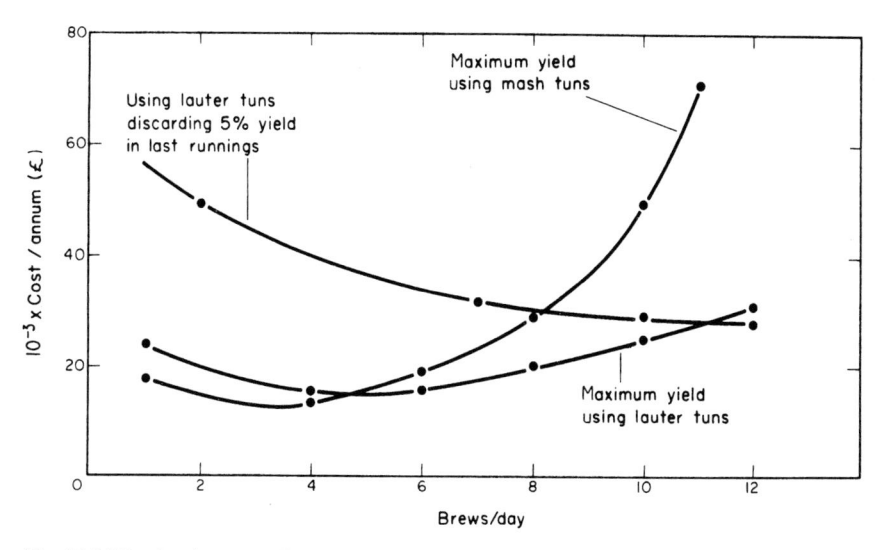

Fig. 11.23 Production costs for different systems of mashing {Royston (1966) [29]}.

Taking a specific example of two extraction devices both dealing with 5 tons, one has a low capital cost (say £25 000) but relatively low extract yield, while the other has a higher initial cost (say £100 000) and higher yield. If it is assumed that the more expensive and efficient plant should begin to justify the extra capital cost after 3 years, £[(100 000 − 25 000)/3] will have to be saved each year in raw materials. This £25 000 approximates at the present time to about 1% of the annual malt bill for a 5 ton mash tun. The cheaper plant can therefore be justified only if its efficiency of extraction differs from that of the expensive device by less than 1%.

The development of new mashing devices which present a much greater area of mash to the straining surface, such as is shown in Fig. 11.15, has great significance in achieving a large number of brews per day. Filtration and run-off is faster. Thus comparing a mash tun, a lauter tun and this new type of masher, the weight of grist per unit area of filtration surface is 476, 143 and 118 kg/m² respectively (9·7, 3·2 and 2·6 lb/ft²). The corresponding rates of extraction (kg/m²/min) are 1·86, 1·03 and 1·48. Thus the new type of masher separates wort at a rate intermediate between a mash tun and a lauter tun. The down-time (when the devices are not actually filtering) is about 3 hr for a mash tun, about 30 min for a lauter tun and about 20 min for the new masher. Under British brewing conditions, between seven and eleven brews/day can be satisfactorily achieved with this new development.

11.10 High-gravity brewing

For many breweries, it makes good economic sense to produce wort of higher gravity than would be expected from the original extract of the beers that are produced. The higher-gravity wort is fermented and dilution occurs at a very late stage in processing, either just before or after the final beer filtration. On the credit side, the brewer uses either smaller vessels or fewer vessels or he brews less frequently. Virtually the same amount of energy is required for altering the temperature of a high-gravity wort or beer compared with a similar volume of normal gravity, so that overall there are large savings of energy per volume of final beer. There are, however, problems in high-gravity brewing, and some of these will receive attention in Volume 2, for instance poorer hop utilization when conventional hopping is employed, and sometimes modified flavour of the final beer. Relevant to the present chapter is that some wort-extraction equipment is more suited than others to yield high-gravity worts without loss of overall extract. Thus the infusion-mash tun is ideal, the lauter tun and mash filter satisfactory, but the Strainmaster is reported to give 2·3% less extract when collecting worts of SG 1034 (8·5°Plato) than of SG 1051 (12·6°Plato).

There are three methods of obtaining high-gravity worts, namely (*i*) adding syrups (of say SG 1150 or 34°Plato) to the copper, (*ii*) discarding the dilute last runnings of wort from the mash tun, lauter tun or mash filter, or (*iii*) using these last runnings either as mashing-in water or as sparge. The first method will be considered in Chapter 15. Last runnings (and press water from spent grains) are of higher pH, lower extract, richer in polyphenols and silica than the earlier worts. They are also richer in certain lipid materials, notably linoleic acid. All these characteristics may be detrimental, and therefore some breweries, concerned more with quality of the final beer than the cost of its production, will discard the last runnings. Still others will treat with active charcoal and reduce the pH to 5·2–5·4 before using these worts. Finally, others, having established that the practice does not significantly devalue the final products, reuse the last runnings without further treatment (and in some cases press water from spent grains). A tank is then reserved for collecting and holding these weak worts at temperatures in the range 55–75°C (131–167°F), in part to prevent microbial attack and in part because the temperature of collection is 65–75°C (140–167°F) and the temperature of reuse will be either 55–65°C (131–149°F) for mashing-in or 65–75°C (140–167°F) for sparging. If the last runnings will not be reused for 48 hr or more, they are normally discarded.

11.11 Spent grains [30]

Examination of spent grains after mashing gives valuable information on the efficiency of milling and mashing. Whole corns should be absent and husks substantially unbroken. The loss of wort through incomplete conversion and retention of wort should not amount to more than 1–2%. It can be assumed that 100 units of weight of malt of, say, 4% moisture will give 60 units of wet spent grains, and when dried the grains will weigh 15 units. Other information is given in Chapter 9. Spent grains are used as a constituent of cattlefood, and the food value of 5 kg of moist spent grains is equivalent to 1 kg of barley. The protein, lipid, cellulose and hemicelluloses present in the grains are readily broken down by microbial action in the gut of cattle. In the future, with the development of suitable commercial hemicellulases and cellulases, it may prove possible to degrade most of the carbohydrate of spent grains and grow food yeast such as *Candida utilis* from the derived pentoses and hexoses. Doubtless, the economies of malting and brewing will become increasingly linked with the economics of producing cattle food (Table 11.9).

TABLE 11.9

Composition and feed-evaluation data of fresh brewers' grains. Results are based on laboratory dried material, and digestibility trials were carried out with sheep. Metabolizable energy is calculated as digestible energy × 0·81. Figures in parentheses are the digestible amounts

	Mean	Range
Dry matter (%)	26·3	24·4–30·0
Crude protein (%)	23·4 (18·5)	18·4–26·2 (13·9–21·3)
Crude fibre (%)	17·6 (7·9)	15·5–20·4 (6·6–10·2)
Ether extract (%)	7·7 (7·7)	6·1–9·9 (5·6–9·2)
Total ash (%)	4·1	3·6–4·5
Digestible energy (mJ/kg dry wt.)	13·8	13·0–14·8
Metabolizable energy (mJ/kg dry wt.)	11·2	10·5–12·0
Gross energy (mJ/kg dry wt.)	21·4	21·1–21·8
DOMD* *in vivo* (%)	59·4	55·2–64·3
DOMD* *in vitro* (%)	48·6	44·8–51·5

* Digestibility of organic matter (dry).

It may therefore become necessary to re-evaluate mashing procedures in order to decide how much extract should be left in spent grains. For instance, in some countries it may be an attractive proposition to cut down mashing times substantially and receive an enhanced price for spent grains. It has become common practice in recent years to add to the spent grains in the lauter tun the spent hop material and trub. Indeed, in some breweries surplus yeast and tank bottoms are dealt with in the same way thereby reducing substantially the volume and concentration of effluent. Spent hops have little food value, but are acceptable to the merchants buying spent grains; in contrast, the nitrogenous content of trub, yeast and tank bottoms enhances the food value. There is, however, some concern about the percentage of yeast cells present, because high levels cannot be tolerated by cattle. Similarly levels of kieselguhr have to be maintained at low levels. Drying grains is expensive and in many countries they are delivered wet without delay to the farms.

REFERENCES

[1] LLOYD HIND, H. (1950). *Brewing Science and Practice*, vol. 2, Chapman & Hall, London.
[2] RENNIE, H. (1967). *Process Biochem.*, **2**, 9.
[3] KIENINGER, H. (1977). *J. Inst. Brewing*, **83**, 72.
[4] ZANGRANDO, T. (1979). *Brewer's Digest*, (4), 32.
[5] DE CLERK, J. (1958). *A Textbook of Brewing*, vol. 1 (translated by K. BARTON-WRIGHT), Chapman & Hall, London, p. 442.
[6] DOUGHERTY, J. J. (1977). *The Practical Brewer*, 2nd edn. (ed. BRODERICK H. M.), Master Brewers Association of the Americas, p. 62.
[7] DIXON, I. J. (1978). *The Brewer*, (5), 171.
[8] SPILLANE, M. H. (1978). *Brewer's Guardian*, (4), 63.
[9] HUG, H. and PFENNINGER, H. B. (1979). *Proc. Eur. Brewery Conv. West Berlin*, 355.
[10] HARRIS, J. O. (1971). *Brewing Technology* (ed. W. P. K. FINDLAY), Macmillan, London, Chapter 1.
[11] ROYSTON, M. G. (1968). *Process Biochem.*, **3**, 16.
[12] HELY HUTCHINSON, M. (1964). *J. Inst. Brewing*, **70**, 12.
[13] ROBERTS, R. H. (1966). *Brewer's Guardian*, **95**, 17.
[14] LAURIDSEN, P. B. (1966). *Tech. Q. Master Brewers Assoc. Am.*, **3**, 34.
[15] LENZ, C. (1965). *Tech. Q. Master Brewers Assoc. Am.*, **2**, 2.
[16] SCOTT, P. M. M. (1967). *Brewer's Guild J.*, **53**, 339.
[17] WHITEAR, A. L. and CRABB, D. (1977). *The Brewer*, (2), 60.
[18] BUTTON, A. H., STACEY, A. J. and TAYLOR, B. (1977). *Proc. Eur. Brewery Conv. Amsterdam*, 377.
[19] VAN WAESBERGHE, J. W. M. (1979). *Brewing Distilling Int.*, **9**(9), 54.
[20] German Patent 144, 146 (1902).
[21] US Patent 2, 127, 759 (1938).
[22] REITER, F. (1962). *Brauwelt*, **102**, 449–451, 614–615, 626–627.
[23] WILLIAMSON, A. G. and BRADY, J. T. (1965). *Tech. Q. Master Brewers Assoc. Am.*, **2**, 79.
[24] SCHULTZE-BERNDT, H. G. (1959). *Brauerei*, **13**, 308.
[25] British Patents 935, 570; 943, 811 and 943, 812 (1963); 963, 699 (1964).
[26] DAVIS, A. D., POLLOCK, J. R. A. and GOUGH, P. E. (1962). *J. Inst. Brewing.* **68**, 309.
[27] HALL, R. D. and FRICKER, R. (1969). *Brewer's Guardian Suppl.* **98**, 15.
[28] EHNSTROM, L. (1977). *Tech. Q. Master Brewers Assoc. Am.*, **14**(3), 159.
[29] ROYSTON, M. G. (1966). *J. Inst. Brewing*, **72**, 351.
[30] GARSCADDEN, B. A. (1973). *The Brewer*, (12), 612.
[31] BROWN, H. T. (1910). *J. Inst. Brewing*, **16**, 112.
[32] HOPKINS, R. H. and CARTER, N. A. (1933). *J. Inst. Brewing*, **39**, 59.
[33] KOLLNBERGER, P. (1979). *Tech. Q. Master Brewers Assoc. Am.*, 101.

Appendix

UNITS OF MEASUREMENT

Two new systems of measurements have been introduced into Britain in the last ten years. Industry and commerce have seen the introduction of metrication to come into line with other European countries. Scientific and teaching establishments have adopted a related but not identical system called Système Internationale (SI for short) illustrated in Tables A.1 and A.2. Both have their place in Brewing Science as have the old systems based on the pound and the foot. Especial difficulties arise because the traditional US measures do not always correspond with the British. For instance the US gallon of water weighs 8 lb, and the constituent eight pints one pound each. The British imperial measure, revised in Queen Anne's reign, 'metricated' the gallon of water so that it weighed 10 lb, each pint weighing 1·25 lb. Measurements of volume such as the bushel or the barrel vary from one region to another. In this book, the barrel will be that used in England, namely 36 gallons. Filled with water this would weigh 360 lb. For rough calculation, it is convenient that the US beer barrel is rather similar to the hectolitre, exceeding it by about 17%. The use of degrees Fahrenheit is rapidly being replaced in the UK by °C (metric) or K (SI) but not in the USA for practical brewing (Table A.3).

With respect to units specifically used in brewing, certain metric units have been selected which are not in the SI, such as litres, hectolitres, metric tonnes and bars, because of their convenience. Other measurements used in brewing are becoming or have become obsolete, for instance Quarters as applied to barley (448 lb) or malt (336 lb). Methods of analysis have changed over the years so that comparisons are difficult. Elsewhere there may be no direct equivalence; a good example is measurement of extract as defined by the Institute of Brewing versus that given by the European Brewery Convention. Set out in Table A.4 are definitions of several SI and metric units plus conversions of non-metric units that are relevant to brewing. Table A.5 gives details of the various systems of measuring specific gravity and extract, while Table A.6 gives equivalents of hot water extracts of malts in the old and new nomenclature.

TABLE A.1
SI derived units

Physical quantity	Name	Symbol	Definition
Energy	joule	J	kg m^2 s^{-2}
Force	newton	N	kg m s^{-2} = J m^{-1}
Power	watt	W	kg m^2 s^{-3} = J s^{-1}
Electrical charge	coulomb	C	A s
Electrical potential difference	volt	V	kg m^2 s^{-3} A^{-1} = J A^{-1} s^{-1}
Electrical resistance	ohm	Ω	kg m^2 s^{-3} A^{-2} = V A^{-1}
Inductance	henry	H	kg m^2 s^{-2} A^{-1} = V A^{-1} s
Luminous flux	lumen	lm	cd sr
Illumination	lux	lx	cd sr m^{-2}
Frequency	hertz	Hz	s^{-1}

where (1) m is the metre　　　　　　　(2) kg is the kilogram
　　　(3) s is the second　　　　　　　(4) A is the ampere
　　　(5) cd is the candela (luminous intensity)　　(6) sr is the steradian (solid angle)
other basic units are: (7) kelvin (K)—thermodynamic temperature and temperature interval
　　　　　　　(8) mole (mol) molecular (or atomic) weight in kg m^{-3}

TABLE A.2
Prefixes for SI units

Fraction	Prefix	Symbol
10^{-12}	pico	p
10^{-9}	nano	n
10^{-6}	micro	μ
10^{-3}	milli	m
10^{-2}	centi	c
10^{-1}	deci	d
10^{1}	deka	da
10^{2}	hecto	h
10^{3}	kilo	k
10^{6}	mega	M
10^{9}	giga	G

TABLE A.3
Comparison of thermometers
Showing the relative indications of the Fahrenheit, Centigrade and Réaumur thermometer scales

	Boiling point	Freezing point
Fahrenheit	212°	32°
Centigrade	100°	0°
Réaumur	80°	0°

Conversion of thermometer degrees

°C to °R, multiply by 4 and divide by 5. °C to °F, multiply by 9, divide by 5, then add 32. °R to °C, multiply by 5 and divide by 4. °R to °F, multiply by 9, divide by 4, then add 32. °F to °R, first subtract 32, then multiply by 4, and divide by 9. °F to °C, first subtract 32, then multiply by 5, and divide by 9.

F	C	R	F	C	R
230	110	88	120·2	49	39·2
221	105	84	118·4	48	38·4
212	100	80	116·6	47	37·6
210·2	99	79·2	114·8	46	36·8
208·4	98	78·4	113	45	36
206·6	97	77·6	111·2	44	35·2
204·8	96	76·8	109·4	43	34·4
203	95	76	107·6	42	33·6
201·2	94	75·2	105·8	41	32·8
199·4	93	74·4	104	40	32
197·6	92	73·6	102·2	39	31·2
195·8	91	72·8	100·4	38	30·4
194	90	72	98·6	37	29·6
192·2	89	71·2	96·8	36	28·8
190·4	88	70·4	95	35	28
188·6	87	69·6	93·2	34	27·2
186·8	86	68·8	91·4	33	26·4
185	85	68	89·6	32	25·6
183·2	84	67·2	87·8	31	24·8
181·4	83	66·4	86	30	24
179·6	82	65·6	84·2	29	23·2
177·8	81	64·8	82·4	28	22·4
176	80	64	80·6	27	21·6
174·2	79	63·2	78·8	26	20·8
172·4	78	62·4	77	25	20
170·6	77	61·6	75·2	24	19·2
168·8	76	60·8	73·4	23	18·4
167	75	60	71·6	22	17·6
165·2	74	59·2	69·8	21	16·8
163·4	73	58·4	68	20	16
161·6	72	57·6	66·2	19	15·2
159·8	71	56·8	64·4	18	14·4
158	70	56	62·6	17	13·6
156·2	69	55·2	60·8	16	12·8
154·4	68	54·4	59	15	12
152·6	67	53·6	57·2	14	11·2
150·8	66	52·8	55·4	13	10·4
149	65	52	53·6	12	9·6
147·2	64	51·2	51·8	11	8·8
145·4	63	50·4	50	10	8
143·6	62	49·6	48·2	9	7·2
141·8	61	48·8	46·4	8	6·4
140	60	48	44·6	7	5·6
138·2	59	47·2	42·8	6	4·8
136·4	58	46·4	41	5	4
134·6	57	45·6	39·2	4	3·2
132·8	56	44·8	37·4	3	2·4
131	55	44	35·6	2	1·6
129·2	54	43·2	33·8	1	0·8
127·4	53	42·4	32	0	0
125·6	52	41·6	30·2	−1	−0·8
123·8	51	40·8	23	−5	−4
122	50	40	14	−10	−8

TABLE A.4

Conversions

METRIC SYSTEM

m = 1·0936 yard = 3·2808 ft; cm = 0·3937 in;
hectare = 2·471 acre; m^2 = 10·764 ft^2; cm^2 = 0·1550 in^2;
m^3 = 1000 dm^3 (or litre) = 33·315 ft^3 = 61024 in^3;
hl = 100 dm^3 (or litre) = 21·998 gal (British) = 26·418 gal (US) = 0·6111 barrel (British)
 = 0·8387 barrel (US) = 0·8522 *beer* barrel (US);
litre = 35·196 fl. oz (British) = 33·815 fl. oz (US);
tonne = 1000 kg = short ton = 0·9842 long ton = 2204·6 lb = 10 doppelzentner = 20 zentner; zenter = 50 kg = 0·984 cwt = 110·231 lb;
g = 0·03527 oz = 15·432 grain.

BRITISH MEASURES

yard = 3 ft = 36 in = 0·9144 m;
in = 2.540 cm = 1000 thou (thousandth of an inch);
thou = 25·4 micron or micrometre;
acre = 4840 yd^2 = 0·4047 hectare;
yd^2 = 0·8361 m^2; ft^2 = 9·290 dm^2; in^2 = 6·452 cm^2;
yd^3 = 0·7646 m^3; ft^3 = 28·317 dm^3; in^3 = 16·387 cm^3;
ton (long) = 20 cwt = 2240 lb = 1016 kg;
lb = 16 oz = 256 dram = 7000 grains = 0·45359 kg;
oz = 28·35 g; grain = 64·80 mg;
gal = 160 fl. oz = 8 pints = 1·201 gal (US) = 4·546 litre = 0·1605 ft.^{3};
pint = 0·5682 litre; fl. oz = 28·412 ml;
butt = 2 hogshead = 3 barrel = 108 gal = 4·9096 hl;
brl = 2 kilderkin = 4 firkin = 36 gal = 1·6365 hl = 1·4 brl beer (US).

US MEASURES

beer brl = 31 gal (US) = 25·81 gal (British) = 1·1734 hl = 0·717 brl British;
standard brl = 31·5 gal (US) = 26·23 gal (British) = 1·1924 hl = 0·729 brl British;
gal = 8 pint = 128 fl. oz = 3·7853 litre = 0·8327 gal British = 231·0 in^3.

BARLEY AND MALT MEASURES

Britain and South Africa: Barley bushel = 56 lb = 25·401 kg;
Barley quarter = 448 lb = 203·209 kg;
Malt bushel = 42 lb = 19·051 kg;
Malt quarter = 336 lb = 152·407 kg;
Australia and New Zealand: Barley bushel = 50 lb; Malt bushel = 40 lb;
US and Canada: Barley bushel = 48 lb; Malt bushel = 34 lb;

USEFUL DATA

1 kcal = 4·186 kJ = 3·968 BTU = 1·1628 Wh = 3088 ft lb;
BTU = 1·055 kJ = 0·252 kcal = 0·2931 Wh = 778·2 ft lb;
Wh = 3·6 kJ = 0·860 kcal = 3·412 BTU = 2655 ft lb;
therm = 105·506 MJ = 29·307 kWh;
standard ton refrigeration per 24 hr = 12000 BTU/hr = 3024 kcal/hr;
atm = bar = 14·70 lb/in^2 = 750.1 mm Hg = 10^5 Nm^{-2};
lb in^{-2} = 6894·76 Nm^{-2} = 0.06895 bar = 703 kg m^{-2} = 27·7 inches water;
lb/gal (British) = 99·76 g/l; lb/gal (US) = 119·8 g/l;
lb/brl (British) = 3.336 g/l; lb/brl (US) = 3·865 g/l;
grain/gal (British) = 14·25 mg/l; grain/gal (US) = 17·12 mg/l;
CO_2 in beer: g/100 ml = 5·06 vol/vol beer;
 vol/vol beer = 0·198 g/100 ml.

TABLE A.5
Specific gravity and extract table

The following table is based on those compiled by Dr Plato for the German Imperial Commission (*Normal-Eichungskommission*) and refers to apparent specific gravities, as determined in the usual manner by weighing in a specific gravity bottle in air or by means of a saccharometer. Cane sugar % wt/vol and % wt/wt represent grams per 100 ml and grams per 100 grams of solution respectively. The percentages by weight in column 6, corresponding with the specific gravities at 60°F given in column 1, were computed by interpolation from Plato's table for true specific gravities at 15°/15°C and 16°/15°C corrected to 60°/60°F and then brought to 60°/60° in air by adding (SG − 1) × 0·00121. The cane sugar weight percentages were converted to volume percentages and the solution divisors calculated. The column headed Plato gives the specific gravities in air at 20°/20°C related to the cane sugar weight percentages and, with the latter, corresponds with the Plato Table commonly used in breweries and laboratories where 20°C is the standard temperature. The column headed Balling similarly gives the specific gravities at 17·5°/17·5°C from the Balling Table corresponding with the same sugar percentages. These specific gravities cannot accurately correspond with those at 60°/60°F and 20°/20°C on account of the errors in Balling's Table. The following densities were used in the calculations:

Water at 15°C/4°C	0·999126
60°F/4°C	0·999035
20°C/4°C	0·998234

SPECIFIC GRAVITY CONVERSION TABLE FOR CANE SUGAR SOLUTIONS

SG 60°F	Brewers' pounds	British units Cane sugar % wt/vol	Solution divisor	SG 20°C	Plato Cane sugar % wt/wt Degrees Brix.	Balling SG 17·5°C	Baumé Modulus 145
1002·5	0·9	0·643	3·888	1·00250	0·641	1·00256	0·36
1005·0	1·8	1·287	3·885	1·00499	1·281	1·00513	0·72
1007·5	2·7	1·932	3·882	1·00748	1·918	1·00767	1·08
1010·0	3·6	2·578	3·879	1·00998	2·552	1·01021	1·43
1012·5	4·5	3·225	3·876	1·01247	3·185	1·01274	1·78
1015·0	5·4	3·871	3·875	1·01496	3·814	1·01528	2·14
1017·5	6·3	4·517	3·874	1·01745	4·439	1·01776	2·48
1020·0	7·2	5·164	3·873	1·01993	5·063	1·02025	2·83
1022·5	8·1	5·810	3·872	1·02242	5·682	1·02273	3·17
1025·0	9·0	6·458	3·871	1·02490	6·300	1·02523	3·52
1027·5	9·9	7·107	3·869	1·02740	6·917	1·02776	3·86
1030·0	10·8	7·755	3·868	1·02989	7·529	1·03027	4·20
1032·5	11·7	8·405	3·867	1·03238	8·140	1·03277	4·54
1035·0	12·6	9·054	3·866	1·03486	8·748	1·03527	4·88
1037·5	13·5	9·703	3·865	1·03736	9·352	1·03775	5·22
1040·0	14·4	10·354	3·863	1·03985	9·956	1·04024	5·55
1042·5	15·3	11·003	3·862	1·04234	10·554	1·04273	5·88
1045·0	16·2	11·652	3·862	1·04481	11·150	1·04523	6·21
1047·5	17·1	12·303	3·861	1·04731	11·745	1·04773	6·54
1050·0	18·0	12·953	3·860	1·04979	12·336	1·05022	6·87
1052·5	18·9	13·604	3·859	1·05227	12·925	1·05269	7·20
1055·0	19·8	14·255	3·858	1·05476	13·512	1·05515	7·52

		British units		Plato		Balling	Baumé
		Cane			Cane		
SG	Brewers'	sugar %	Solution	SG	sugar %	SG	Modulus
60°F	pounds	wt/vol	divisor	20°C	wt/wt	17·5°C	145
					Degrees		
					Brix.		
1057·5	20·7	14·907	3·857	1·05726	14·097	1·05760	7·84
1060·0	21·6	15·560	3·856	1·05975	14·679	1·06005	8·16
1062·5	22·5	16·213	3·855	1·06224	15·259	1·06252	8·48
1065·0	23·4	16·866	3·854	1·06472	15·837	1·06500	8·80
1067·5	24·3	17·519	3·853	1·06720	16·411	1·06747	9·12
1070·0	25·2	18·173	3·852	1·06970	16·984	1·06995	9·44
1072·5	26·1	18·827	3·851	1·07218	17·554	1·07244	9·75
1075·0	27·0	19·482	3·850	1·07467	18·122	1·07494	10·06
1077·5	27·9	20·135	3·849	1·07717	18·687	1·07743	10·37
1080·0	28·8	20·791	3·848	1·07965	19·251	1·07990	10·69
1082·5	29·7	21·446	3·847	1·08213	19·812	1·08237	11·00
1085·0	30·6	22·101	3·846	1·08462	20·370	1·08486	11·30
1087·5	31·5	22·758	3·845	1·08712	20·927	1·08737	11·61
1090·0	32·4	23·414	3·844	1·08960	21·481	1·08986	11·91
1092·5	33·3	24·071	3·843	1·09209	22·033	1·09235	12·21
1095·0	34·2	24·726	3·842	1·09457	22·581	1·09481	12·51
1097·5	35·1	25·384	3·841	1·09707	23·129	1·09730	12·81
1100·0	36·0	26·041	3·840	1·09956	23·674	1·09980	13·11
1102·5	36·9	26·700	3·839	1·10204	24·218	1·10230	13·41
1105·0	37·8	27·360	3·838	1·10454	24·760	1·10480	13·71
1107·5	38·7	28·019	3·837	1·10703	25·299	1·10730	14·00
1110·0	39·6	28·679	3·836	1·10952	25·837	1·10983	14·30
1112·5	40·5	29·339	3·834	1·11200	26·372	1·11235	14·59
1115·0	41·4	30·000	3·833	1·11450	26·906	1·11486	14·88
1117·5	42·3	30·660	3·832	1·11698	27·436	1·11735	15·17
1120·0	43·2	31·321	3·831	1·11947	27·965	1·11984	15·46
1122·5	44·1	31·981	3·830	1·12195	28·491	1·12231	15·74
1125·0	45·0	32·643	3·829	1·12445	29·016	1·12478	16·03
1127·5	45·9	33·305	3·828	1·12694	29·539	1·12729	16·31
1130·0	46·8	33·970	3·827	1·12944	30·062	1·12980	16·60
1132·5	47·7	34·632	3·826	1·13191	30·580	1·13228	16·88
1135·0	48·6	35·295	3·825	1·13441	31·097	1·13477	17·16
1137·5	49·5	35·958	3·824	1·13689	31·611	1·13723	17·44
1140·0	50·4	36·621	3·823	1·13938	32·124	1·13971	17·72
1142·5	51·3	37·285	3·822	1·14186	32·635	1·14221	18·00
1145·0	52·2	37·951	3·821	1·14435	33·145	1·14473	18·27
1147·5	53·1	38·617	3·820	1·14685	33·653	1·14727	18·55
1150·0	54·0	39·284	3·818	1·14934	34·160	1·14980	18·82

TABLE A.6

Equivalence between Institute of Brewing (UK) units of hot water extract

Litre °/kg (20°C) against lb/Qr. (15·5°C; 60°F) in the body of the table. Calculated from the Institute of Brewing Recommended methods of Analysis, (1977), on the basis that for litre °/kg (20°C) the equivalent values in lb/Qr. will be 0·5 greater than those for litre °/kg (15·5°C). Example: 251 litre °/kg (20°C) is equivalent to 84·1 lb/Qr. (15·5°C), 251 litre °/kg (15·5°C) is equivalent to 83·6 lb/Qr. (15·5°C).

	240	250	260	270	280	290	300	310	320	330
0	80·2	83·7	87·2	90·8	94·2	97·8	101·3	104·8	108·3	111·9
1	80·5	84·1	87·5	91·1	94·6	98·1	101·7	105·2	108·6	112·2
2	80·8	84·4	87·9	91·4	94·9	98·5	102·0	105·5	109·0	112·6
3	81·2	84·8	88·3	91·8	95·3	98·8	102·4	105·9	109·3	112·9
4	81·5	85·1	88·6	92·1	95·7	99·2	102·7	106·2	109·7	113·2
5	81·9	85·5	89·0	92·5	96·0	99·5	103·1	106·6	110·1	113·6
6	82·3	85·8	89·4	92·8	96·4	99·9	103·4	106·9	110·4	113·9
7	82·7	86·1	89·7	93·2	96·8	100·2	103·8	107·3	110·8	114·3
8	83·0	86·5	90·1	93·5	97·1	100·6	104·1	107·6	111·2	114·6
9	83·4	86·8	90·4	93·9	97·5	100·9	104·5	107·9	111·5	115·0

INDEX

Abraders, 146–7
Abrasion, 124–5
Acetic acid, 128
Acetone, 128
Acids
  alpha-, 6
  beta-, 6
  organic, in wort, 273–4
  to neutralize alkali, 201, 276
Acrospire (Fig. 2.3), 19
  bolting, 131
  growth, 15, 57
  length, 57, 59
Activated sludge systems, 212–13, 216, (Fig. 7.6) 218
Additives, 122–33
  application times, 54, 150, 152
  effects, 52, 121
  penetration, 44–45
  restrictions, 10
  uses, 122
Adenine, 87
Adenosine, 271
Adjuncts
  advantages, 223, 298
  analysis, 224–30
  application times, 263
  cereal grains, 224–5
    cooked, 232
    flaked, 234–5
    flours, 235–7
    raw, 230–1
  classification, 222
  disadvantages, 299–300
  effects, 224, 298
  extract analysis, 226
  fermentations, 223
  future role, 251

Adjuncts *Cont'd*
  gelatinization of starch, 225, 233–4, 264
  grits, 232–4
  in double mashing, 266–8
  industrial enzymes, 248–51
  machinery needs, 224
  mash-tun, 222, (Table 8.2) 228, 266–8, 275–8, 297, 329
  precooking, 264
  quality assessment, 226
  starches, 237–8
  sucrose, 238–41
  sugars, from purified starch, 242–4
    invert, 241–2
  syrups, malt extract, 244–7
    wort replacement, 247–8
  usage levels, 223
Adsorbants, 9
Aerobacter aerogenes, in water, 202
Air conditioning, 146
Air flow, in pneumatic malting, 120
Air rests, 51–53, 118, 149
Alanine, 271
Ale
  bitter, grist composition, (Table 10.5) 317
  brown, 10
  description, 7
  enzyme survival, 105
  ionic concentration, (Tables 7.4–5) 206–7
  mild, analysis, (Table 5.7) 134
  pale, *see* Pale ale
  processing, 8
  specifications, 137
Aleurone layer
  α-amylase content, 102

Aleurone layer *Cont'd*
  composition, 63
  description, 20-21
  enzyme release, 101, 102
  gibberellin effects on, 101
  lipid content, 89
  minerals content, 91
  physical changes during germination, 58, 62-63
  polyphenol content, 96
  respiration, 100
Alfa-Laval processing, 307
Allantoin, 87-88
Aluminium sulphate
  for sewage coagulation, 212
  for water softening, 200
Amadori rearrangement, 105
American Society of Brewing Chemists (ASBC), 12
Amines
  barley grain content, 88-89
  cereal syrups contents, 247
  effects of kilning on, 105
  sweet wort contents, 271
Amino acids
  barley grain contents, 84-86
  cereal syrups contents, 247
  sweet wort contents, 270
Ammonia
  barley grain contents, 88-89
  for reducing malting losses, 128-9
  worts contents, 271
Amylamine, 271
α-Amylase
  analysis, 259
  for barley flakes, 235
  fungal, 249-50
  high-residual, 231
  importance of respiration for production of, 101, 102
  mash temperature effects on, 282
  starch degradation by, 71-72, 243, 267, 290-3
β-Amylase
  barley grain content, 70-71
  destruction, 104
  for maltose production, 244
  liberation by enzymes, 86, 102
  raw barley content, 231
  soya bean content, 248

β-Amyalse *Cont'd*
  starch degradation by, 71-72, 290-3
Amyloglucosidase *Syn.* glucamylase (AMG)
  for glucose preparation, 243
  for light beers, 10
  fungal, 250-1
Amylolysis, 3, 4
Amylopectin, 67-69
Amylose, 67-69
Anaerobic digestion, of sewage, 213
Analysis of malt, 'Midwestern Lashes', (Table 5.8) 135
Aneurine, 272
Animal feed
  barley grain rejects, 36, 305
  dust, 191, 305
  floaters, 52, 148
  maize by-products, 238
  rootlets, 39, 191
  spent grain, 210, 295, 353
  sugar beet pulp, 240
Anthocyanidin pigments, 93-96
Anthocyanogens
  barley grain content, 62, 92-96
  decorticated grain content 61
  malt content 272
  reduction 277
Anti-foaming agents, from spent grain liquor, 89-90, 274, 296
Apple-jack, 12
Arabinose, 65-67
Arch breakers, 157
Arginine, 271
Aroma
  compounds, 105, 107
  improving, 323
Aromatization, 316
ASBC HWE determination units, 258
Ascorbic acid, 92-93
Aspartic acid, 271
Aspergillus spp. enzymes, 248, 250
Astringency, from polyphenols, 94, 96, 272, 280
Attenuation, 7
Attrition milling, 315-16
'Aurore' barley, effect of gibberellic acid during malting, (Fig. 5.3) 123
Awns, barley, (Fig. 2.1) 16, 17

INDEX

Bacillus spp. enzyme, 244, 248–9, 250, 251
Bacteria
  biochemical reactions, (Table 7.2) 203
  coliform, 202–3
  non-pathogenic, 202
  pathogenic, 202
Bag malting tests, 53
Bagasse, 239
Baker's yeast, 12, 239
Bakery products, 296
Ball mill, 306, 307
Balling, Tables 10, 227
Barley corn, *see* Barley grain
Barley grain
  abrasion, 124–5, 146–7
  aeration, 51–52, 118, 148–9
  after-ripening, 30
  agitation, 129–30
  amino acid content, 84–86
  aspiration, 34
  batch sizes, 146
  biochemistry, 57–107
  bolting, 50–51
  bulk density, 27–28
  carbohydrate content, 65–66
    regularities in 79
    soluble 77–79
  chemical changes during malting, 64–96
  chitting, 15, 43, 45, 51
  cleaning, 29, 34–36, 52, 210
  cold water extract, *see* Cold water extract
  conveying, 37, 170–1
  cooling, 31, 32, 171–3, 178–81
  crushed GA_3-treated, 124
  damage, by heat, 31–32
    by physical stress, 67
    by pumps, 150
  dead, 114
  debranned, 233
  dehusked, 124, 247, 309
  destoning, 151
  discharging from steeps, 150, 155
  dormancy, *see* Dormancy
  dry weight, 114
  dry-casting, 48
  drying, 24, 29–32, 155, 171–90, *see also* Kilning

Barley grain *Cont'd*
  dust dangers, 170–1
  dust for fuel, 188
  evaluation for malting, 1–3, 26–28, 40–41
  extracts, 114–15
  flaked, 234–5
  floaters, 148
  flour, 235, 297, 310
  flow problems, 157
  freezing, 129
  fumigation, 34
  gelatinization, 225, 233–4
  germination, *see under* Germination
  grading, 36–37
  gum content, 72–77
  handling systems, 37
  hardness tests, 59
  hemicellulose content, 72–77
  husk content, 60
  inorganic constituents, 91
  laboratory tests, 40–41
  leaching, 52
  lipid content, 89–90
  loss of dry matter, 52
  low-nitrogen, 298
  mashing, *see* Mashing
  mealy, 20, 27
    water uptake, 42
  metabolic inter-relationships, (Fig. 4.13) 103
  micronized, 232
  milling, 3, 247, 315–16
  modification, *see* Modification
  moisture content, *see* Moisture content
  moisture losses, 152–3
  morphology, 25
  nitrogen content, (Table 2.1) 23, 28, 79–86
  nitrogenous materials, 79–86
  nucleic acid content, 87–88
  organization during malting, 96–104
  oxygen restriction, 129
  pearl, 233, 234
  phenol content, 92–96
  phosphate content, 90–91
  physical changes during malting, 57–64
  physiology, 39–56

Barley grain Cont'd
  piece, 111, 145
  ploughing, 145
  precleaning, 29, 34–36
  pregermination, 44
  pressure in steeps, 150
  purchase for malting, 26–28, 113
  quality, 1–3, 21, 26–28, 113
  raking, 145
  raw, 230–1
    milling, 231
    washing, 231
  reception, 28–29
  relative humidity, 171, 173–6, 180–1
  respiration, 29, 50–52, 99–104
  ripening 'Earl', 25
  roasted, 222, 232, 297
  sampling, 26–27, 40–41, 44, 113–16
  screening, 34
  sections, (Fig. 4.1) 58
  separation, 36
  shrinkage, 173
  size, 42
  solute penetration, 43–44
  starch content, 65–72
  steely, 20, 27
    water uptake, 42
  steeping, *see under* Steeping
  stewing, 105, 175
  storage, (Fig. 1.1.) 2, 24, 29, 32–34
  structure, 19–21
  temperature, *see under* Temperature
  torrified, 222, 232
  transection, 45
  turning
    compartment malting, 153–5, 160
    continuous malting, 164
    floor malting, 145
    paddle, 184
    pneumatic malting, 152
  viability, 28, 44–50
  vitamin content, 91
  washing, 52, 147
  water uptake, 41–43
  water-sensitivity, 46–47, 49–50, 53
  water-vapour pressure, 171, 173–5, 180
  weighing, 37
  wet-casting, 48
Barley plant, botany, 15–21

Barley plant Cont'd
  breeding, 24–26
    for malt quality, 113
  cultivation, 21–24
  diseases, 24
  fertilization, 17–19
  flowering, 17
  four-rowed, 17
  germination, 15
  harvesting, 24
  height, 17
  naked, 19, 61
  pollination, 17
  root formation, 15
  six-rowed, 17
    analysis of malt from, (Table 5.8) 135, 137
    carbohydrate content, (Fig. 4.7) 80
    extract proportion, 79, (Fig. 4.7) 80
    for malt whisky, 140
    high-enzyme malt from, 298
    husk content, 60
    nitrogen-solubility fractions, (Table 4.7) 81
    starch content, 64
  stems, 15–17
  spring, 17, 22
  threshing, 24
  two-rowed, 17
    analysis of malt from, (Table 5.8) 135
    carbohydrate content, (Fig. 4.7) 80
    chemical composition, 64–65
    extract proportion, 79 (Fig. 4.7) 80
    husk content, 60
    starch content, 64
    varietal constants, 113
  uses, 21
  varieties, 22, 24–26, 97
  winter, growth, 17, 22
  yields, 22 (Table 2.1) 23
Barley syrup
  composition, (Table 8.5) 246
  production, 247
Beans, malted, 15
Beer
  adjuncts, *see under* Adjuncts
  African, 12
  alcohol levels, 283

Beer Cont'd
  analysis, (Table 5.7) 134, (Table 9.14) 298
  anti-foam agents, 89–90, 274, 296
  aroma, 105, 107, 316, 323
  calorific value, 269
  carbonation, 8
  chilling, 8
  chill-proofed, 248
  clarification, 180
  consumption, 12
  dispensing, 8
  excise on, 11
  filtering, 8
  flavour, see under Flavour
  foaming, see under Foaming
  from chit malts, 140
  gelatinous precipitates, 295
  gushing, 49, 208–9
  ionic concentration, (Table 7.6) 207
  light, 10, 250
  non-alcoholic, 10
  old Louvain, 141
  pasteurization, 8
  post-fermentation treatment, 8–9
  processing, 8–9
  quality, adjunct effects on, 224
    criteria, 8–9
    nitrogen effects on, 79
  shelf-life, 277, 298, 309
  silica content, (Table 9.12) 295
  staling, 274
  stone, 273, 280
  strength, 10–11
  tank bottoms, 210
  tannin content, (Table 9.12) 295
  types, 10–11
  volatile constituents, 270
  water in, 219–20
  world production, 12
  see also Ale
Beetles, in stored barley, 33
Benzoxazolone, 128
Beta vulgaris, 239
Betaine, 88–89
Bicarbonates, in brewing water, 199–200, 275–6
Biflavanoids, 94
Bifurcose, 78
Bins, storage, 32, 304

Biological Oxygen Demand (BOD)
  in spent grain waste, 296
  tests, 201–2, 213–19
Biotin, 92–93, 272
Biotower, 216
Bisulphite adjuncts, 277
Bitter ale, grist composition, (Table 10.5) 317
Blade, see Acrospire
Block brewhouses, 336–8
Boby drum, 169, 310
Boiler
  blow-down liquid, 210
  corrosion, 201
  water, 201
Bore holes, 196–7
Bottles, washing, 210–11
Bracts, barley, 17
Brandy, 12
Breaks, see Trub
Breweries
  automated, 336
  bungalow, 305, 337
  effluents from, see Effluent
  location, 207
  water usage, 207
Brewing
  confluent, 336
  high-gravity, 8, 210, 224, 247, 341, 352
  industry, 12
  journals, 13–14
  sequence of operations, (Table 1.1) 4
  water, see under Water
°Brix, 227
Bromates, 125–7
  dose effects, 126
  effects, 125
  entry into grain, 44
  survival in malt, 126
  time of application, 54, 150
Bromelain, 248
Brown ale, 10
Brumalt, 138
Bühler-Miag
  mill, 257, 261, (Fig. 10.2) 306, 307
  tower malting, (Fig. 6.9) 159
Bulk handling
  barley grain, 28–29
  malt, 191–2, 304
  syrup, 304

Bulldozer, 153–5
Bungalow brewing, 305, 337
Burton ales, 206–7
Butylamine, 271

Caffeic acid, 92–94
Calcium
  in sparge liquor, 328
  in sweet wort, 275, 276
Calcium bicarbonate, in brewing water, 199, 275
Calcium bisulphite adjunct, 277
Calcium carbonate, in brewing water, 195, 200, 275
Calcium hypochlorite, 52
Calcium oxalate, 273
Calcium pantothenate, 272
Calcium sulphate
  for mash pH reduction, 276
  in brewing water, 195, 199
Calgon, 201
Calvados, 12
Candicine, 88–89
Caramel malt, 105, 138–9
Cara-pils, 139
Carbohydrate
  barley grain content, 65–66, 79
  degradation, 7, 269
  fermentability, 269
  fraction proportions, 79
  insoluble, 114
  malt contents, 268
  mash temperature effects on, 283
  respiratory substrate, 99
  soluble, 77–79
  total content in sample, 79
  wort content, 259, 268–9, 290–3, 297
Carbon dioxide
  extraction, 149
  output, see Respiration
Carbonates, in brewing water, 195, 275–6
Carbonation, 8
Carotenoids, 92
Casella mill, 307
Catchment water, 196
Catechin, 93, 95
Caterpillars, in stored barley, 34
Cattlefeed, see Animal feed
Cellobiose, 71, 76

Cellulose, 72
Centribrew process, 307, 343–4, (Fig. 11.19) 345
Centrifugation
  for hop separation, 6,
  for wort separation, 342
  for yeast separation, 7, 8
Cercosporella herpotrichoides, 24
Cereal
  cooker, 5, 267, 330–1, 337–9
  grain
    adjuncts, 222, 224–5
    cooked intact, 232
    composition, 224–5
    flaked, 234–5
    flours, 235–7
    oil content, 224–5
    raw, 222, 230–1
    refined starches, 237–8
    syrups, 223
Charcoal
  for polyphenol reduction, 277
  for wort treatment, 293–5
'Chevalier', respiration rates, 98
Chilling beer, 8
Chip-sugar, 243
Chit, see Root sheath
Chitting, 15, 43, 45, 51
Chloride, in sweet wort, 275
Chlorination, of water, 204–5
Chlorine demand, 202
Chlorogenic acid, 92, 94
Choline, 271–2
Cider, 11
Citric acid, 273
Citrobacter freundii, in water, 203
Clark degrees, 199
Cleaning
  automated, 210
  barley grain, 29, 34–36, 52, 210
  malt, (Fig. 1.1) 2, 305
Clova malting plant, 155–7
Coagulation
  materials, 6
  sewage, 212
Cognac, 12
Cold trub, 6
Coldewe test, 45
Cold water extract (CWE), 59
  effect of gibberellic acid on, 124

Cold water extract *Cont'd*
  losses, (Fig. 5.1) 112
  preformed soluble sugars in, 268
  ranges, 260
Coleoptile, *see* Acrospire
Coleorhiza, *see* Root sheath
Colloidal silver, for sterilization, 206
Colours of Malt
  gibberellic acid effects on, 124
  husk effect on, 329
  kilning effects on, 133
  polyphenol effects on, 272
  production, 105-6, 107
  sugar effects on, 241
Comparamill, 307-8
Conbrew process, 223, 245
Confluent brewing 336
  HWE determination Congress EBC units, 258
Converters, continuous tube, 342
Conveyors, 37, 170-1
Cooker
  cereal, 5, 267, 330-1, 337-9
  mash 336-7
Cooling
  barley grain, 31, 32, 171-3, 178-81
  malt, 191
  wort, 6-7
Coombes, *see* Rootlets
Copper
  mash, 4, 321, 330-2
    adjuncts in, 222-3, (Table 8.2) 228, 247, 267
    sugars, 241
  wort boiling, 6, 336-8
Copper salts, in sweet wort, 275
Copper sulphate, 128
Corn, *see* Maize
Corn flour, 237-8
Corn-steep liquor, 238
Cossettes, 240
Costs
  effluent, 9, 52-53, 150, 209-10, 213
  energy, 9-10
  grists, (Table 10.4) 316
  mashing, 350-1
  plant, 151, 336
Couch, 111, 145
Couch grass rhizomes, 15

Coumarin
  additive, 128
  in barley grain, 92, 94
Cross-pollination, 26
Cryptolestes ferrugineus, 33
Clums, *see* Rootlets
Cummins, *see* Rootlets
Cyclones, separating, 342
Cylinder, cleaning, 36
Cytolysis, *see* Endosperm degradation
Cytosine, 87

Dams, 196
Decantation, 7
Decoction mashing, *see under* Mashing
Decortication, 45
Deep shaft system, 216
Deionization, 198-9
Demineralization, 198-9
Deoxyribonucleic acid (DNA), 87
Deoxyribose, 65-67
  release, 88
Desalination, 197
Dextrin, 268
Dextrinization equivalent (DE), 230
Dextrose, *see* Glucose
Diacetylreductase, industrial, 249
Diacylglycerols, 273
Diastase
  correlation with respiration, (Fig. 4.12) 100
  in barley grain, 69
Diastatic malt
  analysis, (Table 5.9) 136, 141
  extract, 222, 244-5
Diastatic power, 259
2,4-Dichlorophenoxyacetic acid, 128
Dichromate values, 201
Dietetic ale, 10
Dimethyl sulphide, 106
Dimethylamine, 271
Diseases of barley, 23-24, 29-30
Distillation desalination, 197
Domalt malting plant, 150-1, 164, (Fig. 6.14) 166, 167
Dormancy, 44-50
  breaking, 30-31, 49-50, 120
  causes, 49, 50
  formaldehyde effects on, 120
  profound, 46

Draff, *see* Spent grain
Drilling, direct, 22
Drum
 kilning, 187–8
 malting, 167–70
 roasting, 187
Dry-casting, 48, 150
Drying
 barley grain, 24, 29–32, 155, 171–90
 stages of 175–6
Dublin ales, 206
Dust hazards, 170–1, 236, 305, 315

Ear, barley, (Fig. 2.1) 16, 17
EBC mill, 307
Effluents,
 analysis, (Table 7.9) 215
 BOD, 213–19
 discharge into waterways, 219
 disposal costs, 9, 52–53, 150, 209–10, 213
 from spent grain water, 296
 impurities in, 211
 treatment, 197, 211–19
Eijkman reaction, 204
Elevators, 170–1
Embryo barley, 21
 composition, 62
 crushing, 310
 damage during germination, 104, 128, 129–31
 growth, 102
 lipid content, 89
 minerals content, 91
 physical changes in, 61–62
 respiration, 100
 starch formation, 57
Endosperm barley, starchy
 amino acid accumulation, 86
 composition, 63–64
 crushing, 309, 310
 degradation, 1, 58–59, 102
 hard end, 310
 nitrogen content, (Table 4.8) 82
 pentosan content, 73
Enzymes
 debranching, 70, 72, 250, 290
 deficiencies, 297
 degrading starch 69–72, 278, 290–3
 effect during mashing, 262–3, 278–9

Enzymes *Cont'd*
 for barley syrups, 231
 for raw barley extracts, 231
 for syrup production, 245, 247–8
 for malt whisky, 140
 hydrolytic destruction of, 102, 104–5, 107
 industrial, 249
 production, 21
 release of, 39–40, 58, 59, 101, 103
 inactivation, 275, 278, 291, 323
 industrial, 224, 248–51
 mash temperature effects on, 285, 291
 mash thickness effects on, 288
 proteolytic, addition to beer, 9
 hemicellulose degradation by, 74
 respiration effects on production, 99–104
 supplementary, 277
 temperature effects on production, 98–99, 121
Epithelium barley, scutellar changes during germination, 58
Eremothecium ashbyii, 142
Erysiphe graminis, 24
Escherichia coli, in water, 202–3
Essential oils, of hops, 6
Ethanol production, 7, 50, 99
Ethanolamine, 88–89
Ethylamine, 271
Ethylene, 128
European Brewery Convention (EBC), 12
Evaporative cooling, 171–3
Evaporators, vacuum, 245
Excise duty, 11
Extract, *see* Malt extract *and* Wort extract
Extractors, vibrating-screen, 342
Eyespot, 24

Fans, kiln, 184
 variable-pitch, 189
Farmer's lung, 170
Fatty acids, in wort, 273–4
 sodium salts of, 128
Fermentation
 'hanging', 243
 secondary, 8
 temperature, 8

Fermentation, *Cont'd*
  wort, 6–8, 259, 283–4, 290, 295
Fertilization, barley plant, 17–19
Fertilizers
  nitrogenous, 23
  timing of application, 15
Ficin, 248
Filterbeds
  blocked, 328–30
  travelling, 342
Filtering, beer, 8
Filter, mash, *see* Mash filter
Filtration, mash
  improving, 280
  requirements, 309
  theory, 344–7
  sewage, 212, (Fig. 7.6) 218
Finings, 8
Flaked adjuncts, 234–5
  density, 316
Flavour
  astringency, 94, 96, 272, 280
  beer storage effects on, 8
  bisulphite effects on, 277
  charcoal effects on, 293–5
  chip-sugar effects on, 243
  chloride effects on, 275
  compounds, 7, 105, 107
  dependence on malt characteristics, 104
  diacetyl reductase effect on, 249
  fatty acid effects on, 90
  fuel effects on, 106–7
  green-grain, 106, 247
  husk effects on, 61, 309
  kilning effects on, 40, 105–6, 133
  lactate effects on, 275
  malt extract effects on, 245
  mash temperature effects on, 284
  mineral salts effects on, 208–9
  nitrogen effects on, 79
  nitrogenous compounds effects on, 89
  nucleoside effects on, 88
  of flavan-less beers, 95
  polyphenol effects on, 272
  roasted grains effects on, 232
  spoiled, by chlorinated phenols, 205
    by fatty acids, 274
    by fuels, 106

Flavour *Cont'd*
  by micro-organisms, 33
  striking heat effects on, 323
  sugar effects on, 241–2
  sugar beet effects on, 240
  sugar cane effects on, 239
  sulphate effects on, 275
Floaters, 52, 148
Floor malting, *see under* Malting
Flours, 235–7
  extract, 297
  grinding, 310
Fluorides, in brewing water, 195
Flow problems
  barley grain, 157
  flours, 236–7
  starch powders, 238
Fly, in stored barley, 34
Foaming
  fatty acid effects on, 274
  glycoprotein effects on, 269
  mineral salts effects on, 209
  nitrogenous compounds effects on, 270
  protease effects on, 249
  sulphur dioxide effects on, 244
  wheat flour effects on, 237
Folic acid
  barley grain contents, 92–93
  sweet wort contents, 272
Folinic acid, 272
Formaldehyde
  for reducing polyphenol levels 96, 120 277
  in steep liquors, 49, 119–20, 131
Formol nitrogen, wort content, 259, 269, 284
Fountain, 148
Four-vessel brewhouse, 321–2
Freezing desalination, 197
Fructosans
  barley grain content, 77–8
  wort content, 268
Fructose
  barley grain content, 65–67, 77
  specific rotation, 228, 230
  syrups, 244
  wort content, 268
Fuels in malt kilning
  causing malt deterioration, 106–7

Fuels in malt kilning *Cont'd*
  consumption, (Table 6.2) 181, (Table 6.3) 185, 186, 189–90
  costs, 9, 175, 186
  effects, 133
  efficiency, 189
  for brown malt, 139
  for whisky, 10, 140
  saving, 176, 181
  types, 188–9
Fumaric acid, 273
Fungi, attacking barley, 24, 29–30, 33
Fungicides
  for barley crop, 24
  for steeping, 49
Furnaces
  gas-fired, 188
  oil-fired, 188–9
  solid fuel, 188
Fusaria, 49
Fusarium moniliforme, 122

Galactose, 65–67
Galacturonic acid, 65–67
Galland drum, 168
Gelatin stab test, 204
Gelatinization of starch grains, 225, 233–4, 264
Germination (barley grain), 1, (Fig. 1.1) 2
  biochemistry, 97–104
  changes during, 39–40
  drums, 167
  Eckhardt test, 45
  effects of aerating steeps on, 51–52, 100
  enzyme development during, 98–104
  moisture content requirements, 97–98
  respiration effects, 99–104
  restriction, 129
  temperature requirements, 39, 48, 98–99, 111, 121–2
  tests, 28, 44–48
  water-sensitive grain, 46–47
Germination (barley plant), 15
Germination-kiln boxes, 155–9
Germinative capacity, 44
Germinative energy, 45
Geys malting system, 167

Geyser, 148
Gibberella fujikuroi, 122
Gibberellic acid
  application methods, 124
  application timing 54
  effect during malting, 5 (Fig. 5.3) 123
  effects with bromate, 126–7
  effects with hot water steeping, (Table 5.5) 132
  endogenous, 101
  enzyme stimulation, 123
  for over-steeped grain, 47
  germination encouragement, 100, 101, 102
  grain penetration, 44
  uses, 122–3
Gibberellin, 101–2
Glucamylase, *see* Amyloglucosidase
β-Glucan
  barley grain content, 73–74
  debranned barley content, 233
  degradation, 75–77
  grist content, 295
  raw barley content, 231
  roasted grain content, 232
  wort content, 297
β-Glucanase
  analysis, 259
  addition to mash, 330
  destruction, 104
  fungal, 251
  mash temperature effects on, 285
  release, 102
Glucodifructose, 77–78
Glucosan, 268
Glucose,
  barley grain content, 65–67, 77
  confectioner's, 243
  oxidase, industrial, 249
  specific rotation, 228
  syrup, composition, (Table 8.4) 242
    effects on germinating grain, 125
    maize starch for, 238
  worts content 268
α-Glucosidase
  analysis, 259
  barley grain content, 69–70
  destruction, 104
  release, 102
  starch degradation, 290

Glucuronic acid, 65–67
Glumes, barley, 17
Glutamic acid, 271
Glutelin, 81–82
Glycine, 271
Glycolipids, 89
Glycoprotein foam stabilizers, 269
Grading, *see* Screening
Grain, *see* Barley grain; Cereal grain; Seed grain, Starch grains
Grain whisky, 11
Gramine, 88–89
Gramineae, 15
Grant, 327, (Fig. 11.8) 331
Green malt
  composition, 121
  conveying, 170
  drying, 39, 116, 171–2
  extract adjuncts, 223
  flavour dependence on, 204
  for mashes, 141, 277
  kilning, 39, 167–8, 171–80
  milling, 298, 315–16
  polyphenol content, 96, 277
  syrup, 247
Grinding, *see* Milling
Grist, 296–300
  case, 305, 316
  cleaning, 304–5
  constituents, 15, 223, (Table 10.4) 316, 317, (Table 11.2) 333
  costs, (Table 10.4) 316
  extract, (Table 10.4) 316
  β-glucan contents, 295
  grinding, 333
  hydrator, (Fig. 11.3) 323, 330
  infusion mashing, 3
  moisture content, (Table 10.4) 316
  preparation, 304–19
  specification, 310, 314
  storage, 304–5
  temperature, 323
  weighing, 304
Grits adjuncts, 222, 224, 232–4
  production, 310
Guanine
  barley grain content, 87
  wort content, 271
Guanosine, 271

Gum
  barley grain content, 72–77
  degradation, 74–77
Gushing, 49, 208–9
Gypsum additive, (Table 9.5) 277

Hammer mill, 307
Hand evaluation, of barley grain, 27
Handling
  barley grain, 28–29, 37, 191–2
  malt, 303–5
  starch powders, 238
  wheat flours, 61, 236
Hardness, water, 199–200, 275–7
Hartong number, 261
Haze
  charcoal effects on, 293, 295
  formation, 95, 96
  kilning effects on, 107
  mash pH effects on, 280
  mineral salts effects on, 275
  nitrogen effects on, 79, 270
  oxalate, 280
  polyphenol effects on, 272
  reduction, 89, 120
  wheat flour effects on, 237
Health hazards, 170–1, 305
Heat, exchangers, 190
  output by germinating grain, 99
  pipes, 190
  pump, 190
  waste, causes, 175
  usage for, 190, 191, 197
Hemicellulose
  barley grain content, 72–77
  degradation, 74–77
  roasted grain content, 232
Herbicides, 15
Herniarin, 92, 94
Histidine, 271
Holocellulose, 72
Home brewing
  adjuncts, 223
  hopped wort extracts for, 245
Hop separator, 6, 337
Hop-back, 6
Hop-boil
  calcium ions in, 275
  enzyme inactivation in, 291
  evaporation from, 298

Hop-boil *Cont'd*
  pH, 274, 276
Hops
  addition to wort, 344
  boiling with wort, 5, 95
  essential oils, 6
  extraction of bitterness, 208
  extracts, 6
  resins, 6
  pellets, 6
  powders, 6
  selection, 6
  spent, 210
  utilization rate, 280, 288
Hordein fraction, 81–82
Hordenine
  barley grain content, 88–89
  wort content, 271
Hordeum agriocrithon, 24
H. deficiens, 25
H. distichon, 25
H. intermedium, 25
H. spontaneum, 24
H. vulgare, 25
Horizontal brewing, 337–8
Hot trub, 6
Hot water extract, 40, 255–8
  analysis, 256–8
  definition, 255
  grinding, 112–13, 261–2
  losses, (Fig. 5.1) 112
  prediction, 113–14
  relationship to malt solids and spent grains, 256
  relationship to nitrogen content, 114
Hot-spots in stored grain, 29
Humulones, 6
Husk, of barley, 15, 19
  sieve fraction composition, 60–61
  function, 21
  pentosan content, 73
  physical changes, 59
  silica content, 91
  splitting, 306, 309
  wetting, 314
HWE units, 258
Hybridization, of barley, 26
Hydrochloric acid, for water treatment, 201

Hydrogen peroxide
  for dormancy breaking, 49, 120
  for polyphenol reduction, 277
  test, 44
Hydrostatic pressure, in steeping, 43
p-Hydroxybenzoic acid, 92, 94
Hypoxanthine, 271

Igneous rock, 195
'Imber', cultivar, 26
Immersion, *see* Steep
Indole-production test, 203
Infusion mashing, *see under* Mashing
Inoculation, *see* Pitching
myo-Inositol 91–92, 272
Inositol, sweet wort contents, 274
Insecticide, for stored grain, 31
Insects, in stored barley, 33
Institute of Brewing (IoB), 12
  HWE units, 257–8
International Sugar Degrees, 227
Invertase, 290
Ion-exchange deionization, 198–9
Ions, effects on brewing process, (Table 7.7) 208
'Irish Archer I'
  biology, (Fig. 2.1) 16
  selection, (Table 2.2) 25
Iron salts,
  in brewing water, 195
  in sweet wort, 275
Isoamylase, fungal, 250
Isohumulones, 6
Isokestose, 77–78
Isoleucine, 271

Journals, technical, 13–14

Kaffir beer, 15
Katadyn process, 206
'Kenia', extract yield, (Table 10.2) 309
Kestose, 77–78
Ketol isomerase, 244
Kettle, mash, *see* Copper, mash
Kieselguhr
  filtering through, 273
  spent grain content, 354
Kiln for malt
  continuous, 167, 187, 189
  double box, (Fig. 6.25) 190

Kiln for malt *Cont'd*
  early type, 183
  European, 184–6
  fogging, 175
  furnaces, 188–90
  insulation, 189
  linked, 189
  mechanization, 184
  performance, 181
  semi-continuous, 189
  shallow-loaded, (Fig. 6.22) 182, 183–4
  single floor, (Fig. 6.22) 182, 183–7
  stripping, 191
  three-floor, 176–7, 184
  two-floor, 175, 177, 184, (Fig. 6.23) 185
  types, 183–8
  vertical, 187
  Wanderhaufen, 160–4
Kilning malt (Fig. 1.1) 2
  amino acid destruction during, 86
  chemistry of, 104–7
  cycle duration, 181
  drums, 187–8
  effects of, 40, 133
  effects of varying conditions, 176
  efficiency, 177–8
  enzyme degradation during, 72
  evaporation rates, 174, 178–81
  fuel, *see* Fuel
  liquid, 141
  mathematical analysis, 177–82
  micro-, 115–16
  nitrogen content changes during, 83, 89
  nucleic acid degradation during, 88
  phases, 175–6
  phytase survival during, 91
  purpose of, 40, 171
  sugar content during, 79
  technology, 171–90
  temperature, 133
  variations in, 141
'Klages' analysis of malt, (Table 5.8) 135
Klebsiella spp
  in water, 202–3
  pullulanase from, 250
Knives, for mashing, 325–7, 332–3
Koji, 248
Kolbach index, 260, 269
Kropff malting system, 129–30

Lactate, 275
Lactic acid
  additive, 128, 139–40, 279
  for water treatment, 201, 276
Lactobacillus delbruckii, 279
Lactose, 238
Laevulinic acid, 273
Laevulose, *see* Fructose
Lager
  analysis, (Table 5.7) 134, (Table 5.9) 136
  dark, description of, 10
    ionic concentration, (Table 7.4) 206
    Kropff malting for, 129
  description, 7
  enzyme survival, 106
  grist compostion, (Table 10.5) 317
  pale, analysis, (Table 5.9) 136
    ionic concentration, (Table 7.4) 206
  Pilsner, 10
  specifications, 137
Lagoons, 212
Laminaribiose, 71, 76
Laminarinase, 102
Land races of barley, 25
Lausmann transposal malting system, 164, (Fig. 6.12) 165
Lauter tun
  block brewhouse, 337
  decoction mashing, 4, 321, 330–1
  double mashing, 5, 332
  efficiency, 346–51
  for high-gravity worts, 352
  grist specification, 314
  Strainmaster, 339–41
Lauterbottich, 321
Lautering
  continuous, 342–4
  rapid, 273
  run-off problems, 293
  theory, 344–7
Leaching efficiency, 346
Leaves
  barley, 15
  nitrogen compounds in, 89
Leucine, 271
Light ale
  production, 250
  types, 10
Lignin, 92

Lime
  in steep water, 49, 52
  slaked, for water softening, 199
β-Limit dextrin degradation, 259, 290–1
Linoleic acid, 274
Linolenic acid, 274
Lipase, industrial, 249
Lipids
  barley grain content, 89–90
  starch content, 238
  wort content, 273–4
Lite, *see* Light ale
London ales, 206
Lundin fractions, 288
Lupulones, 6
Lysine, 271
β-Lysolecithin, 274

MacConkey's medium test, 204
Magnesium bicarbonate, in brewing water, 199, 275
Magnesium carbonate, in brewing water, 195, 200, 275
Magnesium sulphate
  for mash pH reduction, 276
  in brewing water, 195, 199
Maillard reaction, 105
Maischbottich, 321
Maischpfanne, 321
Maize
  adjuncts, 222, 224
  flakes, 234
  for double mashing, 5
  for whisky, 11
  gelatinization of starch, 225, 233–4
  grits, 233
    refined, 237–8
  malted, 15
  oil, 238
  starch, 237–8
Malic acid, 237
Malt
  acid, 139–40, 279
  amber, 139, 304
  amino acid content, 84–86
  analyses, 110 (Tables 5.7–9) 134–6, 254–5
  attemperated, 323
  black, 107, 138
  bleaching, 106, 129

Malt *Cont'd*
  blending, 293
  brown, 139
  caramel, 105, 138–9
  carbohydrate content, 65–66
    potentially extractable, 268
    soluble, 77–79
  character alteration, 104
  chit, 140
  chocolate, 138, 304
  cleaning, (Fig. 1.1.) 2, 305
  colour production, 105–6, 107
  coloured, 137–9
  composition, 64–65
    change during kilning, 105
    prediction, 79
  cooling, 191
  crystal, 105, 138–9, 304
  curing, 176
  density, 261–2
  derooting, 111, 191
  description, 1
  diastatic, (Table 5.9) 136, 141
    power, 259
  dispatch, 191–2
  drying, 104, 106, 171–90
  duty on, 11
  enzyme content, 28
  extracts, adjuncts in, 223
    analysis, 245
    bromate effects on, 125–7
    diastatic, 222, 244–5
    gibberellic acid effects on, 123–4
    glucose effects on, 125
    grinding yields, 309
    hot water steeping effects on 131–2
    modification effects on, 309
    pH effects on, 280
    powders, 245
    prediction of quality, 113–15
    preparation, 244
    storage, 245
    syrups, 244–8
    temperature effects on, 121–2
    transport, 245
    used, 141
    yield 40, 50, 306
    see also Hot water extract *and* Cold water extract
  flavour, *see* Flavour

Malt *Cont'd*
  gelatinization, 225
  grading, 304–5
  green, *see* Green malt
  gum content, 72–77
  hand-dry, 141, 176, 191, 298
  handling, 303–5
  hemicellulose content, 72–77
  high-enzyme, 140–1, 298
  hopper, 314–15
  inorganic constituents, 91
  lager, 313–14
  lipid content, 89–90
  liquid, 141, 247
  magpie 189
  mashing, *see* Mashing
  milling, dry, 261–2, 306–14
    for decoction mashing, 330
    wet, 314–16
  modification indices, 259–62
  moisture content, 137, 176–7, 189
  nitrogen content, 137
  nucleic acid content, 87–88
  phenol content, 92–96
  phosphate content, 90–91
  polyphenol content, 272
  production method in UK, 137
  pulverized, 310
  purchasing, 254
  quality, barley grain types for, 21, 40–41, 77
    determination by HWE analysis, 258
    extraction prediction, 113–15, 261
    importance of, 40
    mashing different ages, 113
  relative humidity, 171, 173–6, 178–80
  replacement by adjuncts, 225
  roasted, enzyme destruction in, 105
    flavour compounds in, 107
    production, 138
  silica content, (Table 9.12) 295
  slack, 141, 176, 324
  slaking heat, 324
  special, 137–8, 187
  specific heat, 324
  specifications, 134, 137, 254
  spraying, 314
  starch content, 67–72
  steaming, 314, 323

Malt *Cont'd*
  storage, 304–5
  tannin content, (Table 9.12) 295
  temperature, 324
  transport, 191
  types, 134–42, 176
  vinegar, 11, 140
  vitamin content, 91
  water-vapour pressure, 171, 173–5, 178–80
  weighing, 117
  well-modified, 259–60
  whisky, *see under* Whisky
  wind-, 141
Maltase, 290
Malting, (Fig. 1.1) 2
  advances in, 110
  anaerobic, 99–100, 118, 129
  batch-size, 120
  biochemistry of, 96–104
  box, 153–60
  changes in grain during, 57–107
  compartment, 153–60
  conditions, 110–33
  continuous, 160–7
  development of processing, 110
  double, 133
  drum, 167–70
  effluents, *see* Effluents
  embryo-damaging technique, 104
  equipment, 146–71
    adjunct needs, 224
    deterioration prevention, 153
    economies from, 310
  floor, 110–13, 145–6
  grain, *see* Barley grain
  introduction to, 1–3, 39–40
  Kropff, 129–30
  losses, 39, 116–18
    composition, 111–12
    effects of moisture content on, 97–98
    effects of temperature on, 121
    reducing techniques, 122–33
    total, 43, 64
  mechanized, 145–72
  micro-, 40, 53, 115–16
  phases of kilning, 175–6
  pilot-scale, 116
  pneumatic, 120, 151–3

Malting *Cont'd*
  semicontinuous, 160–7
  sugar use during, 79
  technology, 145–71
  temperature, 48
  tower, 159–60
  water for, *see under* Steep liquor, *and* Water
Maltopentaose, 78
Maltose
  barley grain content, 70–71, 77, 79
  syrups, 243–4
  wort content, 268
Maltotetraose
  barley grain content, 78
  wort content, 268
Maltotriose
  barley grain content, 78–79
  wort content, 268
Maltroom, 305
Maltsters
  brewer, 118
  sales, 118
Manganese, 274
Mannose, 65–67
Manpower costs, 9
'Maris Otter'
  for malting, 22
  selection, 26
Mash
  acidifying, 279
  adjuncts in, 222 (Table 8.2) 228, 266–8, 275–8, 297, 329
  analytic, 255–8
  copper, 4, 267, 321, 330–2
  filter, construction, 333–5
    fineness of grist, 309, 314
    for high-gravity wort, 352
    for wort separation, 4
    high-pressure, 307, 341–2
    vacuum drum, 342
  filtration, improving, 280
    requirements, 309
    theory, 344–7
  flour in, 237
  kettle, *see* Mash copper
  loading depth, 327, 331
  mineral salts effects, 207–8, 274
  mixing, 314, 330

Mash *Cont'd*
  mixing vessel, 321, (Fig. 11.6) 329, 330, 334, 336–8
  oxygenating, 277
  pH, 208, 274, 275, 279–80, 328
  pump, 314
  raw barley in, 231
  thickness effects, 281, 288–90
  tun, 3, 320–1, (*see also* Lauter tun)
    efficiency, 346–51
    for high-gravity worts, 352
    for infusion mashing, 325–8
    pre-1880 law, 309–10
    Strainmaster, 339–41
  vessel, 4
Mashing
  barley extracts, 114–15
  chemical changes during, 40, 262–8
  choice of systems, 347–51
  continuous, 342–4
  decoction, 4, 264–8, 270
    boiling, 275
    double, 265–6, (Table 9.9) 287
    phytate hydrolysis during, 276
    single, 266
    traditional methods, 320, 330–4
    triple, 264–5, (Table 9.9) 287
    unit operations, 321
  double, 4–5, 266–7, 330–4, 339
  effects on worts, (Table 9.9) 287
  enzyme properties during, 278–9
  equipment, efficiency, 348–51
    modern, 336–42
  European, 258, 320
  high-quick, 285–7, 307
  infusion, 3–4
    nitrogen formation during, 270
    phosphate removal during, 276–7
    rapid wort production, 339
    temperature, 262–3, 323–5
    traditional, 323–30
    unit operations, 320
  introduction to, 3–5
  malt mixtures, 293
  malt quality and age effects, 113
  North American, 258
  phytase survival during, 91
  stand period, 327
  stand dissolution during, 290–3
  technology, 320–54

Mashing *Cont'd*
  temperature achievement, 320
  temperature effects, 278–88, 291–2
  temperature-programmed, 5, 247–8 263, (Fig. 9.8) 283, 285, (Table 9.9) 287, 334
  time effect, 280–8
  traditional British, 320–1
  underletting, 324–5
Mashing-in water, 314
Massecuite, 239
Master Brewers' Association of America (MBAA), 12
Maturation, beer, 8
'Mazurka', milling, 308
Mealworm, in stored barley, 33–34
Melanoidins, 105
Mesaconic acid, 273
Metamorphic rock, 195
Methanol, 128
Methionine, 271
Methyl Red test, 203
Methylamine, 271
N-Methyltyramine, 88–89
Microkilning, 115–16
Micromalting, 40, 53, 115–16, 256, 300
Micro-organisms
  beer contamination with, 284
  for sewage treatment, 212–13, 216
  fungal, 248–9
  growth factor source, 142
  in barley grain, 33, 49
  in steep liquor, 46, 119–20, 202–3
  on rootlets, 131, 155
  oxygen competition by, 60, 99
  sterilization of, 204–6
Microprocessor regulation, 189
Micropyle, 19
  solvent penetration, 44
Mild ale, 10
  analysis, (Table 5.7) 134
Mildew, 24
Milk stout, 238
Millet, 12
Mill
  six-roll, 313
  five-roll, 313
  four-roll, 311–12
Milling
  dry, 306–14

Milling *Cont'd*
  capacity, 309
  energy measurement, 308
  extract recovery, 261–2
  flour, 235
  green malt, 298
  introduction to, 3
  malt, 306–16
  raw barley, 231
  wet, 314–16
Millival unit, 199
Millroom, 305
Mineral salts, barley grain content, 91
  mashing content, 195, 274–5
  removal, 198–200
Modification, changes during, 39–40, 58–59, 63
  embryo damage during, 104
  guides to progress of, 101
  indices, 259–62
  over-, 124
  temperature effects on, 121
  'two-way', 125
Mogden formula, 213
Moisture content, of barley grain
  at steep out, 118
  determination, 27
  effects on respiration, 29
  increase during steeping, 41–43, 53–54
  regulating, 97–98
  temperature effects on, (Table 2.6) 31, 32, (Fig. 5.2) 119
Moisture content, of grists, (Table 10.4) 316
Moisture content, of malts, 137, 141, 176–7
  effect on surface area, 173
  pneumatic malting, 120
  raised, 189
  relationship to relative humidity, 173–5
  storage effects, 304
  temperature effects, 176
Molasses, 239
Mollier diagram, 178–9
Monoacylglycerols, 273
Moulds, on barley grain, 49
  on malting plant, 153
  on rootlets, 155

MTI malting plant, (Fig. 6.8) 158
Müger kiln, 181, 186–7
Multistage flash distillation, 197
Munich lagers, 206
Mushroom compost, 296
Mycotoxins, 49
Naphthalene acetic acid, 128
Nicotinic acid, barley grain content, 92–93
  wort content, 272
Nitric acid, 128
Nitrogen, barley grain content, (Table 2.1) 23, 28, 79–86, 137
  coagulable, 260
  effect on malt extract, 113
  formol, 259, 269, 284
  index of modification, 84, 260
  oxides, (NOX) in fuel gases, 106–7, 188
  permanently soluble, 83–84, 260, 269, (Fig. 9.7) 282
  soluble, mash temperature effect on, (Fig. 9.10) 285
  ratio, 84, 260
  total α-amino, 260, (Fig. 9.11) 286
  total content, 111–12
  total soluble, *see* Total soluble nitrogen
  wort content, 259, 260, 269–71, 284–7, 293–4, 297
Nitrosamines, accumulation of, 104, 142
  carcinogenic, 106–7, 188
  reduction, 129
Non-alcoholic beer, 10
Nordon steep, 151
Nordon-van Waesberghe process, 307, 341–2
Nucellar tissue, 62–63
Nucleases, 88
Nucleic acids, barley grain content, 87–88
  mash temperature effects on, 283–4
  sweet wort content, 270–1
Nucleosidases, 88
Nucleosides, 88
Nucleotides, barley grain content, 88
  wort content, 270–1

Oats
  adjuncts, 222, 224
  flakes, 234

Oats *Cont'd*
  husks, 329
  malted, 134
Oil content, of adjuncts, 224–5, 226
  maize, 238
Oleic acid, 274
Ophiobolus graminis, 24
Original gravity, 11
Oryzaephilus surinamensis, 33
Osborne method of protein fractionation, 81
Oxalic acid, 273
α-Oxoglutaric acid, 273
Oxygen, for dormancy breaking, 49
  uptake by grain, 50, 99
Ozone, for water sterilization, 120, 206

Palate fullness, nitrogenous effects on, 79, 270
Pale ale, 10
  analysis (Table 5.7), 134
  ionic concentration, (Table 7.4) 206
  malting method, 137
Pale lager, analysis, (Table 5.9) 136
  ionic concentration, (Table 7.4) 206
Palmitic acid, 273
Pantothenic acid, 92–93
Papain, 248
Particle size control, 306
Pasteurization of beer, 8
Pasveer ditch system, (Fig. 7.6) 218
Pearl barley, *see under* Barley
Pearl starch, wet-milled, 237–8
Pearling, 60
Peas, malted, 15
Peat, for flavouring whisky, 107, 140, 188
Pentosan, barley grain content, 73–74
  degradation, 74–75
Pentoses, 269
Pepsin, 248
Peptidases, barley grain content, 86
  mash temperature effect on, 285
  wort content, 270
Peptides, 270
Pericarp, 21
  functions, 60–61
  polyphenol content, 96
Permanently soluble nitrogen (PSN), 83–84, 260, 269 (Fig. 9.7) 282

Permanganate value (PN), 201–2
Pesticides, for stored barley, 33–34
  timing of application, 15
Pests attacking barley, 23–24, 29, 33–34
Petri dish test, 45, 53
pH
  hop-boil, 274
  mash, 208, 274, 275–6, 279–80
  wort, 209, 293
Phenolic compounds, barley, 95
Phenols, barley grain content, 92–96
  chlorinated, 205
  malt whisky content, 140
Phenylalanine, 271
Phlobaphenes, kilning effects on, 106
  wort content, 272
Phosphatase, 102
Phosphate, barley grain content, 90–91
  spent grains content, 277
  sweet wort content, 274, 293–4
Phosphatidylcholine, 88–89
Phospholipids, barley grain content, 89–90
  wort content, 274
Phosphoric acid, for mash acidification, 279
  for reduction of malting losses, 128
  for water treatment, 201
Phosphorylase, 69
Phytase, mash temperature effect on, 185
  survival, 91
Phytate, hydrolysis, 91, 276
  sweet wort content, 274
Phytic acid, 90–91
Piece (malt), 111, 145
Pilsen lager, analysis, (Table 5.9) 136
  water composition effects, 206
Pilsner lager, 10
Pin milling, 315–16
Pin-disc mill, 306–7
'Piroline', analysis of malt, (Table 5.8) 135
Pitching, 7
Plato tables, 10, 227
Plough, 153
'Plumage-Archer', crude protein fractions, (Fig. 4.8) 82
  selection, 26
Pneumatic malting, 120, 151–3
Pol, 227

Pollution, environmental, 10, 219
  tests for, 202–4
Polymers, 93, 95
Polyphenols, barley grain content, 92–96
  effects on beer quality, 270
  reduction, 120, 277
  wort content, 272–3, 293, 297
Polysaccharides, 72–77
Popp malting plant, 160, (Fig. 6.10) 161
Porter, 10
Potassium bromate
  effects, 125–7
  grain penetration, 44
  used with GA_3, 126–7
Potassium dichromate, for BOD test, 201–2
Potassium metabisulphite, for dechlorination, 205–6, 277
Potassium permanganate, for BOD test, 201–2
Potato, flakes, 15
  gelatinization, 225
  starch, 15, 222, 237, 238
Precleaning, 29, 34–36
Pregermination, 44
Premasher, 330
Present gravity, 11
Proanthocyanidins, *see* Anthocyanogens
'Proctor', moisture content, 41
  nitrogen content, (Fig. 4.10) 84
  prediction equation, 114
  selection, 26
  varietal constants, 113
Proline, barley grain content, 86
  wort content, 247
Protease, addition to mash, 330
  barley grain content, 86
  correlation with respiration, (Fig. 4.12) 100
  effects, 249
  fungal, 251
  inhibitors, in raw barley, 86, 231
  mash temperature effects on, 285
  stability, 290
Protein, crude content, 79–80
  effects on beer quality, 270
  in endosperm, 63, (Table 4.8) 82
  index of modification, 260
  interactions with polyphenols, 94–95
  precipitation, 208

Protein *Cont'd*
  rest, 285
Proteolysis, during mashing, 3, 4, 270
  mash temperature effects on, 288
Pullulanase, analysis, 259
  fungal, 250
  starch degradation, 290
Pumps, 314
  damage by, 150
Purines, 88
Pyridoxal, 272
Pyridoxamine, 272
Pyridoxin, 272
Pyrimidines, 88
Pyrrolidine, 271
Pyruvic acid, 273

Quality control, 204

Rachis, barley, 17
Raffinose, barley grain content, 77–79
  wort content, 297
Rakes, *see* Knives
Recirculation, 189
Relative humidity (RH), 171, 173–6, 178–81
Reservoirs, storage, 212
Resin, hop, 6
  regeneration, 199
Respiration, 99–104
  effects of moisture content on, 29, 50–52, 97–98
  losses, 111–12, 117
  restriction, 129
Resteeping, 54, 131, (Table 5.6) 133
Retorrification, 316, 323
Reverse osmosis desalination, 197
Rhizopus spp enzyme, 248, 250
Riboflavin
  barley grain content, 92–93
  production, 142
  wort content, 272
Ribonucleic acid (RNA), 87
Ribose
  barley grain content, 65–67
  release, 88
Rice
  in double mashing, 5
  gelatinization, 225, 233–4
  grits, adjuncts, 222, 224, 233

Rice *Cont'd*
  proportion, 331
  malted, 15
Roasting, drums, 187–8
  enzyme destruction during, 105, 107
Roller mill, 307, 308, 316
Root sheath, 15
  appearance of, 57
Rootlets
  appearance of, 57
  bromate effect on, 125
  composition, 142
  importance of embryo for, 100
  killing, 131
  losses, 111–12, 117
  nitrogen content, 88–89
  protein content, 142
  removal, 39, 111, 191
  temeperature effects on growth, 121
  uses, 141–2
  vitamin content, 142
Roots, 15
  nitrogen content, (Table 4.8) 82, 89
Rouser, 148–9
Ruck, 111, 145
Rum, 239
Rye, adjuncts, 222
  damage through handling, 61
  malted, 15, 134

Saccharum officinarum, *see* sugar cane
Sacks, 191–2
Saladin box, 152, 153–60
Salmonella spp, in water, 202
Salts, content in brewing water, 195–6
  removal, 197
Sampling, for germinative capacity, 44
  for grain evaluation, 40–41
  for grain purchasing, 26–27
  for malting quality, 113–16
  malt extract, 261
Saturne malting system, 167
Scale removal, 201
Schmitz process, 266
Schock-Gusmer brewhouse, (Fig. 11.14) 339
Schönfeld germination test, 45
Schönjahn germination test, 45
Screens, grain, 34–37
  malt, 304–5

INDEX 383

Screens *Cont'd*
  inclined, 343-4
  mesh width, (Table 10.3) 310, 311
  sewage, 211
Screw-conveyors, wort separating, 342-3
Scutellar epithelium, 58-59
Scutellum
  amino acid accumulation in, 86
  α-amylase content 102
  enzyme release 101, 102
Seck mill, 310
Sedimentation tanks, 212
Sediments, in brewing water, 194-5
Seed bed, 22
Seed grain
  dressed, 22, 23
  germination, 15
  inferior, 23
  planting depth, 22
  quality, 21-22
  selection, 25-26
  sowing, 22
  structure, (Fig. 2.1) 16, (Fig. 2.2) 18, 19-21
Seedlings, winter killing, 22
Selection, of barley seed, 25-26
Serine, 271
Sewage, *see* Effluent
Shigella dysenteriae, in water, 202
Sieves, *see* Screens
Silica, barley grain content, 91
  brewing water content, 198
  sweet wort content, 274, 293-5
Silos, storage, 32, 304
Sitophilus granarius, 33
Skimmings, 7, 117
Sludge removal, 201
Small-scale malting, *see* Micromalting
Soda, for water softening, 200, 201
Sodium bromate
  effects, 125-7
  grain penetration, 44
Sodium carbonate, for water softening, 200, 201
  in steep water, 52
Sodium citrate bacterial test, 203
Sodium hydroxide, in steep water, 49, 52
Sodium hypochlorite, for chlorine demand test, 202
  for water sterilization, 204-5

Sodium octanoate additive, 128
Sodium phosphate, for water treatment, 201
Softening systems, 199-201
Soil, for barley growing, 22-23
Solek tunnel maltings, 164, (Fig. 6.13) 165, 167
Soluble nitrogen ratio (SNR), 84
  effect of gibberellic acid on, 124
Solutes, effects of, 52
  penetration into grain, 43-44, 59
Sorghum, adjuncts, 222, 224
  gelatinization, 225, 233-4
  grits, 233
  malted, 12, 15
Soya beans, 248
Sparge liquor, hardness, 328
Sparge-arms, 4, 325
Sparges, contactor, 343-4
Sparging, 293-5
  decoction mashing, 332
  infusion mashing, 328
Specific gravity, *see under* Wort
Specific rotation, 227
Spend safe, 327
Spent grains, 4, 40
  analysis, 295-6
  composition, 353
  dried, 295-6
  lipids in, 273
  mineral salts in, 274
  phosphate content, 277
  pressing liquor, 210, 295, 341
  relationship to extract, 256, 257
  value of, 352-4
Spent hops, 210, 353
Spikelet, 17
Spire, *see* Acrospire
'Spratt-Archer', selection, 26
Spray-steeping, 53-54, 118, 149, 150-1
Sprouts, *see* Rootlets
Starch, adjuncts, 222, 225
  barley grain content, 65-72
  conversion products, 223
  degradation, 67-72, 278, 282, 290-3
  formation in embryo, 57
  formation in endosperm, 63-64
  granular, 238, 297
  hydrolysis, 242
  malt contents, 268

Starch *Cont'd*
  roasted grain content, 232
  lipids, 89
  liquefaction, 231, 233, 243
  potato, 15
  refined, 237–8
    sugars from, 242–4
  resistance to amylolysis, 293
  run-off problems, 237, 293, 297
Starchy endosperm, *see* Endosperm
Steam condensates recovery, 210
Steel's masher, 322–3
Steep
  aeration, 50–52
    germination encouragement by, 100
  equipment, 147–51
    conical-bottomed, 147–50
    flat-bed, 149–51, 157
    hopper-bottomed, 150
    self-emptying, 145
  liquor, 52–53
    alkaline, 52, 118
    BOD, 217–19
    changing, 54
    concentrated, 128
    corn, 238
    disposal, 150
    effects on germination, 49
    nitrogen content, 219
    pH, 218–19
    sparging, 50, 118
    temperature, 118
Steep-germination-kiln boxes, 155–9
Steeping 1, (Fig. 1.1) 2, 39, 47
  air-rest, 51–53, 118, 149
  discharging, 150, 155
  floor malting, 110
  flush-, 55
  hot water, 4, 131–2
  in running water, 54
  incremental, 53–54
  loss of dry matter, 52, 111–12, 117
  multiple, 54, 131, (Table 5.6) 133
  prolonged, 47–48, 51
  schedules, 53–55
  spray-, 53–54, 118, 149, 150–1
  traditional, 53
  variations in conditions, 118–20
  warm, 53
  water usage, reduction, 210

Steeping *Cont'd*
  water uptake during, 41–43, 47
Steinecker brewhouse, (Fig. 11.13) 338
Stems, 15–17
Sterilization, water, 204–6
Sterols, 92, 273
Stewing, 175
Storage, barley grain, 1, (Fig. 1.1) 2, 24, 32–34
  malt, 304
  syrup, 243, 245, 304
Stout, 10
  grist composition, (Table 10.5) 317
  milk, 238
  oats for, 134, 234
  roasted barley for, 232
  rye for, 134
Strainmaster mash tun, (Fig. 11.15) 340, 352
Straw adjunct, 329
Street, 160, 164
Streptococcus faecalis, in water, 202
Striking heat, 323–4
Succinic acid, 273
Sucrose, adjunct, 223, 238–42
  analysis, 226–7
  barley grain content, 71, 77, 79
  from sugar beet, 240
  from sugar cane, 239–40
  wort content, 268
Suction, 7
Sugar, adjuncts, 223
  analysis, 226–30
  barley grain content, 65–67, 77–79
  beet, 239, 240
  brewing, 240
  candy, 241
  cane, 239–40
  copper, 6, 241
  density, 230
  for secondary fermentation, 8
  from purified starch, 242–4
  granulated, 241
  invert, 241–2
  primings, 241–2
  reduction during kilning, 105
  refining, 240
  specific rotation, 227
  spectra, 230
  wort content, 268–9

Sulphates, in brewing water, 195, 199–200
  in sweet wort, 275
Sulphur dioxide, for germinating grain, 128
  for kilning, 106, 129, 188–9
  for syrups, 244
Sulphuric acid, for decortication, 45
  for reducing malting losses, 128
  for mash acidification, 279
  for water treatment, 201, 276
Surface film, on grain, 43, 47–48, 54
Syrup, brewing, 240–1
  copper, 6, 223
  density, 230
  extract analysis, 226–30
  from enzyme converted grain, 247–8
  from purified starch, 242–4
  high-fructose, 223, 244
  high-glucose, 243
  high-maltose, 243–4
  refining, 244
  storage, 243, 245, 304
  transport, 245
  types, 223
  uses, 141

Take-all, 24
Tankers, 192
Tannin, for scale removal, 201
  laboratory brews content, (Table 9.12) 295
  precipitation, 208
  wort content, 272
Tanninogens, 94–96
Tapioca, starch adjunct, 222
Temperature, ambient, 181–3
  brewing water, 323–4
  control in steeps, 148
  curing, 176
  decoction mashing, 264–7
  double mashing, 267
  drying barley, 171, 174–5
  effects on enzymes, 291
  effects on water uptake of grain, 43, (Fig. 5.2) 119
  floor malting, 111
  gelatinization, 225
  germination, 39, 48, 98–99, 121
  grist, 323

Temperature *Cont'd*
  infusion mashing, 262–3, 323–5
  in-grain, 178
  kilning, 133
  mashing, 278–88, 291–2, 320, 323
  malt, 324
  pneumatic malting, 120, 152, 153
  reducing, 189
  sparging, 293, 328
  stored barley, 29–31, 33–34
  striking heat, 323–4
Temperature-programmed mashing, *see under* Mashing
Testa, composition, 61
  solvent penetration, 44
Testinic acid, 52, 61
Tetrazolium tests, 28, 44
Thiamin, barley grain content, 92–93
  wort content, 272
Thousand corn dry weight (TCW), 27
Threonine, 271
Thunaeus test, 44
Thymine, barley grain content, 87
  wort content, 271
Tides, *see* Steep
Tillers, 15–17
α-Tocopherol, 92–93
Total nitrogen content (TN), 111–12
  relationship to HWE, 114
Total soluble nitrogen (TSN), 83
  barley grain contents, (Fig. 5.1) 112
  mash temperature effects on, (Fig. 9.11) 286
  sweet wort contents, 269–71
  wort contents, 259, 260
Transglucosylase, fungal, 250
Triacylglycerols,
  barley grain contents, 89–90
  wort contents, 273
Tribolium castaneum, 33
T. confusum, 33
Trickling bed system, 212
Trichoderme viride enzyme, 248
Trimethylamine, 271
Triticale, adjunct, 222
Trogoderma granarium, 33
Trub
  cold, 6
  formation, 95
  hot, 6

Trub *Cont'd*
  pressing, 210
  separation, 344
  use, 353–4
Tryptophan, 271
Tun, *see under* Lauter *and* Mash
Turners, 153–5
Two-high mill, 311–12
Two-vessel brewhouse, 321
Tyramine
  barley grain contents, 88
  wort contents, 271
Tyrosine
  barley grain contents, 85, 88
  wort contents, 271
'Tystofte Prentice', selection, 25
Ultraviolet light for sterilization, 206
Underback, 328
Underletting, 324–5, 329
Upperback, 328
Uracil
  barley grain contents, 87
  wort contents, 271
Urea nitrate, 128

Valentine tube, 327
Valine, 270, 271
Valley bottom mash tun, 339
Vanillic acid, 92, 94
Versene, 201
Viability, 28, 44–50
Vibrio cholerae, in water, 202
Vienna-type lager, (Table 5.9) 136
Vinegar, brewing, 330
  malt, 11, 140
Vitamins
  barley grain contents, 91–2
  rootlet contents, 142
  wort contents, 271–2
Voges-Proskauer test, 203
Vorlaufvehrfahren, 327

Wanderhaufen malting plant, 160, (Fig. 6.11) 162–3
Water, *see also* Moisture content
  beer content, 219–20
  brewing
    aeration, 200
    boiling, 200
    costs, 9, 53

Water *Cont'd*
  deionization, 198–9
  desalination, 197
  economy, 209–11
  effluent, *see* Effluent
  geology, 194–5
  in beer, 219–20
  International Standards, (Table 7.3) 205
  ionic concentration, (Tables 7.4–6) 206–7
  ozonized, 120, 206
  purity, 201–6
  quality, 195
  softening, 199–200, 275
  supply, 195–7
  temperature, 323–4
  treatment, 201
  types, 206–9
  usage, 217
  mashing-in, 314
  physical properties, (Table 7.10) 220
  reuse, 196–7, 210, 211
  steep, *see* Steep liquor
  uptake, of barley grain, 41–43
    grain size effects on, 42
    temperature effects on, 43
Water-sensitivity, of grain, 46–47
  causes, 47
  effects of warm steeping on, 53
  overcoming, 49–50
Water-vapour pressure, 171, 173–5, 178–80
Weevils, in stored barley, 33
Weissbier, 15, 134, 296–7
Weizenbier, *see* Weissbier
Wells, 197
Wet-casting, 48, 150
Wheat grain
  adjuncts, 222, 230
  damage through handling, 61
  flaked, 234
  flour, 235–7, 297
  gelatinization, 225, 233–4
  malted, analysis, (Table 5.9) 136
    for Weissbier, 15, 134, 296–7
  milling, 235–6
  roasted, 222, 232
  starch, 237, 238
  torrified, 222, 232

INDEX

Whirlpool tank, 6
Whiskey (Irish), raw barley for, 230
Whisky, flavour production, 107
  grain, 11
  malt, 11, 140
Wind-malt, 141
Wine, 11
Winkler kiln, 177–8, 181, 186–7
Wort
  aeration, 6
  analysis, (Table 9.14) 298
  attenuation, 293
  bacteria, 203
  boiling, 6, 275–6
  bright, 328
  carbohydrate content, 259, 268–9, 290–3, 297
  clarity, 330
  cloudy, 330, 333
  collection, 293
  composition, 263
    from unmalted adjuncts, 297
    mashing effects on, 263
  cooling, 6–7
  copper, 6, 336–8
  effect of gibberellic acid, 124
  excise on, 11
  extenders, 242
  extract, efficiency, 328 (Fig. 11.22) 350
  hopped, 223, 245
  fermentation, 6–8, 259, 283–4, 290, 295
  first, 328
  from green malt, 298–9
  high-gravity, 210, 224, 247, 341, 352
  ionic concentration, (Table 7.5) 207
  last runnings, 203–5, 328, 352
  lipids in, 273–4
  losses, 353
  nitrogen content, 259, 260, 284–7, 290, 297
  nitrogen-diluents, 234
  organic acids content, 273–4
  oxygenation, 6
  pH, 209, 280, 293

Wort *Cont'd*
  polyphenol content, 272–3, 297
  quality analysis, 256–62
  recycling, 327
  replacement syrups, 247–8
  run-off, 280, 327–30, 347
  second, 328
  separation, 342–4
  specific gravity, 10–11, 255–8, 328
  steam-stripping, 106, 298
  sweet, analysis, (Fig. 9.14) 294
    clarity, 330
    mash thickness effects on, (Table 9.11) 289, 290
    mineral salts content, 274–5
    nitrogenous compounds in, 269–71
    separation, 3–4, 40
    turbidity, 107, (Fig. 11.3) 334
    viscosity, (Fig. 5.1) 112, 288, 297
    vitamin content, 271–2
    wheat flour effects on, 237
Wurzepfanne, 321

Xanthine, 271
Xylobiase, 75
Xylobiose, 71, 74
Xylose, 65–67

Yeast, baker's 12, 239
  bottom, 7
  for other beverages, 11–12
  fatty acid effects, 274
  flocculation, 280
  growth encouragement, 6–7
  in cattlefood, 354
  mineral salts effects on, 208–9, 275
  overgrowth, 284
  selection criteria, 7–8
  separation, 7
  surplus, 210
  top, 7

Ziemann brewhouse, (Fig. 11.13) 338
Zinc salts, in sweet wort, 274
Zymomonas test, 204